Understanding Geometry for a Changing World

NCTM yearbooks focus concerted attention on timely issues by viewing them in depth, from multiple perspectives, through interdisciplinary lenses, and across various grade bands.

Understanding Geometry for a Changing World

Seventy-first Yearbook

Timothy V. Craine
Seventy-first Yearbook Editor
Central Connecticut State University
New Britain, Connecticut

Rheta Rubenstein
General Yearbook Editor
University of Michigan—Dearborn
Dearborn, Michigan

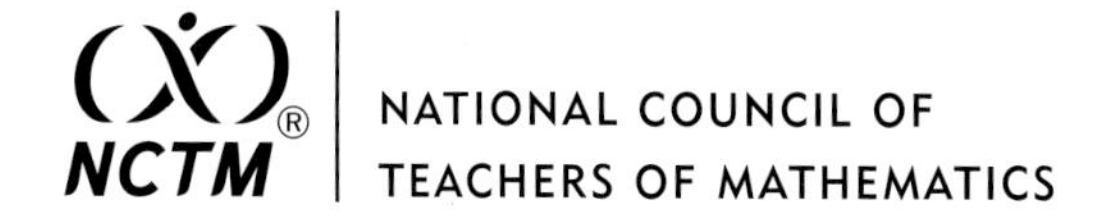

Library of Congress Cataloging-in-Publication Data

Understanding geometry for a changing world / Tim Craine, seventy-first yearbook editor ; Rheta Rubenstein, general yearbook editor.
p. cm. — (Yearbook ; 71st)
ISBN 978-0-87353-619-6
1. Geometry—Study and teaching. I. Craine, Timothy. II. National Council of Teachers of Mathematics.
QA461.U53 2009
516.0071—dc22
2008051331

The National Council of Teachers of Mathematics is a public voice of mathematics education, providing vision, leadership, and professional development to support teachers in ensuring mathematics learning of the highest quality for all students.

Printed in the United States of America

Contents

Contents of the CD-ROM

Preface

A yearbook gives an organization an opportunity to take stock, at one moment in time, of the status, concerns, understandings, and expectations regarding a specific topic. Such is true of this Seventy-first Yearbook of the National Council of Teachers of Mathematics (NCTM). Only three of the seventy yearbooks published by NCTM prior to this one have had "geometry" in their titles. The following look at previous yearbooks highlights several changes in how our profession has attended to geometry over nearly eight decades.

NCTM's Fifth Yearbook, *The Teaching of Geometry* (Reeve 1930), said in its preface that the yearbook was intended to supplement and assist a committee appointed by NCTM and the Mathematical Association of America to study the feasibility of a combined one-year course in plane and solid geometry. Many articles in that yearbook advocate such a course; others suggest that to prepare students for this course in the tenth year of high school, groundwork in "demonstrative" geometry (i.e., informal proofs) should be laid in grades 7, 8, and 9.

The basic framework for the Fifth Yearbook was the assumption that the tradition of teaching some variation of Euclid's geometry should be continued. The major departure from Euclid's approach was to allow some propositions that Euclid proved (such as the equality of vertical angles) to be taken as postulates. In one article, George Birkhoff and Ralph Beatty of Harvard University argued for a system of postulates based on measurement of distance and angle (1930, p. 92). These eventually became the ruler and protractor postulates incorporated into many textbooks from the 1950s onward (Sinclair 2008, p. 60).

By the time the Thirty-sixth Yearbook, *Geometry in the Mathematics Curriculum* (Henderson 1973), appeared, a major upheaval in mathematics education had occurred with the introduction of "new math" curricula. No longer was it assumed that geometry should be taught as a one-year course in synthetic, Euclidean geometry. Instead, the core of that yearbook was a series of articles proposing a variety of ways to organize the high school geometry curriculum. These included a modification of the synthetic approach; courses based primarily on coordinates, transformations, or vectors; developing affine properties (e.g., incidence and parallelism) prior to introducing distance and angle measure; an eclectic approach; and spreading geometry over six years of an integrated program.

Almost every article focused on identifying an appropriate set of axioms for the approach it discussed. However, one author reported studies indicating that few students completing a formal course in geometry could reliably identify axioms, definitions, and theorems (Brumfiel 1973, p. 102):

> What can I conclude? Students of 1954 who studied an old-fashioned, hodge-podge geometry had no conception of geometric structure. Students of today who have studied a tight axiomatic treatment also have no conception of geometric structure.

He cautioned, "We need to listen to students and learn what they really think. If we do listen, what we hear will provide useful guidance as we experiment in the years ahead with various approaches to the teaching of geometry" (ibid).

Ironically, aside from this observation, the Thirty-sixth Yearbook focused largely on curriculum with virtually no discussion of what students "really think" or how they come to understand geometry. In a departure from its predecessor, however, it included two articles that dealt with informal geometry (one for grades K–6, the other for grades 7–12) and contained sections on the place of geometry in modern mathematics and on the preparation of teachers.

The most recent previous yearbook on geometry was *Learning and Teaching Geometry, K–12* (Lindquist 1987). The opening article in that yearbook, "The van Hiele Model of the Development of Geometric Thought" (Crowley 1987), represented a fundamental shift away from an almost exclusive focus on content and curriculum to a consideration of issues related to students' learning.

In contrast with its predecessors, the 1987 Yearbook placed little emphasis on axiomatics, although a few articles offered suggestions for helping students learn how to construct proofs. Rather, geometry as a vehicle for problem solving was highlighted, in keeping with the *Agenda for Action* (NCTM 1980) and anticipating the centrality of problem solving in the later *Standards* documents (NCTM 1989, 2000). One entire section of the yearbook was devoted to "activities," including those appropriate at the elementary and middle school levels. Another section emphasized geometry's relationship to other branches of mathematics including algebra, calculus, probability, and combinatorics, anticipating the Connections Standard. As in 1973, the yearbook closed with two articles on the preparation of teachers.

Only two of the twenty articles in the 1987 Yearbook dealt with computers. One of them advocated giving increased attention to such topics as matrices, parametric equations, and homogeneous coordinates that are applied in computer graphics (Smart 1987). The other described how the program Logo with turtle graphics could be used to enrich the secondary school mathematics curriculum (Kenney 1987). At the time the yearbook was written, interactive geometry software lay in the future.

A lot has changed in the past twenty-two years. First and foremost, all mathematics education has been influenced by the Standards movement (NCTM 1989, 2000). In *Principles and Standards for School Mathematics* (NCTM 2000), geometry is given continual emphasis throughout all grade levels; in fact, the graph on page 30 (NCTM 2000), reproduced on the cover of this yearbook, suggests that geometry is the one content standard that should receive relatively constant attention from prekindergarten through grade 12.

Research on students' learning of geometry has continued to inform curriculum developers. Some textbooks, for example Serra's *Discovering Geometry* (1989), were guided by the van Hiele model. Projects supported by the National Science Foundation produced curricula at all grade levels aligned with the Standards and emphasizing developmentally appropriate activities, real-world applications, and the integration of algebra and geometry (Sinclair 2008, pp. 80–81).

Perhaps the most significant change was the development and dissemination of interactive geometry software, specifically such products as Cabri (Laborde and Bellemain 2005) and The Geometer's Sketchpad (Jackiw 1991). That change is reflected in this yearbook, in which nine of twenty-three articles refer to interactive geometry software as a tool for teaching and learning geometry.

Parallel with these developments in our understanding of students' learning and the availability of new tools for teaching, the field of geometry itself has experienced a revival. After languishing during of the early twentieth century as a field peripheral to mainstream mathematics (Sinclair p. 46), in the latter half of the century geometry again emerged as an area of research. A conference on "Geometry's Future" held in 1990 assessed the implications of this development on grades K–12 and university curricula (Malkevitch 1992).

Nevertheless, the state of the geometry curriculum remains unsettled much as it was twenty-two years ago. At that time Usiskin (1987, p. 20) observed, "There is lack of agreement regarding not just the details but even the nature of geometry that should be taught from elementary school through college" (Usiskin 1987, p. 20). This volume contains numerous articles with insights about teaching and learning but few that take a more global curricular perspective. A need still exists for a detailed discussion in the mathematics education community on what school geometry ought to be. We hope that the insights provided by the articles in this yearbook will contribute to that discussion.

This yearbook is divided into three parts. The first, "Expanding Visions of Geometry" attempts to bring us up to date in the work that today's geometers do. It focuses on topics in geometry that are current but not traditionally part of the curriculum. The section opener by Editorial Panel member Malkevitch gives an overview of problems contemporary geometers are working on. Schattschneider engages us in detective work to establish the existence of exactly seven different types of frieze patterns. Handa, James, and Mattman share the beauty and intrigue of the well-known Möbius strip, its extension to Möbius tori, and its applications in art and architecture. Iseri shows how to make the concept of curvature accessible to school students. Camp and Hauenstein reveal how fractal geometry can be used to model the structure of plants. Finally, O'Rourke poses an engaging question about folding polygons that is simple to state but remains a challenge.

Articles in the second section, "Learning Geometry," give important attention to the ways students perceive shape, location, angle, and other geometry concepts and processes. Editorial Panel member Battista opens the section with a summary of current research on learning school geometry, including an explanation of the van Hiele model, which appears in several subsequent articles. Yu, Barrett, and Presmeg describe research on how interactive geometry software affects students' ability to reason about categories. Browning and Garza-Kling report on studies of how school students and preservice teachers develop the concept of angle. Sack and van Niekerk share activities to develop children's spatial visualization abilities. Driscoll, Egan, DiMatteo, and Nikula detail a professional development program based on identifying and fostering students' "geometric habits of mind."

The third section, "Teaching Geometry for Understanding," brings us to the actual teaching of geometry. The section opener by Editorial Panel member Paniati relates one geometry teacher's evolution as a practitioner of discovery learning. DeVilliers, Govender, and Patterson share perspectives on how teachers can develop and give to students a nuanced view of definitions. Casa and Gavin then give examples of how to develop elementary school students' understanding of definitions for quadrilaterals. Hollebrands and Smith provide an overview of research on the use of interactive geometry software. Contreras and Martinez-Cruz collaborate in two distinct articles to show how interactive geometry software can be used to help students become better problem solvers and to invent their own theorems. In a similar vein, Quesada reports on discoveries students have made using interactive geometry software and implications for professional development. Blair and Canada demonstrate how one carefully chosen, open-ended problem can lead to a very rich exploration. Flores shows how interactive computer-generated figures can be used to develop and show connections among area formulas.

In the one article that discusses an "integrated" approach to the secondary school geometry curriculum, Wilson discusses the advantages and challenges of structuring a course around a set of carefully chosen problems. Davis shows how a traditional geometry lesson can be redesigned to promote students' more active involvement. The final article by Todd looks to the future, in which software integrating algebra and geometry may further extend teachers' capacity to use technology effectively to stimulate students' thinking.

At the outset I mentioned the three previous yearbooks (Reeve 1930; Henderson 1973; Lindquist 1987) that specifically refer to geometry in their titles. In addition, in the Thirteenth Yearbook, *The Nature of Proof* (Fawcett 1938, 2001), Harold Fawcett described an experiment in which a class of high school students constructed their own system of geometry—undefined terms, definitions, postulates, and theorems—from scratch. Many of the articles in the third section of this Seventy-first Yearbook emphasize the power that students gain when they are encouraged to "invent" their own mathematics—much in the spirit of Fawcett.

The preparation of this yearbook would not have been possible without the work of an outstanding editorial panel. Each person brought his or her unique perspective, and together they functioned as an effective team. The following individuals served as editorial panel members:

Michael Battista, Ohio State University

Earlene Hall, Detroit Public Schools, Detroit, Michigan

Joseph Malkevitch, City University of New York—York College

James Paniati, Northwestern Regional High School, Winsted, Connecticut

Ann Spinelli, Bristol Public Schools, Bristol, Connecticut,

I would also like to acknowledge the contribution of the editorial and production staff at the NCTM headquarters in Reston, Virginia. Ann Butterfield acted as project manager, and David Webb served as copyeditor. Randy White was responsible for the cover design and many of the figures that appear in this yearbook. David Barnes assembled the material that appears on the accompanying CD.

Above all, I would like to thank Rheta Rubenstein, University of Michigan—Dearborn, general editor, whose wisdom and guidance were invaluable.

Timothy V. Craine
Central Connecticut State University
Seventy-first NCTM Yearbook Editor

REFERENCES

Birkhoff, George D., and Ralph Beatley. "A New Approach to Elementary Geometry." In *The Teaching of Geometry,* Fifth Yearbook of the National Council of Teachers of Mathematics (NCTM), edited by William David Reeve, pp. 86–95. New York: Teachers College, Columbia University, 1930.

Brumfiel, Charles. "Conventional Approaches Using Synthetic Euclidean Geometry." In *Geometry in the Mathematics Curriculum,* Thirty-sixth Yearbook of the National Council of Teachers of Mathematics (NCTM), edited by Kenneth B. Henderson, pp. 95–115. Reston, Va.: NCTM, 1973.

Crowley, Mary L. "The van Hiele Model of the Development of Geometric Thought." In *Learning and Teaching Geometry K–12,* 1987 Yearbook of the National Council of Teachers of Mathematics (NCTM), edited by Mary Montgomery Lindquist, pp. 1–16. Reston, Va.: NCTM, 1987.

Fawcett, Harold. *The Nature of Proof.* Thirteenth Yearbook of the National Council of Teachers of Mathematics (NCTM). New York: Teachers College, Columbia University, 1938. Reprint 2001.

Henderson, Kenneth B., ed. *Geometry in the Mathematics Curriculum.* Thirty-sixth Yearbook of the National Council of Teachers of Mathematics (NCTM). Reston, Va.: NCTM, 1973.

Jackiw, Nicholas. The Geometer's Sketchpad. Software. Emeryville, Calif.: Key Curriculum Press, 1991.

Kenney, Margaret J. "Logo Adds a New Dimension to Geometry Programs at the Secondary Level." In *Learning and Teaching Geometry K–12,* 1987 Yearbook of the National Council of Teachers of Mathematics (NCTM), edited by Mary Montgomery Lindquist, pp. 85–100. Reston, Va.: NCTM, 1987.

Laborde, Jean-Marie, and Franck Bellemain. Cabri II. Software. Temple, Tex.: Texas Instruments, 2005.

Lindquist, Mary Montgomery, ed. *Learning and Teaching Geometry K–12.* 1987 Yearbook of the National Council of Teachers of Mathematics (NCTM). Reston, Va.: NCTM, 1987.

Malkevitch, Joseph. *Geometry's Future*. 2nd ed. Lexington, Mass.: Consortium for Mathematics and Its Applications, 1992.

National Council of Teachers of Mathematics (NCTM). *An Agenda for Action: Recommendations for School Mathematics of the 1980s*. Reston, Va.: NCTM, 1980.

———. *Curriculum and Evaluation Standards for School Mathematics*. Reston, Va.: NCTM, 1989.

———. *Principles and Standards for School Mathematics*. Reston, Va.: NCTM, 2000.

Reeve, William David, ed. *The Teaching of Geometry.* Fifth Yearbook of the National Council of Teachers of Mathematics (NCTM). New York: Teachers College, Columbia University, 1930.

Serra, Michael. *Discovering Geometry.* Berkeley, Calif.: Key Curriculum Press, 1989.

Sinclair, Natalie. *The History of the Geometry Curriculum in the United States.* Charlotte, N.C.: Information Age Publishing, 2008.

Smart, James, R. "Implications of Computer Graphics Applications for Teaching Geometry." In *Learning and Teaching Geometry K–12,* 1987 Yearbook of the National Council of Teachers of Mathematics (NCTM), edited by Mary Montgomery Lindquist, pp. 32–36. Reston, Va.: NCTM, 1987.

Usiskin, Zalman, "Resolving the Continuing Dilemmas in School Geometry." In *Learning and Teaching Geometry K–12,* 1987 Yearbook of the National Council of Teachers of Mathematics (NCTM), edited by Mary Montgomery Lindquist, pp. 17–31. Reston, Va.: NCTM, 1987.

Part I

Expanding Visions of Geometry

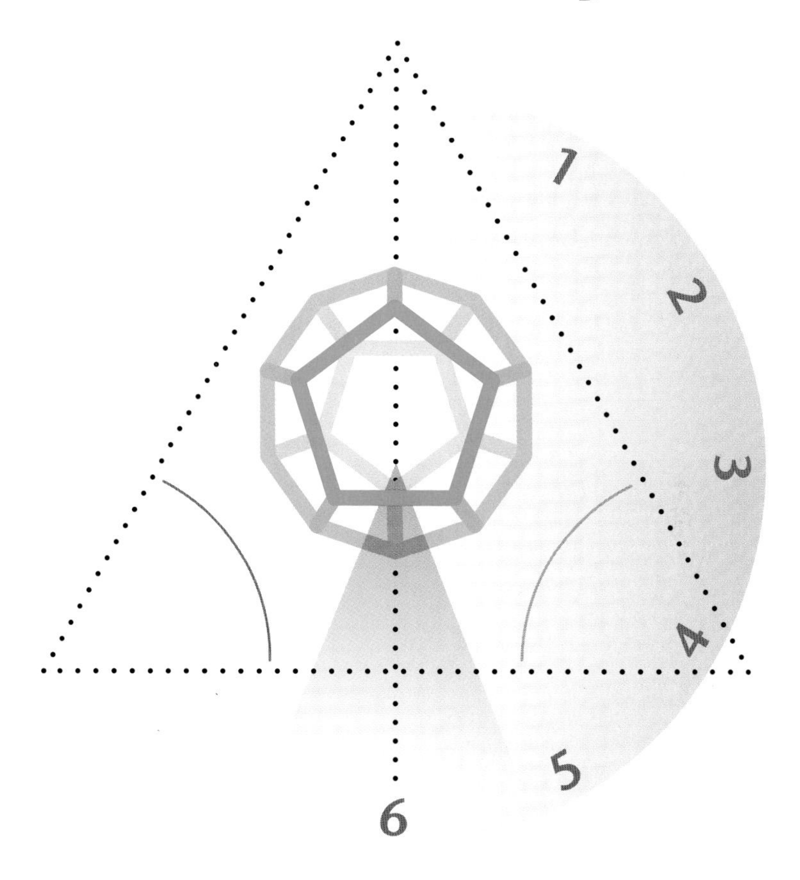

What Is Geometry?

Joseph Malkevitch

THE WORLD of geometry is changing. To illustrate the scope of this change, let us examine six problems that illustrate some questions of interest to contemporary geometers. The ideas on which these problems are based and their solutions—for those that have been solved, and not all of them have—can be explained in terms no more complicated than the geometry that has been part of the grades K–12 curriculum in the past. In what follows, I try to trace the transition between geometry as it developed over the past three millennia and some of the many developments in the field that have sparked the imagination of researchers in recent years. I hope to show that the geometry of today and tomorrow can be compelling, intriguing, and engaging to learn and to teach.

Some Geometrical Problems

Problem Situation 1: Guarding an Art Museum

Suppose that an n-sided polygon, such as the one in figure 1.1 where n is 20, represents the floor plan of an art museum or a bank. Guards or sensors are to be placed at some of the vertices so that if a "point intruder" is detected in the interior of the polygon, an alarm will sound. An interior point I is visible from a point P on the boundary, where P is not necessarily a vertex, if the line segment from P to I contains no points in the exterior of the polygon. A polygon with a set of locations for the sensors is said to be well guarded if every point in its interior is visible from at least one sensor, and hence any intruder in the interior will be detected.

Fig. 1.1. A non-self-intersecting, twenty-sided polygon

1. What is the smallest number of sensors located at vertices needed for the particular polygon in figure 1.1 to be well guarded?
2. The number found in question 1 above might not be sufficient for every twenty-sided polygon. What is the largest number of sensors located at vertices that might be required for some twenty-sided polygon to be well guarded?
3. Find a formula (depending on n) for how many guards (sensors) located at vertices of any n-sided polygon are needed to guarantee that an intruder into the interior of the polygon will be detected. Furthermore, devise a method for determining at which vertices of the "art gallery" the guards should be placed.

Problem Situation 2: Graphing Calculator Screen

The reason that graphs of lines on a calculator or computer screen sometimes look jagged is that the lines are represented by tiny squares called *pixels*. Which of the pixels should be "lit up" to represent a particular straight line that is to be displayed on the screen?

More specifically, suppose that the origin of an x-y coordinate system, (0, 0), represents the pixel in the lower-left corner of figure 1.2. Which cells shown should be lit up to represent the line $y = 2x + 1$? Obviously the pixels representing (0, 1), (1, 3), and (2, 5) should be included. However, if the line is to look connected, more pixels are needed. How do we determine which ones? How is this problem solved in general?

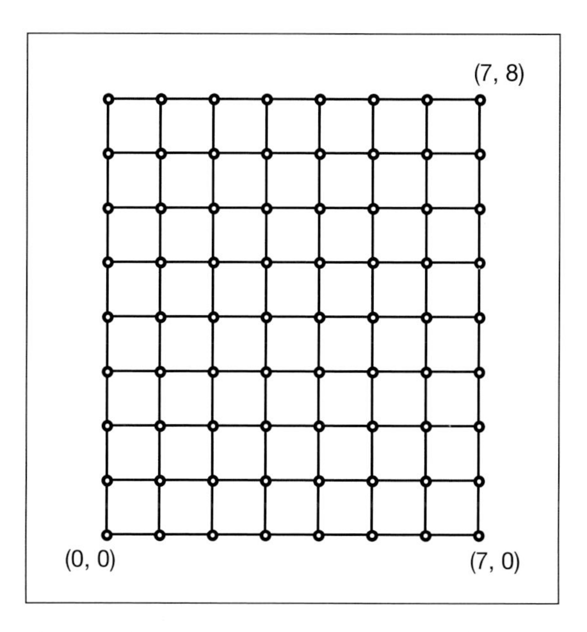

Fig. 1.2. Each pixel square is represented by the coordinates of its lower-left corner.

Problem Situation 3: Folding Boxes

Many polygonal nets will fold to a $1 \times 1 \times 2$ box (rectangular prism). Two such nets with fold lines are shown. Observe that one of the nets (fig. 1.3) contains all four 1×2 rectangular faces, whereas for the other net (fig. 1.4), two of

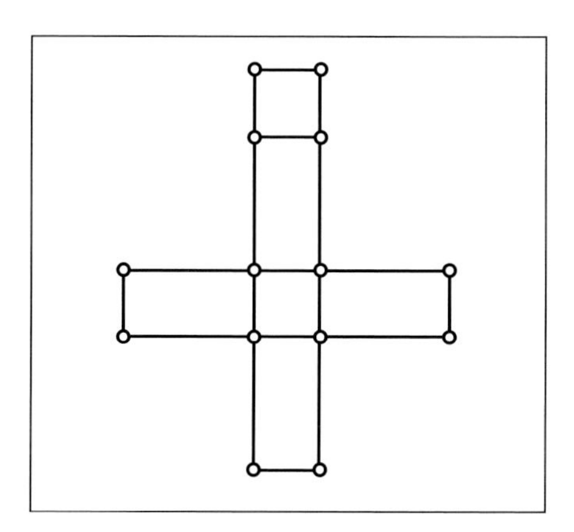

Fig. 1.3. A polygon that can be folded to a $1 \times 1 \times 2$ box

the 1×2 rectangular faces of the box are formed by gluing together two 1×1 squares. In the diagram in figure 1.4 we actually have a polyomino (a figure that is the union of 1×1 squares adjoined edge to edge, in this instance with a hole) rather than a polygon. The gray square is not part of the panels to be folded, and the two 1×1 squares that touch at point P are not attached there, but separated when the shape is folded into the $1 \times 1 \times 2$ box. The question arises, How many nonequivalent polygonal shapes can be folded to form a $1 \times 1 \times 2$ box? What about an $n \times m \times p$ box?

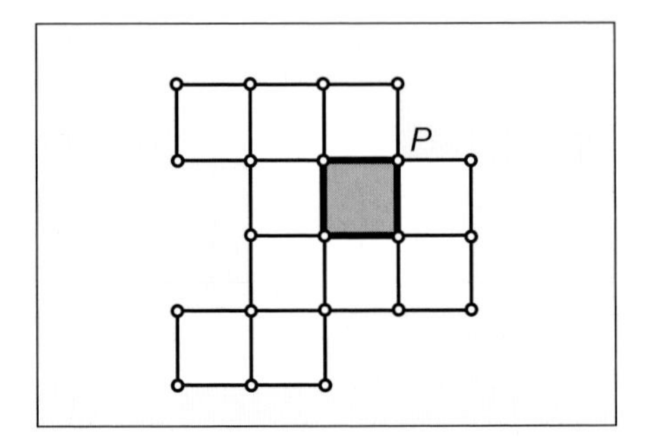

Fig. 1.4. A polyomino that can be folded to a $1 \times 1 \times 2$ box

Problem Situation 4:
Representing Three-Dimensional Polyhedra in the Plane

In problem situation 3 above, a three-dimensional polyhedron was represented by a two-dimensional *net* of polygons that cover the surface of the polyhedron when the net is folded. In this problem, a polyhedron is represented by a *graph* that shows only the vertices and edges and how they are connected.

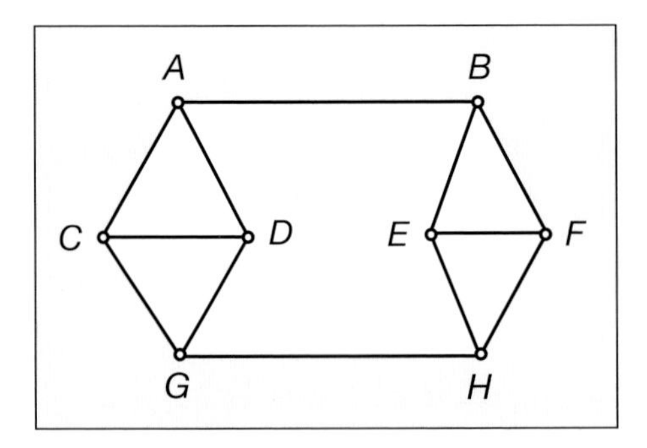

Fig. 1.5. Does this graph correspond to a convex, three-dimensional polyhedron?

1. Does a convex, three-dimensional polyhedron exist whose pattern of vertices and edges corresponds to the graph shown in figure 1.5? Such a polyhedron would have six faces—two hexagons (*ABEHGD* and *ABFHGC*) and four triangles.
2. Does a nonconvex polyhedron exist that this diagram represents?
3. Is there a general way to determine whether a given graph of vertices and edges represents a polyhedron?

Problem Situation 5:
Robot Arm

Geometry is being put to work at the International Space Station in the form of robotic arms to help the astronauts do their work. The reader can get some flavor of what is involved by considering a simplified problem.

The diagram in figure 1.6 shows a robot arm with three links, which are confined to move in the plane, one link of length 1 ($\overline{PA}$), one of length 4 ($\overline{AB}$), and one of length 2 ($\overline{BE}$). The arm is pinned down at *P* and can rotate about that point, but the two links at *A* are free to rotate an unrealistic 360 degrees with respect to each other, as are the two links at *B*. A pen is located at *E*, the "end effector," which can place an ink dot (or do some work, say, on a space station) at point *E*.

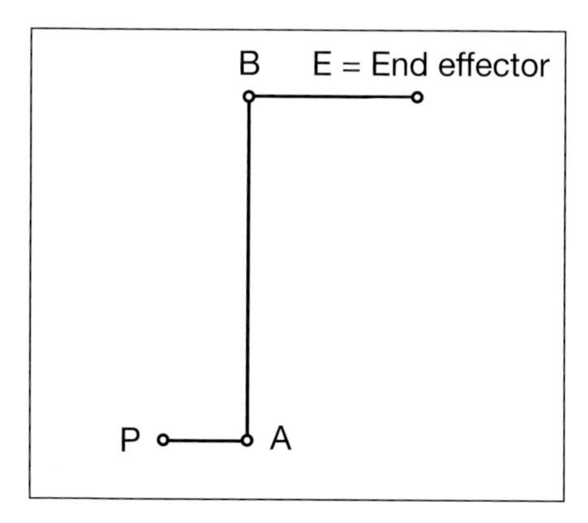

Fig. 1.6. A robot arm pinned at *P* but free to swivel at *A* and *B*

Suppose that *P* has coordinates (0, 0) and that the positive *x*-axis lies along $\overline{PA}$.

1. What are the coordinates of *A*, *B*, and *E*?
2. How far from *P* is the end effector *E* for the position of the robot arm shown?

3. Can one "reach" (e.g., place an ink dot) at (5, 6) using this arm?
4. Show that a dot can be placed at (1, 6) by indicating the angles between the links of the robot arm that accomplish this task. Use the convention of measuring angles in the counterclockwise direction from the positive x-axis. For the arm shown the settings are ($\overline{PA}$, 0°) at P; ($\overline{AB}$, 90°) at A; and ($\overline{BE}$, 0°) at B. If the arm is set at 0° at P, A, and B, then E will be at the point (7, 0).
5. Exactly what is the set of points where this robot arm can place a dot? A first guess might be the interior of the circle of radius 7 with center at the origin, but this answer is not quite correct.

Problem Situation 6:
Equidecomposability

Tangrams (fig. 1.7) are a collection of polygonally shaped regions that can be assembled to make a variety of different polygonal shapes. Although the perimeter of these shapes can vary, all shapes made using the set of seven tangram pieces have the same area. This observation is related to the general fact that if one cuts a polygonal region R into a finite number of polygonal pieces, all polygonal shapes one can assemble with the pieces made from R will have the same area.

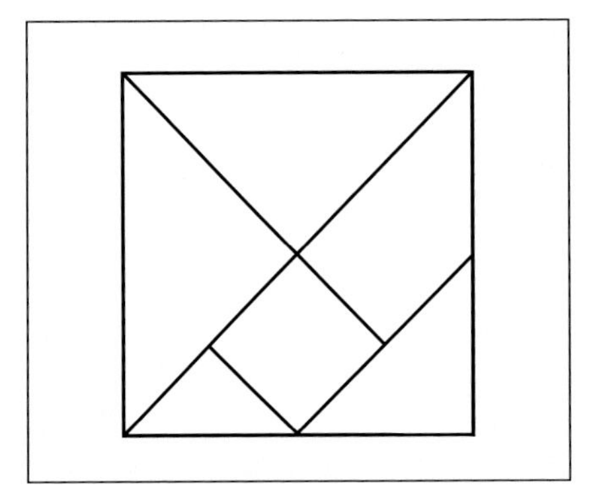

Fig. 1.7. The tangram pieces

However, what about the converse? Given any two polygons $P1$ and $P2$ with the same area, is it possible to cut $P1$ into a finite number of pieces and assemble these pieces in the style of a jigsaw puzzle to form $P2$? Two polygons that have this property are said to be *equidecomposable*.

Remarkably, the answer to this question is yes. To see that the problem involves some not totally transparent issues, we might consider the question of whether a rectangle whose sides have length $\sqrt{2}$ and $\sqrt{8}$ can be cut into a finite number of pieces and reassembled to form a square of side 2.

One approach to the general case is to show that any polygon can be subdivided into triangles, and each of these triangles can be shown to be decomposable into pieces that can be assembled into a rectangle with one of its sides having length 1. The rectangles formed from each of the triangles can then be glued together along the sides of length 1 to form a longer rectangle also with a side of length 1. Because *P1* and *P2* have the same area, they are both decomposable to the same rectangle with a side of length 1. When two polygons are each equidecomposable to a third polygon, they are equidecomposable to each other.

The analogous question in three dimensions—Can one decompose a three-dimensional polyhedron *Q1* into a finite number of polyhedral pieces to form a polyhedron *Q2* of the same volume?—is not universally true! For example, one cannot decompose a regular tetrahedron of volume 1 into a cube of volume 1. However, in the twentieth century the discovery was made that a ball (sphere together with its interior) can be decomposed into a finite number of disjoint parts and reassembled to form two spheres identical with the original one, a phenomenon known as the Banach-Tarski paradox.

The Geometric Tradition

Geometry's history helps us understand its progress and future. The word *geometry* is of Greek origin. *Geo* refers to the earth or land, and *metry,* to measure. The roots of the subject thus seem to be in physics. On the one hand, geometry can foster insight into empirical questions of the kind that concerned the Egyptians, who needed to keep track of the consequences of the Nile River's obliterating field boundaries in the spring or to help in constructing the pyramids. On the other hand, Euclid's *Elements*, undoubtedly one of the great landmarks in all intellectual history, was designed as a deductive system based on ideal notions of points, lines, and planes. To this day we do not know whether Euclid thought of his work as a compendium of "facts" about the world (i.e., physics) or whether he thought of geometry as a "game with rules," in which, beginning with a collection of simple assumptions, other results followed using the laws of logic.

Thus we think of geometry both as a description of physical space and as an axiomatic system. Throughout history, the development of geometry has exhibited this duality. Geometric researchers have both tried to understand the nature of the space in which we live along with the matter that makes it up (e.g., general relativity) and have investigated the many different kinds of geometry we now know are possible.

Geometry as a subject flowered during the Greek period up through the third century CE, after which progress in obtaining geometrical insights slowed down.

Meanwhile, Greek geometry was kept alive in the Middle East while developing separately in India and China (Joseph 2000). By the time of the Renaissance, geometry had been revived in Western Europe.

A new great stimulus to geometry came during the Renaissance in Italy and Germany. As a response to the challenges of building new palaces and cathedrals and of drawing realistic paintings of architectonic settings, the foundation of what is now called *projective geometry* was laid. If we take the points and lines of Euclidean geometry and modify them, we can construct "real projective geometry." Here is the idea. We partition all lines in the plane into classes, in which each class consists of all lines with a given direction. Then all the lines within the same class are parallel. To each line in one of these classes L, we add a single "point" L_∞ (shown with a small circle in fig. 1.8). We think of all such points as lying on a new "line," a "line at infinity," corresponding with the horizon in perspective drawing (fig. 1.8).

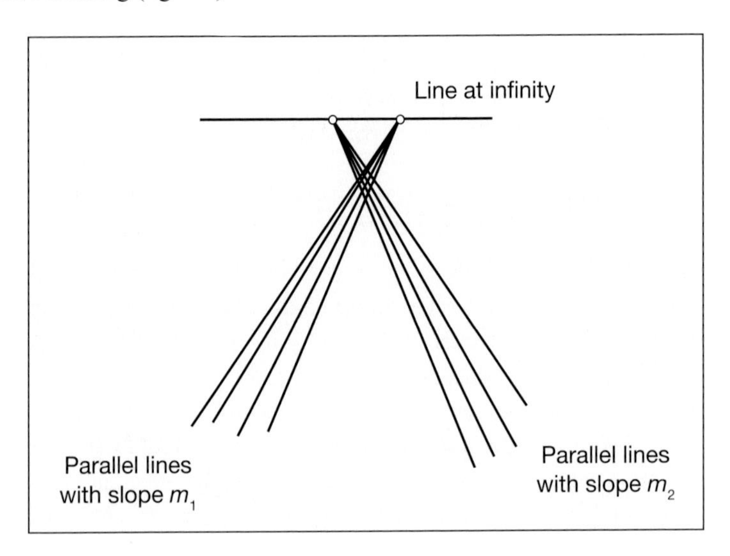

Fig. 1.8. Classes of mutually parallel lines are modified by adding a single point "at infinity" belonging to each of the parallel lines. All such new points are placed on a single line "at infinity."

Projective geometry consists of all the points and lines of Euclidean geometry with the addition of the new points (L_∞ for each set of parallel lines L) and the "line at infinity." We now easily see that no parallel lines exist in this geometry, because two projective lines arising from Euclidean lines that were not parallel naturally meet at a Euclidean point. However, two lines that were originally in the same class of parallels now meet at the "new" projective point we added to each of these lines. Finally the "line at infinity" intersects every other line at one of the new projective points. Although the formal development of projective geometry awaited a future day, the Renaissance artists had seen beyond Euclid and raised the possibility that the "physics" of space might actually be projective, although this observation is only implicit in the work of the time.

The fact that different kinds of geometry can be considered candidates for the geometry of physical space was the result of an intellectual revolution culminating in the early nineteenth century. This revolution resulted from the failure of attempts to show that Euclid's fifth postulate—in recent times often stated that if P is a point not on a line l, there is a unique line through P that is parallel to l—could be proved from Euclid's other axioms. In the early twentieth century, David Hilbert provided a rigorous axiomatization of the geometry Euclid was describing in *Elements*. Hilbert's work added to and summarized the work of others, including Gauss, Bolyai, Lobachevsky, and Reimann, and included a demonstration that the fifth postulate is independent of the others, so geometries rejecting it were sound.

As a result we have an abundance of geometries today—Euclidean, hyperbolic, elliptic, projective, finite—each based on its own set of axioms. These geometries are typically studied in college geometry courses.

Practicing geometers, however, do not typically go back to axioms for doing their research. They have a "toolkit" of theorems and methods that serve as a launch pad for proving new results. These investigations might come up in the course of designing some special effects for a Hollywood motion picture, devising a new compression system for high-definition television, creating a reliable face-recognition software package, or proving an extension of known results about "guarding polygons."

Why Has the Word *Geometry* Taken on a Broader Meaning? The Concept of Distance

Many of the theorems that grew out of the great Euclidean tradition deal with distance. When points are represented by coordinates, the distance is computed using the distance formula, $d = \sqrt{(x_2 - x_1)^2 + (y_2 - y_1)^2}$, which is based on the Pythagorean theorem. Different ways exist, however, to measure the distance between two points. Suppose that in an urban area all streets run north and south or east and west, forming a large square grid with blocks one tenth of a mile long. Suppose that the hospital for this community is located at (0, 0) and an accident occurs at (50, 120), how far must the ambulance go to get from the accident to the hospital? If the ambulance is a helicopter, it can fly as the crow flies, and the Pythagorean theorem helps us see that the Euclidean distance is 130 units, or

13 miles. However, if the ambulance is a vehicle with wheels confined to the urban road system, then 130 is not the correct answer. The distance will instead be 170 units, or 17 miles.

A geometry based on this metric, or distance function, is sometimes called *taxicab geometry*. In this geometry we have a radically different notion of what a circle is. Consider the set of points in the coordinate plane whose distance from the origin, in this system, is one unit. This "taxicab circle" takes on the shape shown in figure 1.9.

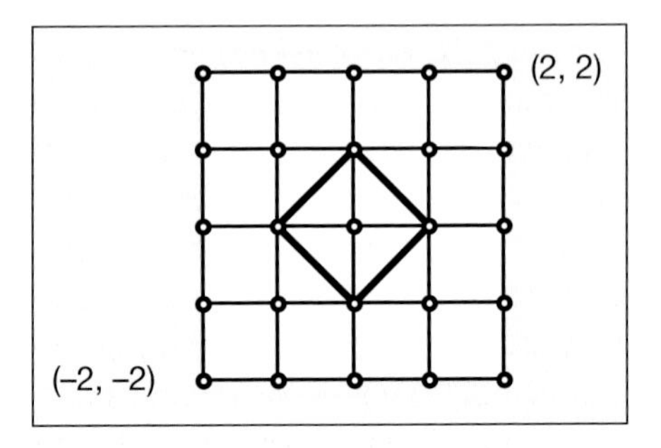

Fig. 1.9. A "circle" in taxicab geometry

Next let us examine another notion of "distance." When a young child looks at the two leaves in figure 1.10, the child may accurately report that the one on the left is an oak leaf whereas the one on the right is a maple leaf. This identification is accomplished despite the fact that two leaves from the same tree are never identical. We have little insight into how children or adults achieve this amazing feat. Describing the difference between the shapes of these two leaves, such as over the phone to an intelligent alien in another star system who understands English perfectly, goes beyond our linguistic skills. To get the idea across, we would send pictures and hope our new alien friends could "see."

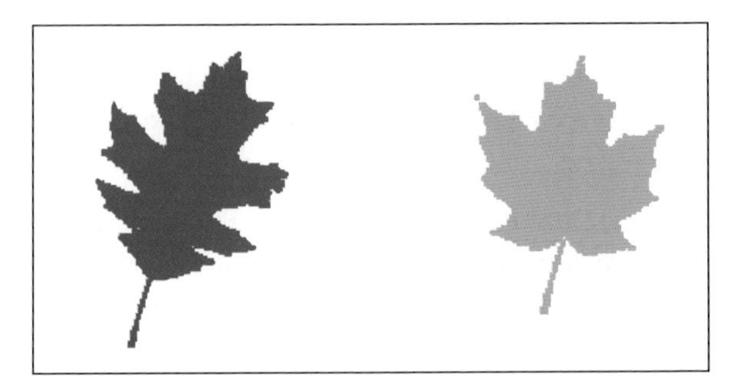

Fig. 1.10. An oak leaf and a maple leaf

Computer programs, however, can now take the coordinates of essential points on a leaf and use a rather complex algorithm to compute the "distance" of the given leaf from a prototypical oak or maple leaf. The program then compares the distances and "recognizes" the leaf as being a maple leaf or an oak leaf on the basis of these calculations. The ideas necessary to tell the "distance" between leaves from different species of trees might well be of use in defining the "distance" between two faces that would be the basis for software used to identify people uniquely.

Yet another application of distance is found in a word processing program's spell checker. A spell checker is designed with this purpose in mind: if a word keyed by the user does not appear in the word processor's dictionary, the software should determine which words in the dictionary to suggest as possibilities for what the user meant to write.

Geometric thinking can help us understand what is involved. Given two strings of the same length whose letters are from some common alphabet, we can compute the *Hamming distance* between them. The Hamming distance is found by lining up the two strings and counting the number of columns (positions) at which they differ. For example, for the two strings 11101 and 10100 in a binary alphabet, the Hamming distance is 2 because the strings differ in columns 2 and 5. Similarly with the English alphabet, the Hamming distance between "vowel" and "towel" is 1.

We may think of all the strings that are at most k units from a given string C as making up the "interior" of a Hamming circle of radius k, center at C. If a keyer makes only one mistake when keying a string, then the word that was meant to be keyed should be in the dictionary and lie on the Hamming circle of radius 1. Thus, if we make the simplified assumption that the only error in a typed word is that some letter was keyed incorrectly, a simple spell checker can be designed as follows.

- Step 1. Check the keyed word to see whether it appears in a dictionary of legal words that might have been keyed.
- Step 2. If the keyed string does not appear in the dictionary, generate a list of all strings in the dictionary that are Hamming distance 1 from this string and indicate this list to the keyer.

The user can now select the intended word from the list of proposed "legal words." Thus if one keys SMEAK (which is not a word in English), one gets the list SMEAR, SNEAK, SPEAK, and STEAK.

This approach is naive because mistakes can also occur by keying an extra letter, not keying enough letters, or using a phonetic spelling for the intended word. Thus, the user might enter the string "fonetic" into the word processor in the hope that although this string is one letter shorter than the word intended, the tool might suggest the correct spelling. In fact, on my computer when I enter the string "fonetic," the spell checker helpfully suggests that I meant to type "phonetic." This response means that the spell checker, in addition to using ideas related to Hamming distance, also has a way of looking for strings in its dictionary that are "phonetically" close to the string entered.

Conclusion

Do new theorems remain to be found in the spirit of those that Euclid codified in *Elements* thousands of years ago? Certainly. Some geometers are still working on geometrical questions of this type. Are geometers examining the structure of the axioms of geometric systems? Again the answer is yes. Yet most of the progress made in mathematics is not that old questions that have gone unresolved are being answered, but rather that new questions that have not been thought of before are being addressed. Modern geometers for the most part are asking and answering different kinds of questions from the ones that occupied Euclid and geometers up to the start of the twentieth century.

So, what is geometry? Years ago an informal definition might have been that it was the branch of mathematics devoted to the study of shapes and space. Now, however, a more apt definition might be "the branch of mathematics that studies visual phenomena." No brief set of words will fully do the job, but the foregoing situations and discussion support this broader view (Linquist 1987; Malkevitch 1992; Mammana and Villani 1998).

Coda: "From The Book"

Perhaps the reader is hoping that the problems posed at the start will all be resolved. I will leave it to the reader to think more about, and consult the references to get further insight into, these questions. But the developments concerning the first problem are so elegant that perhaps I will be forgiven for letting the cat out of the bag.

Paul Erdös (1913–1996), the famous prolific Hungarian combinatorist, had a metaphor that the most elegant ways to demonstrate some mathematical result were kept in a special book by God. When a proof was particularly elegant, Erdös would exclaim of the proof, "Straight from The Book!" Here is a proof of what has come to be called the Klee-Chvátal Art Gallery theorem. The problem is due to Victor Klee; the first proof, to Vašek Chvátal; and the proof that I outline below, "from The Book," is by Steve Fisk.

The fact that any plane polygon can be triangulated is well known. So we first begin by triangulating the art gallery. The result for a simpler polygon than in figure 1.1 is shown in figure 1.11, in which $n = 8$. Next we easily see that we can color the vertices of the triangulated polygon with exactly three colors. We can start by labeling the vertices of any triangle in the triangulated polygon with the colors *a, b,* and *c*. If we color each vertex with *a, b,* and *c*, then only one way exists to continue the coloring so that if two vertices are joined by a line segment, they get different colors.

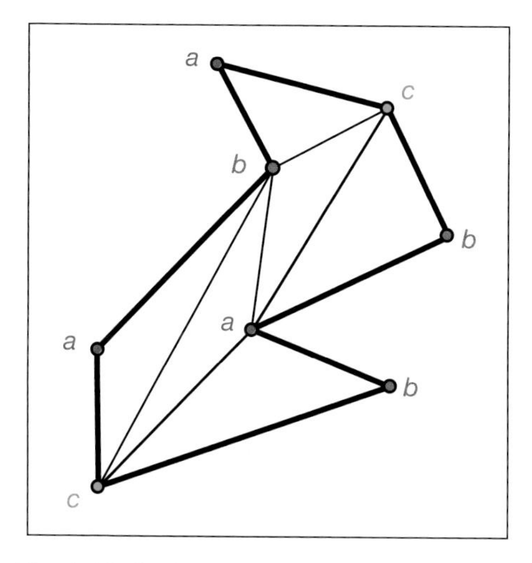

Fig. 1.11. An 8-gon, which can be guarded by two guards, who can be placed at the vertices labeled *c*.

Many triangulations may be possible for a given polygon, but once the triangulation is done, essentially only one way exists to three-color the vertices. Next we place guards at those vertices colored with a color that appears least often. In our example, if we put guards at the two vertices that are colored *c*, we are certain that guards at these two vertices "see" the complete interior of the polygon. In the general case, the number of guards required can never be more than the floor function $\left\lfloor \frac{n}{3} \right\rfloor$. The floor function of an integer k is k, whereas the floor function of a number between k and $k + 1$ is k. We can easily see that for any n, some polygon with n sides exists that requires the floor function of $n/3$ guards. The foregoing argument shows that we never need more than this number of guards. As the reader can see, this proof is, indeed, "from The Book!"

Readers are encouraged to learn more by consulting works in the reference list. Art-gallery (guarding) problems are discussed in O'Rourke (1987), Urrutia (2000), and Goodman and O'Rourke (2004). Problems involving nets, folding, and unfolding are presented in Demaine and O'Rourke (2007). The robot-arm questions and robotic problems are discussed in Meyer (2006) and Goodman and

O'Rourke (2004). The reader can learn more about different notions of distance in Krause (1975) and Gusfield (1997), whereas equidecompatibility is treated in Hartshorne (2006).

Geometers are at work trying to discover other truths in that Book that are still hidden from us. Engaging students in these emerging problems and efforts toward their solutions should enrich both teaching and learning geometry.

REFERENCES

Demaine, Erik, and Joseph O'Rourke. *Geometric Folding Algorithms: Linkages, Origami, and Polyhedra*. Cambridge: Cambridge University Press, 2007.

Goodman, Jacob, and Joseph O'Rourke, eds. *Handbook of Computational Geometry*. 2nd ed. Boca Raton, Fla.: Chapman and Hall/CRC, 2004.

Gusfield, Dan. *Algorithms on Strings, Trees, and Sequences: Computer Science and Computational Biology.* Cambridge: Cambridge University Press, 1997.

Hartshorne, Robin. *Geometry: Euclid and Beyond*. New York: Springer, 2000.

Joseph, George G. *The Crest of the Peacock: Non-European Roots of Mathematics*. Princeton, N.J.: Princeton University Press, 2000.

Krause, Eugene F. *Taxicab Geometry*. Menlo Park, Calif.: Addison Wesley Publishing Co., 1975.

Lindquist, Mary, ed. *Learning and Teaching Geometry, K–12.* 1987 Yearbook of the National Council of Teachers of Mathematics (NCTM). Reston, Va.: NCTM, 1987.

Malkevitch, Joseph. *Geometry's Future*. 2nd ed. Lexington, Mass.: Consortium for Mathematics and Its Applications, 1992.

Mammana, Carmelo, and Vinicio Villani. *Perspective on the Teaching of Geometry for the Twenty-first Century*. Dordrecht, Netherlands: Kluwer Academic Publishers, 1998.

Meyer, Walter. *Geometry and Its Applications.* 2nd ed. New York: Elsevier Academic, 2006.

O'Rourke, Joseph. *Art Gallery Theorems and Algorithms*. Oxford, U.K.: Oxford University Press, 1987.

Urrutia, Jorge. "Art Gallery and Illumination Problems." In *Handbook of Computational Geometry*, edited by Jorg-Rudiger Sack and Jorge Urrutia, pp. 972–1027. Amsterdam, Netherlands: Elsevier Science Publishers, 2000.

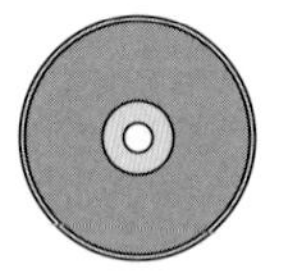

Blackline masters of the Art Gallery and Folding Boxes problems are found on the CD-ROM disk accompanying this Yearbook.

Enumerating Symmetry Types of Rectangle and Frieze Patterns:

How Sherlock Might Have Done It

Doris Schattschneider

One topic in the geometry curriculum in both middle and high school is *symmetry*, that aesthetic component of designs and patterns that can be defined mathematically. Mathematicians say that a design or pattern in the plane *has symmetry* or *is symmetric* (an attribute) if a rigid motion (a distance-preserving transformation, also called an *isometry*) exists that moves individual points in the design to new positions in the design while mapping the whole figure onto itself. Thus the motion transforms the design so that it is superimposed exactly on itself and some, but not all, points may stay fixed. After the transformation is performed, the design looks as though nothing at all has happened. The transformation that leaves the design unchanged is also called *a symmetry of the design.* Both the attribute and the action share the same name. In figure 2.1, the design in the rectangle has reflection symmetries and 180-degree rotation symmetry; the frieze pattern it can generate has these symmetries and others, including translation symmetries.

Fig. 2.1. A rectangle with a symmetric design and a periodic frieze pattern generated by it

Students first learn to recognize symmetries of designs by looking at many examples, testing the designs with mirrors, folding them, turning them around, rotating them back and forth, or manipulating traced or acetate copies of the designs on top of the original. Later they learn careful definitions of the four possible isometries of the plane: reflection, rotation, translation, and glide reflection. The definitions of these transformations can be given either synthetically or analytically. For example, a reflection in a mirror line m in the plane can be defined as the transformation that sends each point P not on m to its image P' where m is the perpendicular bisector of the segment PP', and keeps P fixed if P is on m ($P = P'$). The analytic definition of a reflection is given as a function of the coordinates of points in the plane; for example, a reflection in the line m with equation $y = x$ is defined by the rule $(x, y) \to (y, x)$. Reflections and rotations that leave the origin, (0, 0), fixed can be defined by multiplication of coordinate vectors by 2×2 matrices.

The study of these special transformations and the symmetry of designs enriches the study of functions, of synthetic geometry, and of analytic geometry, and it connects geometry directly with an important component of artistic design. In this article, I illustrate by asking and answering several questions, first about rectangles with a design, and then about frieze patterns.

- How many different symmetries does a rectangle have? What are they?
- What happens when you combine two of those symmetries, performing the transformations one after the other?
- How can symmetry be used to differentiate (or classify) designs?
- How many "essentially different" decorated rectangles are possible? First we need to define "essentially different."
- What about frieze patterns? Answer questions similar to those above.

Symmetries of a Decorated Rectangle

Discovering the symmetries of an arbitrary rectangle (one that is not a square) is an easy exercise that can be done at any level; having a paper rectangle to manipulate makes it easier. The identity transformation leaves every point of the figure fixed, so it is a symmetry of the rectangle. *Just three nonidentity symmetries of the rectangle exist* (assume it is positioned with one edge horizontal; see fig.2.2):

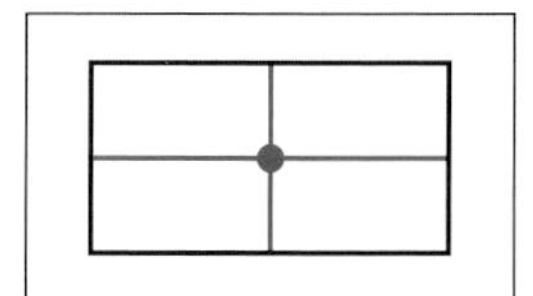

Fig. 2.2. Rectangle with center of rotation symmetry and lines of reflection symmetry

- reflection in the horizontal line bisecting two sides (H),
- reflection in the vertical line bisecting two sides (V), and
- 180-degree rotation (half-turn) about the center (R).

Many students want to believe that reflection in a diagonal is a symmetry of the rectangle, but it is not. To see this fact, simply fold a rectangle (that is not a square) along a diagonal. The rectangle does not fold onto itself.

Although folding and rotating are convincing ways to verify that H, V, and R are symmetries of a rectangle, formal proofs can be given using the definitions of reflection and rotation. (Here, the coordinate arguments are especially simple when the rectangle is centered at the origin.) To see that no other nonidentity symmetries of the rectangle are possible, note that any isometry mapping the rectangle onto itself must map the center of the rectangle onto itself—so only reflections and rotations leaving the center fixed can be symmetries of the rectangle. The facts that edges must be mapped to edges and that length cannot be changed leave only the three possibilities listed.

Next consider a rectangle with a simple decoration in it, where the frame of the rectangle and everything in its interior is considered to be part of the decoration. What are the possible symmetries of such a figure? Any symmetry of a decorated rectangle must superimpose the rectangle and its decoration exactly on the original figure, so the only possible symmetries are the identity, any of the three symmetries listed above (H, V, or R), or some combination of these. Figure 2.3 shows an example of a decorated rectangle with no nonidentity symmetry; we say the design d is *asymmetric*.

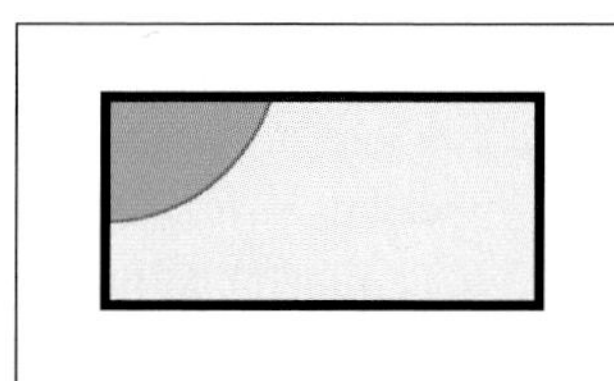

Fig. 2.3. Asymmetric design d

Three other designs, shown in figure 2.4, each have exactly one symmetry: *H*, or *V*, or *R*. In fact, these isometries can also be used to create each design from the asymmetric design *d* in figure 2.3. For example, if we take the design *d*, perform reflection *H*, and superimpose the image on the original design, we obtain the first decorated rectangle in figure 2.4. We appropriately name this design *H*(*d*). The decorated rectangles *V*(*d*) and *R*(*d*) are obtained in a similar manner from *d*, applying the transformations *V* and *R*, respectively.

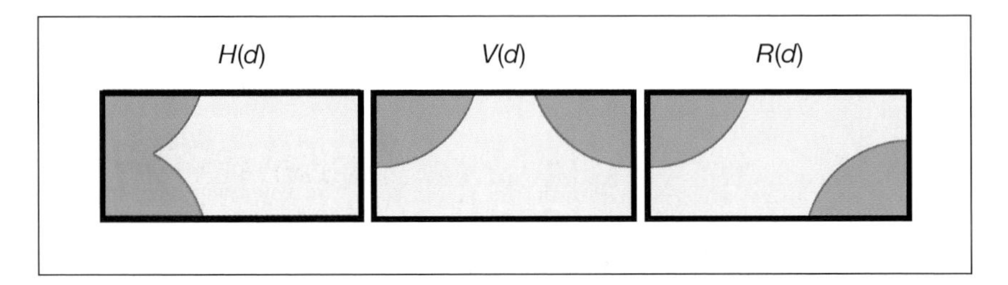

Fig. 2.4. Three designs having exactly one symmetry: *H*(left), or *V*(center), or *R*(right)

Is it possible for a decorated rectangle to have exactly two of the three possible symmetries of the rectangle? *No*—if it has two of these symmetries, then it must have all three because the composition of any two produces the third. The following theorem can be discovered by using interactive geometry software to carry out the two-part composite transformations on a simple asymmetric design, or by using overlaid acetate transparencies. It can be proved carefully using the property that isometries preserve distance and angles, arguing on congruent triangles (Yaglom 1962), or it can be proved by analyzing the composite transformations in analytic form (Martin 1982).

Theorem 1: (1) The composition of *V* and *H* (in either order) is *R*.
(2) The composition of *V* and *R* (in either order) is *H*.
(3) The composition of *H* and *R* (in either order) is *V*.

Theorem 1 implies that a decorated rectangle is either asymmetric, or has exactly one nonidentity symmetry, or has all three symmetries, *H*, *V*, and *R*, as illustrated in figure 2.5. The design in figure 2.5 can be obtained by acting on the design *d* first by one of these symmetries, and then acting on the transformed image by another of the symmetries.

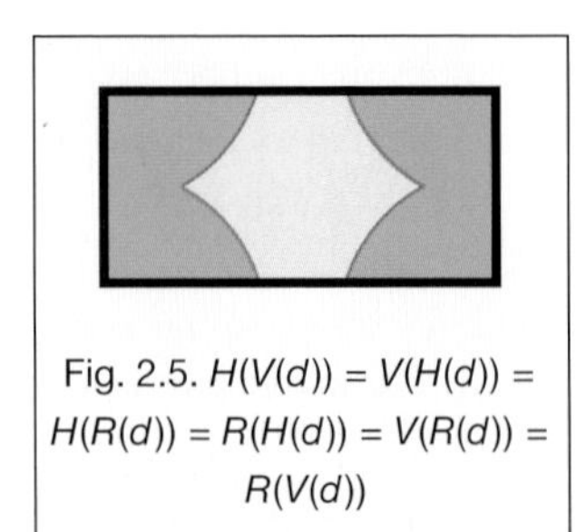

Fig. 2.5. *H*(*V*(*d*)) = *V*(*H*(*d*)) = *H*(*R*(*d*)) = *R*(*H*(*d*)) = *V*(*R*(*d*)) = *R*(*V*(*d*))

Mathematicians, as well as most scientists, like to classify objects according to certain attributes. One method of classification of designs and patterns is to sort them according to their symmetries. Two designs are declared to be of the *same symmetry type* if they have the same symmetries. So, for example, any decorated rectangle that has all three symmetries, *H, V,* and *R,* would be of the same symmetry type as our example in figure 2.5. The flags of Jamaica and Israel in figure 2.6 have the same symmetry type as the design in figure 2.5.

Fig. 2.6. Top: flag of Jamaica; bottom: flag of Israel

We want to answer the question "How many essentially different decorated rectangles are possible?" We will say that two designs are "essentially the same" when they have the same symmetry type. So our question is already answered! Five essentially different decorated (nonsquare) rectangles are possible; these are illustrated in figures 2.3, 2.4, and 2.5. Note that this classification according to symmetries is very crude; it gives no detailed information about the artistic qualities of the design, or about the shape of the rectangle. Within each of the five classes of symmetry types are an infinite number of possible shapes and decorations of rectangles.

We have considered only designs that are decorated rectangles; most often designs are not framed in a rectangle. But any design that can be contained in a finite region of the plane and is either asymmetric, or has only one reflection symmetry, or one half-turn symmetry, or two reflection symmetries with their mirror lines perpendicular (and hence a half-turn symmetry as well) can always be positioned in a rectangle in which those symmetries coincide with *H, V,* and *R*. All such patterns are classified as being one of the five symmetry types for decorated rectangles. Figure 2.7 shows a few such designs; all are highway signs, the last one found in Bethlehem, Pennsylvania.

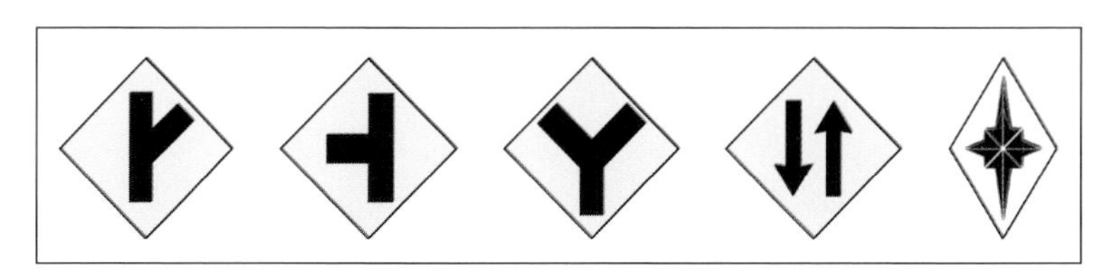

Fig. 2.7. Highway signs that illustrate the five symmetry types for decorated rectangles

We note that the collection of all symmetries of a design or pattern is known to form what mathematicians call a *group*—that is, the collection has an algebraic structure in which compositions of its elements satisfy rules similar to those of addition of integers, with the exception that composition is not always commutative. In college-level mathematics, symmetry groups are typically studied in abstract algebra. With this terminology, designs and patterns are often classified according to their symmetry group. In many instances, mathematicians do not differentiate between decorated rectangles with only H symmetry and those with only V symmetry—after all, each has only one reflection symmetry. The geometric classification (by symmetry types) and the algebraic classification (by symmetry groups) differ in recognizing that H and V are different reflections.

Symmetries of Frieze Patterns

Next let us consider frieze (or strip, or border) designs and patterns. These are defined as designs contained between two parallel lines that extend infinitely in both directions, like an infinitely long flowered ribbon (see fig. 2.8). Portions of such patterns are found in most cultures: framing book pages, paintings, doors, and ceilings; edging porches, garments, furniture, jewelry, and pottery; crowning buildings; wrapping around gears, pulleys, and waists.

Fig. 2.8. A classic frieze pattern

In analyzing such patterns, we can never see more than a finite portion, but for our mathematical classification by symmetries, the whole infinite patterned strip is considered (including the top and bottom edges that are the parallel lines enclosing it). We will orient such an infinite strip so that its edges are horizontal, and call the line equidistant from the two edges its *midline*. As with decorated rectangles, we consider the edges of the strip and the undecorated area enclosed by the strip as part of the design.

Since a horizontal strip formed by two parallel lines may be considered a rectangle with fixed height and infinite width, the symmetries of a rectangle are also symmetries of the strip:

- reflection in the midline, which we denote as H;
- reflection in a line perpendicular to the edges of the strip; and
- half-turn about a point on the midline.

Students can easily discover these symmetries by manipulating a strip of paper or tape; ask them to describe the precise locations of the mirror lines of the reflections and centers of half-turns. What is so different from the collection of all symmetries of the rectangle? The infinite length makes all the difference. An infinite number of half-turns are symmetries of the strip, not just one—each point on the midline can be chosen as a half-turn center. And similarly, an infinite number of reflection symmetries have their mirror lines perpendicular to the edges of the strip. Students will discover that the infinite length of the strip also allows such transformations as translations that slide the strip along itself—more on these transformations subsequently.

We already know that five symmetry types of decorated rectangles are possible, so at least those five types of decorated strips exist. These kinds of patterns are indeed just like decorated rectangles that are very long and skinny. Figure 2.9 shows some examples of such designs, all created using variations of the function $f(x) = x/(1 + x^2)$.

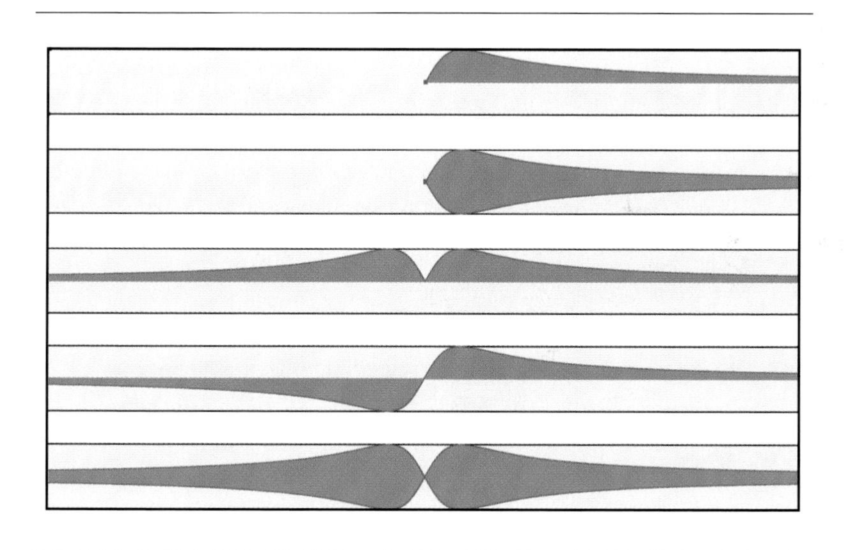

Fig. 2.9. Decorated infinite strips of the same five symmetry types as decorated rectangles

Frieze patterns with only a finite number of symmetries are often not counted in the classification of all frieze patterns by symmetries, but because they occur with great frequency in architectural and furniture decoration, we include them here.

Unlike a rectangle, an infinite strip bordered by two parallel lines can be translated horizontally (in the direction of its edges) by any amount to be superimposed on itself. An infinite number of these translation symmetries of the strip are possible. And these can be combined with any of the reflection symmetries and half-turn symmetries of the strip to produce new symmetries. So a frieze pattern would seem to have an infinite number of different symmetries! In classifying frieze patterns according to their symmetries, however, we will not try to count actual individual symmetries but only the *kinds* of symmetries the pattern has. To do so, we need to know what possible symmetries result when any two symmetries are combined, that is, performed (nonstop) one after the other.

Although many students are familiar with three isometries—translation, rotation, and reflection—a fourth, lesser-known isometry, called *glide reflection*, exists. It is defined as the composition of a translation and a reflection in a mirror line parallel to the direction of the translation. Thus, when a translation symmetry of an infinite strip is combined with the reflection symmetry H, a glide reflection results. Figure 2.10, left side, shows how a glide reflection acts on a single rectangle—the arrow is the glide vector (the translation vector of the glide reflection), and the horizontal red line is the "glide mirror" in which the translated rectangle is reflected. Here the glide mirror is below the rectangle, and so the rectangle and its glide-reflected image are staggered like footprints. When the glide mirror is the midline of the rectangle and the glide vector is the width of the rectangle, then the glide-reflected image of the rectangle adjoins the rectangle. Repeating this glide reflection produces the frieze pattern on the right side of figure 2.10; this pattern has glide reflection symmetry, in which panels alternate up-down-up-down.

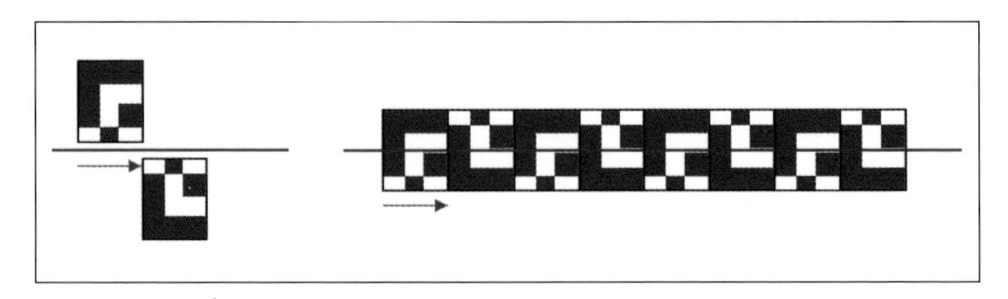

Fig. 2.10. Producing a frieze pattern with glide-reflection symmetry

Although many other combinations of symmetries can be considered, it turns out that the reflections, half-turns, translations, and glide reflections we have discussed are the only possible symmetries of an infinite horizontal strip—no others are possible. The general results are in the following theorem. As with the symmetries of a rectangle, these results can be discovered or demonstrated by performing the two-part transformations on a simple asymmetric design with computer software, or by drawing the sequence of images of a design on graph paper, or by using the analytic definitions of the transformations.

Theorem 2: The transformations in table 2.1 are symmetries of an infinite horizontal strip. A "vertical reflection" is a reflection in a line perpendicular to the edges of the strip, and H denotes a reflection in the horizontal midline of the strip. A "half-turn" is a 180-degree rotation about a point on the midline of the strip. Table 2.1 shows all possible compositions of two of the nonidentity symmetries of the strip: for each of the fifteen rows, the transformations given in the row in the three columns Transformation X, Transformation Y, and Transformation Z fill in the following sentence:

The composition of an X and a Y (in either order) is a Z.

Table 2.1
Compositions of Symmetries of an Infinite Strip—the Composition of an X and a Y (in Either Order) Is a Z

Transformation X	***Transformation Y***	***Transformation Z***
translation	translation	translation or identity
translation	H	glide reflection
translation	vertical reflection	vertical reflection
translation	half-turn	half-turn
translation	glide reflection	glide reflection or H
H	H	identity
H	vertical reflection	half-turn
H	half-turn	vertical reflection
H	glide reflection	translation
vertical reflection	vertical reflection	translation or identity
vertical reflection	half-turn	glide reflection or H
vertical reflection	glide reflection	half-turn
half-turn	half-turn	translation or identity
half-turn	glide reflection	vertical reflection
glide reflection	glide reflection	translation or identity

A more precise statement of theorem 2 can be given, supplying much more detailed information about the composite transformation in each case, such as the length of the translation vector, the location of half-turn center, or the location of mirror line for a vertical reflection. A nice way to do so is to make the midline of the strip a coordinate axis. The choice of origin and unit length can be arbitrary. Then the following notations can describe precisely the symmetries of the strip:

H: reflection in the midline $y = 0$

V_c: reflection in the vertical line $x = c$

R_c: half-turn about the point $(c, 0)$ on the midline

T_c: horizontal translation a directed distance of c units

G_c: glide reflection that is the composition of H and T_c

Table 2.2 uses this notation to display all twenty-five possible compositions of these symmetries; each cell's entry is the result of first performing the transformation in a column, followed by a transformation in a row. The entries in the cells of table 2.2 can be verified either by observation using interactive geometry software or by careful synthetic or analytic proof.

Table 2.2
Compositions of Pairs of Symmetries for a Frieze Pattern

$\circ$	H	V_a	R_a	T_a	G_a
H	I	R_a	V_a	G_a	T_a
V_b	R_b	$T_{2(b-a)}$	$G_{2(b-a)}$	$V_{b-a/2}$	$R_{b-a/2}$
R_b	V_b	$G_{2(b-a)}$	$T_{2(b-a)}$	$R_{b-a/2}$	$V_{b-a/2}$
T_b	G_b	$V_{a+b/2}$	$R_{a+b/2}$	T_{a+b}	G_{a+b}
G_b	T_b	$R_{a+b/2}$	$V_{a+b/2}$	G_{a+b}	T_{a+b}

Just as notation for the composition of two functions $f(g(x)) = (f \circ g)(x)$ indicates that g is the first function to act on the input x (and then f acts on the output $g(x)$), the notation $X \circ Y$ means that transformation Y acts first, followed by transformation X. For example, $H \circ R_a$ denotes the composition of a half-turn with center at coordinate a on the midline, followed by a reflection in the midline; this composition yields V_a, a reflection in the vertical line $x = a$. In table 2.2, the identity transformation is denoted I, and it can be represented by a rotation of 360 degrees or a translation T_0 through a distance of 0 units; also, H can be represented as G_0, a glide reflection whose translation part is through a distance of

0 units. One important property of these compositions is that the entries in the table depend, in most cases, on the order in which the transformations are performed. For example, $V_b \circ V_a = T_{2(b-a)}$, whereas $V_a \circ V_b = T_{2(a-b)}$; these translations have the same magnitude but opposite directions. This fact is not apparent in the general summary given in theorem 2.

We are almost ready to answer the question "How many essentially different frieze patterns are possible?" Because such patterns can have an infinite number of symmetries, we say that two frieze patterns have *the same symmetry type* if they have the same *kinds* of symmetries. Two frieze patterns will be considered to be essentially the same if they have the same symmetry type.

If a frieze pattern does not have translation symmetry, then we know that it must be one of the five types we have illustrated in figure 2.9. If a frieze pattern has translation symmetry, then a translation occurs through a distance $c > 0$ that will slide the pattern horizontally along itself until the whole pattern matches up exactly; applying that translation again must also match up the whole pattern, and so on, again and again. So if a frieze pattern has translation symmetry, it consists of the repetitions of a decorated rectangle whose height is the height of the strip and whose width is $c > 0$. This rectangle repeats again and again, side-by-side, to fill out the whole pattern. Such a frieze pattern is called *periodic*. In almost all instances, it will have a translation symmetry whose translation distance $c > 0$ is a minimum; this minimum distance is called the *period* of the pattern. A simple example of a periodic frieze pattern with period 2π is the graph of the sine function.

Usually frieze patterns that do not have a minimum period (such as a pattern consisting of a few horizontal stripes) are excluded from consideration as periodic patterns, and we will also make that exclusion.

From theorem 2, we can see that if a frieze pattern has glide reflection symmetry, then it also has translation symmetry because the pattern must be superimposed on itself each time the glide reflection is applied, and the composition of two glide reflections is a translation. In fact, table 2.2 shows that if the glide reflection G_a that is a composition of the translation T_a and H is applied to the pattern twice, the resulting translation is T_{2a} (which is a symmetry of the pattern).

Periodic frieze patterns are easily produced by choosing a decorated rectangle and repeating it by translation or by glide reflection. Figure 2.11 displays seven frieze patterns. The first three and the fifth are produced by the repeated translation of the decorated rectangles d, $V(d)$, $R(d)$, and $H(d)$, respectively, in figures 2.3 and 2.4. The fourth pattern is produced by repeated glide reflection of decorated rectangle d in figure 2.3, and the sixth pattern is produced by repeated glide reflection of decorated rectangle $R(d)$ in figure 2.4. The last pattern can be obtained by repeated translation of the decorated rectangle in figure 2.5 or by repeated glide reflection of that rectangle.

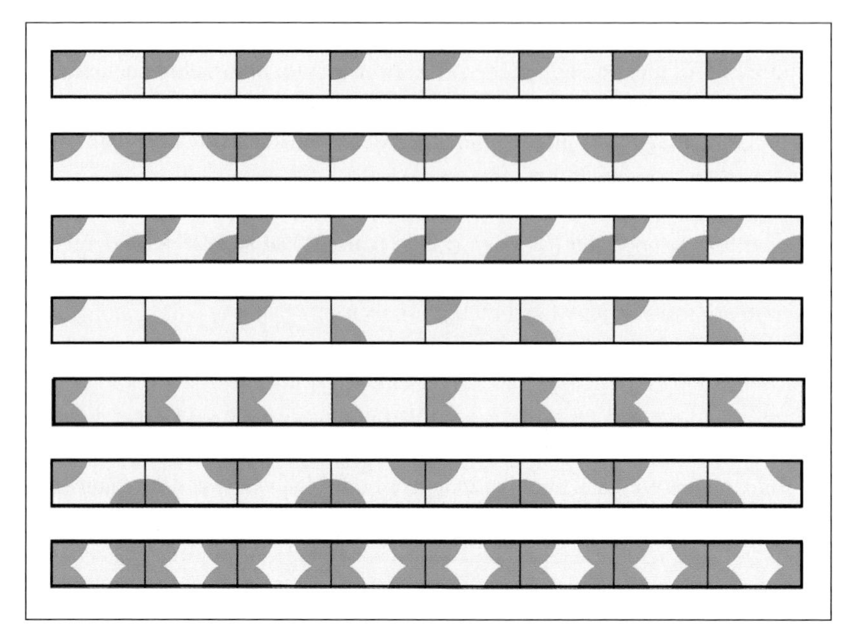

Fig. 2.11. Seven symmetry types of periodic frieze patterns

Enter Sherlock

We want to determine how many different symmetry types of periodic frieze patterns exist. The task seems hopeless because such patterns have an infinite number of symmetries. However, we can approach the problem as Sherlock Holmes might have done: *first, consider all the possibilities*. We know that any periodic frieze pattern has translation symmetries, and it might also have some of the four other possible kinds of symmetry: *H*, vertical reflections, half-turns, or glide reflections. (We note that with the exception of *H*, if a periodic frieze pattern has one of these symmetries, then it has an infinite number of symmetries of that same kind.) So we can make a table of all possible sets of symmetries that a frieze pattern can possibly have. In addition to translations, a periodic frieze pattern might have just one other kind of symmetry, or two other kinds, or three other kinds, or all four other kinds. In table 2.3, a check mark (✓) means that the pattern has symmetries of that kind, and a blank means that it does not have any symmetries of that kind. Each row in the table gives a possible complete collection of symmetries of a periodic frieze pattern.

Note that the last column of table 2.3 is labeled "POSSIBLE?" To paraphrase Sherlock, *one should always look for possible alternatives, and eliminate some of them*. We have indeed provided for all possible alternatives in the table—the rows represent every possible collection of symmetries that might be the complete list of symmetries of a periodic frieze pattern. But now we need to eliminate some of them. To quote Sherlock, "*when all other contingencies fail, whatever remains, however improbable, must be the truth*" (Doyle 1917; [emphasis added]).

Table 2.3
The Sixteen Possible Collections of Types of Symmetries for a Periodic Frieze Pattern

Row	Translations	*H*	Vertical Reflections	Half-Turns	Glide Reflections	POSSIBLE?
1	✓					
2	✓	✓				
3	✓		✓			
4	✓			✓		
5	✓				✓	
6	✓	✓	✓			
7	✓	✓		✓		
8	✓	✓			✓	
9	✓		✓	✓		
10	✓		✓		✓	
11	✓			✓	✓	
12	✓	✓	✓	✓		
13	✓	✓	✓		✓	
14	✓	✓		✓	✓	
15	✓		✓	✓	✓	
16	✓	✓	✓	✓	✓	

Theorem 2 can now be used to eliminate those rows of table 2.3 whose collection of symmetries cannot be *all* the symmetries of a periodic frieze pattern. For example, the collection of symmetries in row 6 cannot be all the symmetries of a periodic frieze pattern, because if the reflection H in the midline and a vertical reflection are both symmetries of the pattern, then the composition of these two symmetries produces another symmetry of the pattern—namely, a half-turn—and half-turns are missing from this row. So row 6 is eliminated; we put NO in the last column of that row. Applying this technique to each of the remaining 15 rows will show that all but seven rows are eliminated, and as Sherlock said, what remains must be the truth. The rows that remain after the elimination procedure are 1, 3, 4, 5, 8, 15, and 16. Since we have, in figure 2.11, examples of seven frieze patterns that have the remaining collections of symmetries as their complete set of symmetries, we know that these seven types are possible, and our elimination shows that no other types exist. This result proves theorem 3.

Theorem 3: *In the classification of frieze patterns by symmetry type, there are exactly five types of nonperiodic patterns and seven types of periodic patterns.*

The seven types of periodic frieze patterns have been given various labels; one frequently used system is based on notation used by crystallographers. Each label consists of two symbols, and is assigned to a periodic frieze pattern according to the following rules:

- The first symbol is *m* if the pattern has vertical reflection symmetry, otherwise the first symbol is 1.
- The second symbol is *m* if the pattern has midline reflection (horizontal reflection) symmetry.
- If the pattern has no midline reflection symmetry but does have glide reflection symmetry, the second symbol is *g*.
- If the pattern has no midline reflection symmetry or glide reflection symmetry but does have half-turn symmetry, the second symbol is 2.
- If the pattern has no midline reflection symmetry, glide reflection symmetry, or half-turn symmetry, the second symbol is 1.

Thus the seven labels to identify the symmetry type of a periodic frieze pattern are (in order of appearance in table 2.3 and also in figure 2.11) the following: *11, m1, 12, 1g, 1m, mg,* and *mm*.

Many other interesting questions about periodic frieze patterns can be pursued; here are a few, along with suggested references that have some answers:

- How can you fold a paper rectangle and cut out parts of it so as to produce a "paper snowflake" that has symmetry of any of the five symmetry types for a decorated rectangle? Can you fold and cut out parts of a long paper strip to produce a periodic frieze pattern of any of the seven symmetry types? (Servatius 1997)
- How can you produce a frieze pattern of interlocked congruent tiles that are more interesting than rectangles or parallelograms? (Schattschneider 1986)

- For each of the seven symmetry types of periodic frieze patterns, can you find a smallest region of the pattern that will produce the whole pattern when the symmetries of the pattern act on that region? (Such a region is sometimes called a *fundamental domain,* or a *generating region.*) What is the least number of symmetries of the pattern that will generate the whole pattern by repeatedly acting on this minimum region? (Lee 1999; Schattschneider 1986)
- How can you color a periodic frieze pattern with two colors so that some symmetries of the uncolored pattern interchange the two colors while the remaining symmetries do not change any colors? Note that our frieze pattern in figure 2.8 is a two-colored *mg* pattern in which the black and white portions are exact copies of each other. Translation symmetries and vertical reflection symmetries preserve all colors, whereas half-turn and glide reflection symmetries interchange colors. How can two-colored frieze patterns be classified according to these "color symmetries"? (Schattschneider 1986; Washburn and Crowe 1988)

In addition to some of the references given here, many resources contain wonderful illustrations of frieze patterns for analysis and inspiration. A particularly rich resource is the Dover Pictorial Archive series. The subject of symmetry, with its many connections with geometry, art, and creativity, provides a wealth of ideas to explore in the classroom and beyond.

Acknowledgment: I first saw the elimination proof of theorem 3 in *Symmetries of Culture* by Dorothy Washburn and Donald Crowe (1988), in their Appendix 2, and was delighted to see such a clear and simple argument.

REFERENCES

Doyle, Sir Arthur Conan. "The Adventure of the Bruce-Partington Plans." In *His Last Bow*. London: John Murray, 1917.

Lee, Kevin D. *Kaleidomania! Interactive Symmetry.* Software and Activity book. Emeryville, Calif.: Key Curriculum Press, 1999.

Martin, George E. *Transformation Geometry: An Introduction to Symmetry*. New York: Springer-Verlag, 1982.

Schattschneider, Doris. "In Black and White: How to Create Perfectly Colored Symmetric Patterns." *Computers and Mathematics with Applications* 12B, no. 3/4 (1986): 673–95. Also in *Symmetry: Unifying Human Understanding,* edited by István Hargittai, pp. 673–95. New York: Pergamon, 1986.

Servatius, Brigitte. "The Geometry of Folding Paper Dolls." *The Mathematical Gazette* 81, no. 490 (1997): 29–36.

Washburn, Dorothy K., and Donald W. Crowe. *Symmetries of Culture: Theory and Practice of Plane Pattern Analysis*. Seattle: University of Washington Press, 1988.

Yaglom, Izaak Moseivich. *Geometric Transformations I*. New Mathematical Library no. 8. Washington, D.C.: Mathematical Association of America, 1962.

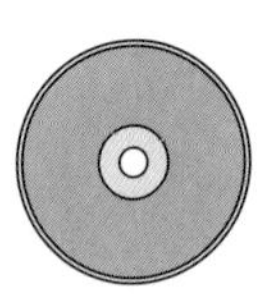

Blackline masters of Frieze Pattern worksheets 1 and 2 are found on the CD-ROM disk accompanying this Yearbook.

Möbius Concepts: Strips and Tori

Yuichi Handa
David A. James
Thomas Mattman

THE STUDY of the twisting and turning of topological objects holds a real fascination for both students and teachers. Among the objects that have the most interesting and instructive twists are the members of the family of Möbius strips and Möbius tori. A broad fascination with these mathematical objects extends beyond the classroom and into the realms of architecture, science, and art,

For example, a beautiful pedestrian and cycling Möbius Band Bridge planned in Bristol, England (illustrated in fig. 3.1), will both fascinate and perplex those who cross over it. At the microscopic scale in the realm of science, chemists have synthesized a Möbius band at the molecular level by joining the ends of a twisted double-stranded strip of carbon and oxygen atoms, and in 2002 physicists produced a Möbius crystalline ribbon of niobium selenide to help in the exploration of topological effects in quantum mechanics. In the area of art, a number of sculptors including Max Bill and John Robinson (see fig. 3.2) have experimented with Möbius strips and Möbius tori to visually express such abstract, philosophical concepts as infinity and eternity.

Courtesy of Hakes Associates

Fig. 3.1. Möbius Band Bridge

In the classroom, Möbius strips and Möbius tori also offer ample opportunities for fascination and inquiry. The explorations that we describe in this article allow students from elementary school to high school to develop their mathematical maturity and to question and improve on their preconceived notions about sidedness and edgeness. This is to say that the study of Möbius strips and Möbius tori offers an exciting avenue for extending the existing geometry curriculum.

According to the Geometry Standards in NCTM's *Principles and Standards for School Mathematics* (2000, p. 41), grades K–12 students are engaged in "Analyz[ing] characteristics and properties of two- and three-dimensional geometric shapes and develop[ing] mathematical arguments about geometric relationships...." The explorations we present address this strand of the Geometry Standards. In fact, students will be analyzing the characteristics and properties of shapes that arise from both two-dimensional and three-dimensional objects.

Our presentation divides into two parts: Part 1 contains explorations of Möbius strips appropriate for elementary and middle school students, and Part 2 presents Möbius tori concepts suitable for high school students.

Courtesy of the Bradshaw foundation

Fig. 3.2.
John Robinson,
"Eternity"

Part 1: Explorations of Möbius Strips for Elementary and Middle-School Students

We begin with an activity that is appropriate for all elementary and middle school students. The students begin with long strips of paper—about one to two feet long and one to two inches wide. As depicted in figure 3.3, students are instructed to curl the strip around (without any twists), joining the ends with tape or glue so that they form fairly wide "cylinders." If the students have already studied polyhedra, they may even be able to identify these figures as cylinders (e.g., "the sides of a soda can," "a simple, closed surface that is bounded by two congruent circles that lie in parallel planes," or something along such lines).

With a second strip, students are told to give one "half-twist" to the strip before joining the ends (see fig. 3.4). With a third strip, students give two "half-twists" forming yet another type of figure (see fig. 3.5), and so on, until students have made anywhere from five to six different such "loops," each with one more "half twist" than the previous one. Some of the later "loops" may require longer strips to accommodate the many half-twists.

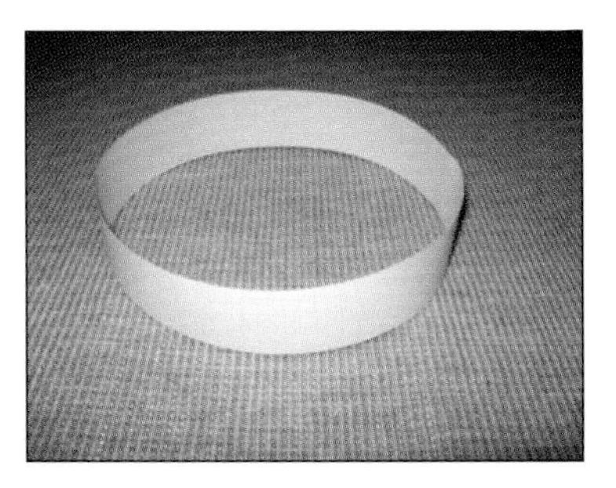

Fig. 3.3. Zero half-twists

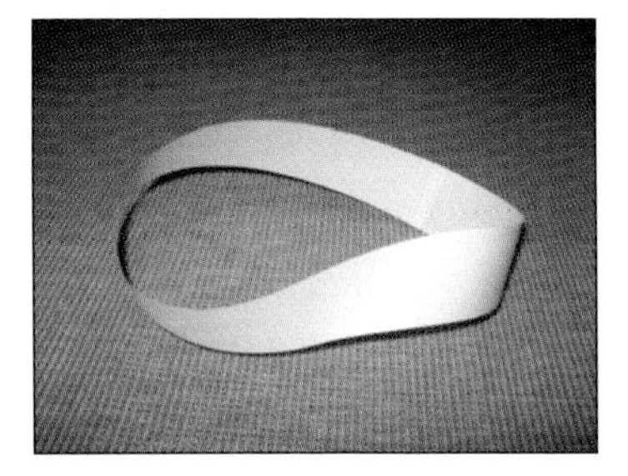

Fig. 3.4. One half-twist

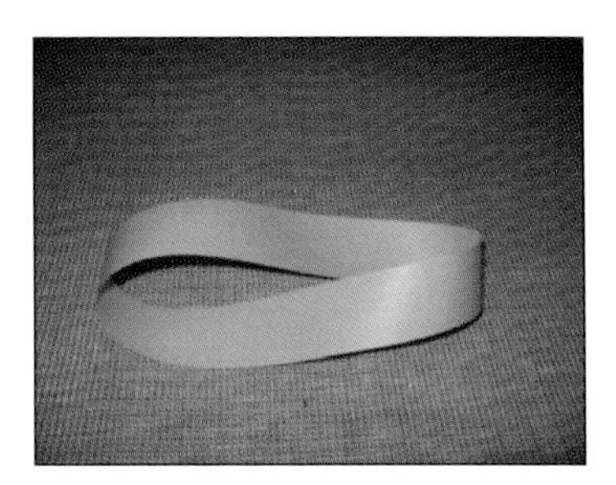

Fig. 3.5. Two half-twists

Exploration 1

This first exploration helps students begin classifying and characterizing these different "loops," using topological distinctions. For elementary school students, the teacher can instruct the students to color the "inside face" and "outside face" of each of the "loops." The important instruction will be to start at some point and to continue coloring until one has arrived back to where one started. When the student has finished coloring one of the faces (or "surfaces"), then the student can continue with shading the other (if there is another!). For demonstration, a helpful tactic may be to use an unlooped strip of paper, showing how the strip can be colored with two colors. Once all the "loops" have been shaded, the teacher can ask the students whether they notice a pattern. For advanced students, the teacher could simply ask the students to think of ways in which they might categorize the different "loops" (i.e., the coloring could be optional), offering the hint of counting the number of "faces" if students are stymied.

The fact should arise that all the "loops" formed with an odd number of half-twists can be colored with only one color, whereas the ones containing an even number of half-twists require two colors. Some students may phrase this outcome as "The odd-number twists have only one face, and the even-number twists have two faces."

As an optional extra task within this initial exploration, the teacher may also suggest that students color the "edge(s)" of all the loops in the same fashion that the "faces" were colored. What students can discover is that all the loops with an odd number of half-twists, in addition to having only one face, also have only *one continuous edge.* This fact will be helpful in the second exploration.

At this point, the teacher can introduce the mathematical terminology that describes these two types of "loops." Those with only one "face" are all called Möbius strips, named after German mathematician August Ferdinand Möbius. Even though the "loop" with *one* half-twist looks different from those with *three* and *five* half-twists, they are all considered, from a "topological" point of view, to be Möbius strips because they all have only one continuous "face." Technically speaking (but couched intuitively), these different-looking Möbius strips are *topologically equivalent* to one another because you can take any one of them (say, the *three* half-twisted Möbius strip), cut it somewhere, and reattach the cut pieces in the exact same orientation as they were before the cut, to get a different Möbius strip (say, the *once* half-twisted Möbius strip). As an aside, the teacher may also mention the interesting fact that some conveyor belts are made of very large Möbius strips so that not just half the surface gets all the wear, leading to longer-lasting belts. In addition, continuous-loop recording tapes exist that use the Möbius concept to double the available playing time.

Likewise, all the "loops" with *two faces* are considered to be in the same class of topological objects. In fact, *from a topological point of view*, they are all called "cylinders!" Some of the students may object, "But these other ones don't look like the cylinders we've studied!" The topologically minded teacher could respond, "Yes, but we now have a bigger and better definition for a cylinder that won't let us be deceived by *how it looks*!"

After all, referring to all these loops as cylinders is no different from using just one name, Möbius strip, to describe all the one-faced loops. Knot theory provides another example of the rich potential that comes from such a topological mindset. The simplest examples of knots are the trefoil (see fig. 3.6) and the circle, or what mathematicians call "the unknot" (see fig. 3.7). Topologically these two (and indeed, all knots) are equivalent; they're all circles! The aspect that makes them different is the way that they are placed in three-dimensional space. The study of knots offers another avenue for extending the geometry curriculum. Although we do not pursue it further here, we recommend *Why Knot? An Introduction to the Mathematical Theory of Knots* (1999) by Colin Adams as a starting point for elementary school students, whereas *The Knot Book: An Elementary Introduction to the Mathematical Theory of Knots* (2004) by the same author includes many exercises suitable for more advanced students.

In addition, our expanded "definition" of a cylinder is on some level much more student-friendly than a previous one ("A simple, closed surface that is

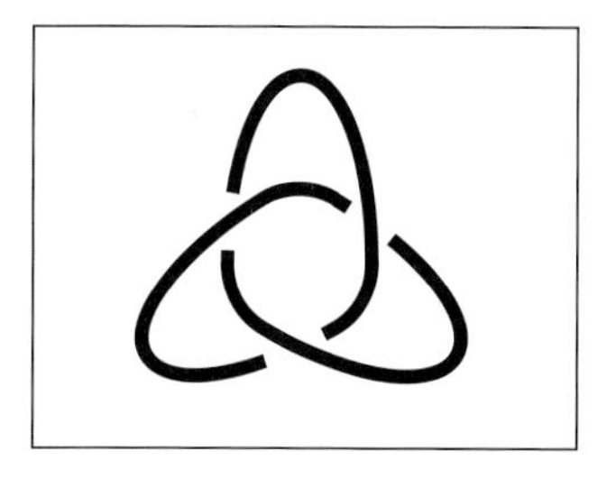

Fig. 3.6. Trefoil knot

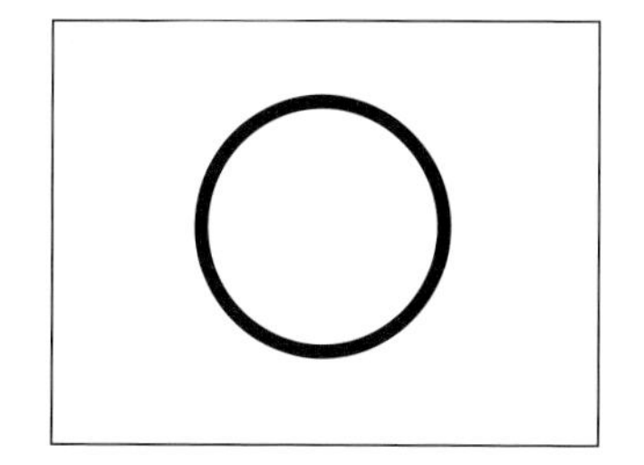

Fig. 3.7. Circle (or "unknot")

bounded by two congruent circles that lie in parallel planes") because it is based on what the students have just made with their own hands. The new definition for the cylinder is that it is a rectangle (or rectangular strip) *in space* whose ends have been joined right side up, and not upside-down (see fig. 3.8; notice the arrows on the ends). A Möbius strip, by contrast, is a rectangle in space whose ends have been joined against their original orientation (see fig. 3.9).

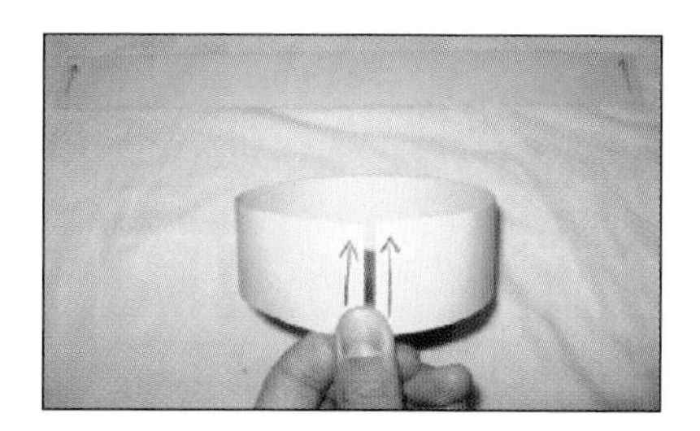

Fig. 3.8. Cylinder with ends joined right-side up

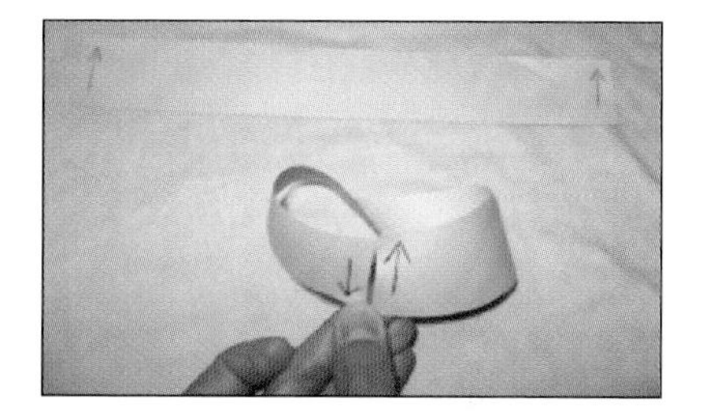

Fig. 3.9. Möbius strip with ends upside-down

We would note that this idea of forming surfaces (such as that of the cylinder) by joining the edges of a rectangle is an interesting way of extending ideas taught in grades K–8 school geometry. In particular, the torus, Klein bottle, sphere, and projective plane can all be formed by joining the edges of a rectangle in different ways. Two very nice books by Jeffrey Weeks (2000, 2001) aimed at upper-elementary through high school students (or interested elementary school teachers) explain these ideas in an accessible fashion.

Exploration 2

The second exploration involves cutting these cylinders and Möbius strips along their lengths (see figs. 3.10 and 3.11 for a partial cutting of each). It allows students to continue classifying geometric objects, but this time, in regard to the resulting figures after the lengthwise cuts are made.

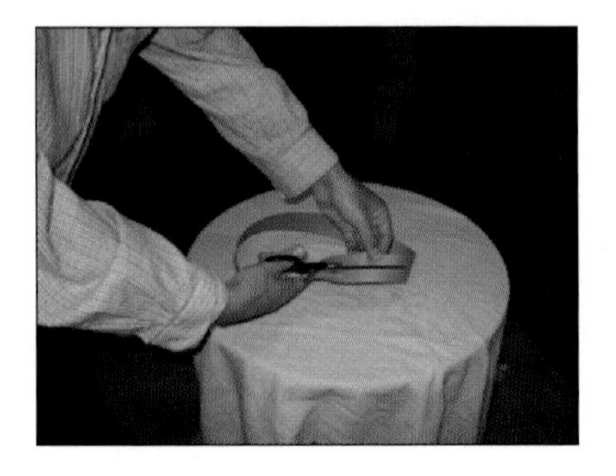

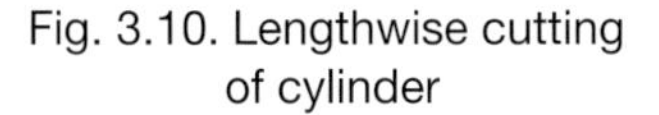
Fig. 3.10. Lengthwise cutting of cylinder

Fig. 3.11. Lengthwise cutting of Möbius strip

To add to the surprise and anticipation for the students, the teacher might ask the students, "What do you think will happen?" Students can then fill in a chart such as the one in figure 3.12, where they first try to predict what will happen, and then check what actually happens, as they cut their Möbius strips and cylinders.

No. of 1/2-twists	What do you predict will happen?	What actually happens?
0	I'll get 2 cylinders ...	I was right ... but also 1/2 as tall.
1		
2		
3		
4		
5		

Fig. 3.12. Chart for keeping track of the result of lengthwise cuts on Möbius strips and cylinders

Some middle school students may have easily anticipated that the "no twist" cylinder would end up becoming two cylinders that are half as wide. Conversely, most students will very likely be surprised by what happens to the Möbius strips, in addition to all the cylinders past the first one. Once all the new figures have been cut out, the students' task will be again to try to characterize the difference between what happens to the Möbius strips and the cylinders. The answer is that all the Möbius strips will result in a longer—and sometimes knotted—but single "loop," whereas all the cylinders will result in two "loops" in various arrangements. This result is somewhat surprising, although not unexplainable.

The mathematical reason for this result may be a bit advanced for the early elementary school student but very much within reach of older students. It comes about from the fact that all Möbius strips have only *one continuous edge*—something learned in the first exploration. When the lengthwise cut is made, the left and right side of where the scissors cut become the other continuous edge. Students can "see" this by coloring the original edge one color, and coloring the center strip where the scissors will cut with another color. The resulting figure after the cut will have the two colors for the edges, which, incidentally, tells us that the resulting figure is *not* a Möbius strip, because it will have two edges.

The cylinders, however, already have two edges. The cut introduces two extra edges, leading to two separate "loops." So cutting the Möbius strips leads to one longer loop, whereas cutting the cylinders leads to two loops.

Exploration 3

The first chart can be extended for some students by introducing the idea of making more than one revolution of a lengthwise cut. Thus, for the cylinder, a student can make two lengthwise cuts (see fig. 3.13 for a partial cutting along two lengthwise cuts). For the Möbius strips, the equivalent type of action will not involve two cuts, but one long cut that winds around the Möbius strip twice. The technique for this cut will be to start one-third of the way from the edge (see fig. 3.14) and continue around the Möbius strip twice until one reaches the starting place.

Fig. 3.13. Two lengthwise cuts on a cylinder

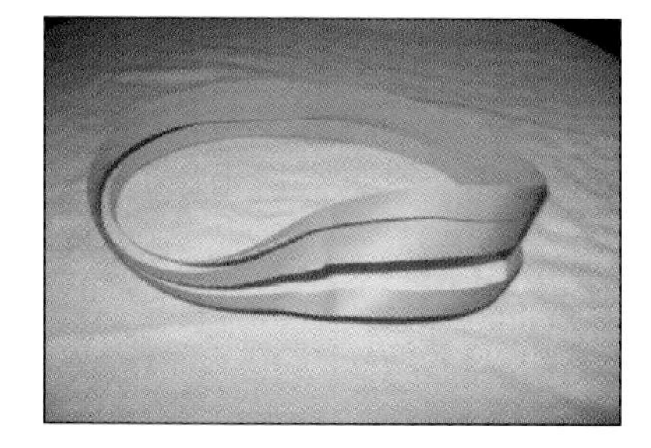

Fig. 3.14. Two lengthwise cuts on a Möbius strip

A chart similar to the one from the second exploration (see fig. 3.15) can be filled in by the students.

We leave it to students to discover what hidden patterns may emerge. We would mention that the result from these different cuttings is what mathematicians call *paradromic rings,* which are studied by knot theorists (see Ball and Coxeter [1987, pp. 127–28]), including one of the authors of this article.

	Width of bands formed by cuts		
No. of 1/2-twists	**1/2 width of strip**	**1/3 width of strip**	**1/4 width of strip**
0			
1			
2			
3			
4			
5			

Fig. 3.15. A chart for tracking the changes with different numbers of cuts

Part 2: Möbius Concepts for Secondary School Students—the Möbius Torus

We recall that a tall, rectangular piece of paper can be looped around to make a cylinder, or it can be given a half-twist before connecting to form a Möbius strip (fig. 3.16). Similarly a tall, *three-faced* prism can be bent around and its top connected to its bottom to form a doughnut with three faces. But if the prism is given a one-third twist before connecting the top to the bottom (see fig. 3.17), then although the resulting object, called a *Möbius torus,* seems at first glance to have three faces, it actually has only one face.

To show this outcome, one could try to paint only one track. After painting once around, one arrives back near the starting point but is surprised to find oneself on an adjacent track. Completing the track requires looping three times around the doughnut to arrive back at the starting point, and by this time one finds the entire surface has been completely painted with the original single color. So the prism really has only one face. The Möbius strip in figure 3.16 and the Möbius torus in figure 3.17 are just the $n = 2$ and $n = 3$ members of a family, where n is the number of faces of the starting prism.

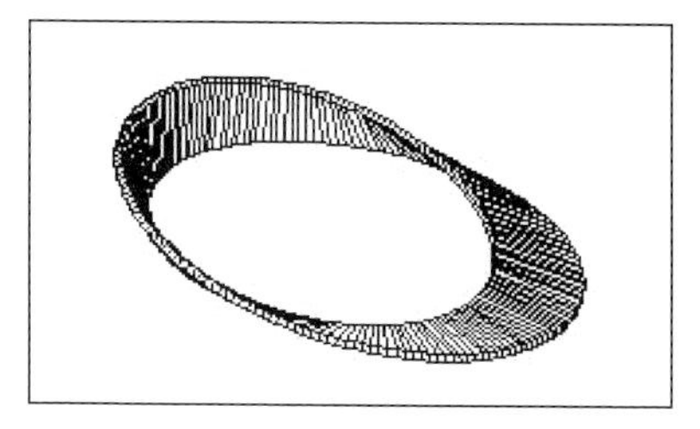

Fig. 3.16. Möbius strip

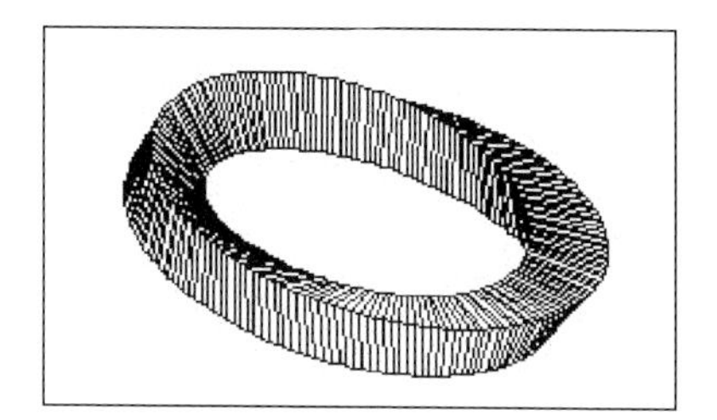

Fig. 3.17. Möbius torus

If one starts with a prism having a square cross section ($n = 4$) rather than a triangular cross section, things are even more interesting. In that instance, if one connects top to bottom without twisting, a four-faced doughnut results, but if one rotates by a one-fourth turn before connecting, then the apparently four-faced object actually has only one face. So far, no surprises emerge beyond what we observed earlier for $n = 3$. But if one rotates by a two-fourths, or one-half, turn, a two-faced object is produced. If one rotates by a three-fourths turn, a one-faced object is again produced. And if one rotates by a complete turn before connecting, a four-faced object results, just as was true for the original no-rotation instance—one loop around the track gets you back to your starting point.

The member with $n = 8$ is the *Möbius torus* or *Möbius doughnut* (as shown in fig. 3.19), made from adjoining the ends of an eight-faced prism (fig. 3.18).

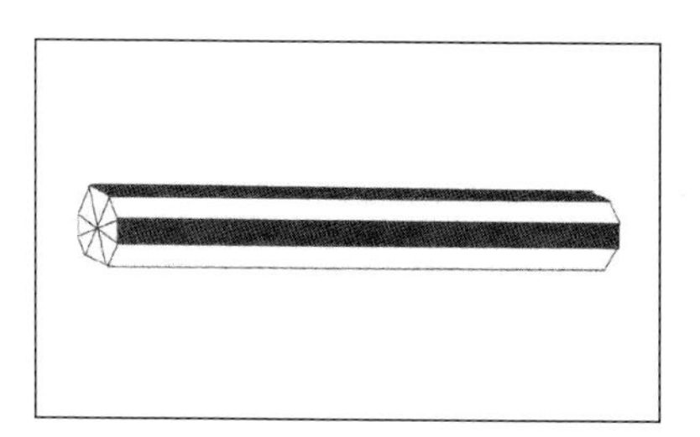

Fig. 3.18. Eight-faced prism

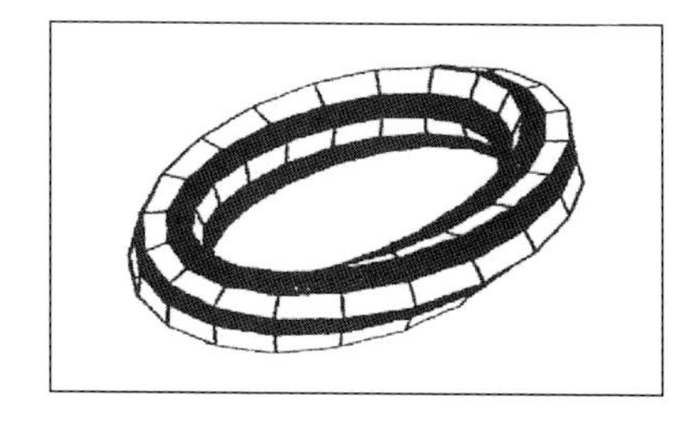

Fig. 3.19. Möbius torus with $n = 8$

Exploring only by touch an enormous Möbius torus like that of figure 3.19 would be both a puzzling and intriguingly satisfying experience for a group of blindfolded adventurers. Part of their instructions would be to refrain from crossing over the dangerous sharp edges between the tracks. The intrigue would come from various levels of surprise. The first surprise would come when, after traveling for a while, the explorers arrive back within earshot of the people they left tending the fire at their starting point. They would correctly conclude that this object they are exploring must circle back on itself in some way. They then return home the way they came, and make further plans. The next day they ask their friends on adjacent tracks to start out at the same time they do, and after looping once around, they have a group discussion and conclude that this object is a many-faced doughnut. But on closer inspection they would realize things are more complicated, because setting off in lane 1 leads to returning in lane 2 after one loop. This puzzle is resolved when they realize that after eight times around the loop (for the torus in fig. 3.19), lane 1 leads back to lane 1. Thus, this object is a combination of a once-around global reconnecting with an eight-times-around local reconnecting—a mystery and a resolution.

We next introduce some notation by way of an example. If one starts with a six-faced prism, and connects the top end to the bottom while twisting k-sixths of a turn, we denote the result as a $(6, k)$ Möbius torus. A (6, 0) torus has six faces. A (6, 1) torus has one face. A (6, 2) torus has two faces. A (6, 3) torus has three faces. A (6, 4) torus has two faces. A (6, 5) torus has one face. And a (6, 6) torus has six faces.

M. C. Escher and Fair Racetracks

Said the ant to its friends: I declare!
This is a most vexing affair.
We've been 'round and 'round
But all that we've found
Is the other side just isn't there.

—Cameron Brown (from Pickover [2006, p. xiii])

It is interesting to compare M. C. Escher's well-known illustration of ants racing around a Möbius strip (see fig.3.20) with the concept of six ants racing around a (6, 1) Möbius torus. On the torus, each ant starts in one of the six lanes on the start line. After once around the torus, each ant is back to the start line, but ant 1 is where ant 2 started, ant 2 where ant 3 started, and so forth. The race continues, and ends when each ant arrives back at its starting position, which takes

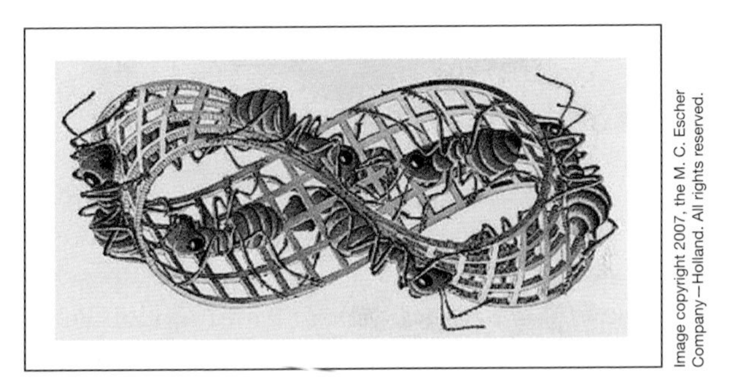

Fig. 3.20. Möbius Strip II, M.C. Escher

six loops around the torus. Each ant has run over exactly the same path as every other ant but has never needed to switch lanes or swing out to pass another ant to accomplish this feat—reflecting the ultimate in a fair racetrack. The ants also are racing side-by-side, so the competitive aspect is still present for the spectators and the racers.

Exploration 4

In this interesting exploration, students are asked to discover connections between the count of the number of faces of various Möbius tori and the number theory concepts of division, greatest common divisor, prime numbers, and modular arithmetic. The students are subdivided into groups, and each group is challenged to discover a pattern that gives the number of faces on a (n, k) Möbius torus. They should be instructed that their answer will have something to do with the numbers n and k, and with the relationship these numbers have with each other. To start them off, the teacher might point out that the $(6, k)$ Möbius torus has k faces for $k = 1, 2,$ and 3 but not for $k = 4$ and 5. Students should study the (6, 4) Möbius torus carefully enough to see why it has two faces rather than four faces, and why the (6, 5) torus has only one face. Each group can study various other (n, k) Möbius tori of their own choosing. The choices of $n = 8$ and 10 and 12 with various k's are particularly revealing.

This area is rich for explorations. Some groups will finish more quickly than others, and these groups should be challenged to apply their result to show that for the special case of an (n, k) torus where n is a prime and $1 \le k < n$, the only possibility is the one-faced torus. Another advanced exploration is to show that $\varphi(n)$ of the (n, k) Möbius tori with $1 \le k < n$ have only one single face. Here the Euler phi-function $\varphi(n)$ is the number of integers 1, 2, …, n that are relatively prime to n.

Students acquainted with the concept of modular arithmetic should give several examples of each of the two following statements and explain why each statement is true:

1. If $j \equiv k (\bmod n)$, then the (n, j) torus has the same properties as the (n, k) torus.
2. The $(n, n - k)$ torus has the same properties as the (n, k) torus. This result is related to *chirality,* the properties of a left-hand twisting compared with those of a right-hand twisting.

Exploration 5

Students are asked to discover a formula for the number of times one must loop around the (n, k) Möbius torus to arrive back at the starting point. They can be asked to combine this result with their answer to Exploration 4 to conclude that the product of the number of faces and the number of times one must loop to arrive back at one's starting point equals n.

Exploration 6

In this exploration students are asked to discover what the relationship must be between n and k if they wish to color a (n, k) Möbius torus with just two colors (like a barber pole), or just three colors, or just m colors.

The Möbius Strip and the Möbius Torus in Art and Architecture

The Möbius concept arises in various manifestations in our world. In U.S. life, we encounter daily the ubiquitous recycling symbol, one version of which is equivalent to a Möbius strip. Several graphic illustrations of M. C. Escher involve the Möbius strip, including the previously mentioned one showing six red ants endlessly traversing such a strip. In the area of sculpture, one of the first artists to incorporate a Möbius strip was Max Bill in 1935. More sophisticated sculptures exploring the use of one-faced bands appear in the work of Brent Collins. Using the Möbius concept in a different context, an art critic has analyzed the use of visual illusion in the landscape painting of the Australian Modernists (International Conference on the Arts in Society).

We next move from applications of the Möbius strip to those of the Möbius torus. The Möbius torus has made its appearance in art and architecture quite recently, only since the mid-1980s. A strikingly beautiful crystal art sculpture designed by Peter Drobny was presented by President William Clinton to the Crown Prince and Crown Princess of Japan (see fig. 3.21). Other impressive renditions of a (3, 1) Möbius torus have been designed by Robert Wilson, Helaman Ferguson, and John Robinson and can be found with a World Wide Web search of each artist's name. Keizo Ushio has designed upwards of 200 attractive sculptures based on the Möbius strip and Möbius torus (see fig. 3.22). That these

Photograph courtesy of Steuben Glass

Fig. 3.21. Peter Drobny, Steuben Glass

Photograph courtesy of Keizo Ushio

Fig. 3.22. Keizo Ushio with Möbius sculpture

Möbius-related objects clearly project a real mystique for both their creators and their viewers is captured by sculptor John Robinson's comment, "By changing one of the edges to meet another, I found I was left with only one edge and one band.... It was a magical moment when I realized that I had created ETERNITY" (Mathematics and Knots/Edition Limitee).

A Möbius prism is simply a Möbius torus that has been snipped and straightened out while retaining the twist. The (n, k) Möbius prism has for centuries appeared in the form of a barber pole—the more stripes, the bigger is n; the more slanted the stripes, the bigger is k. The barber of one of the authors proudly displays his (8, 4) red-and-white Möbius prism outside his shop. Candy canes are another appearance in the popular art realm. Spiral staircases may be Möbius prisms in the instance of two or more intertwined sets of steps. Leonardo da Vinci designed a famous double-intertwined staircase for the Chambord chateau in France, and another incredibly beautiful example is found in Rome at the Vatican Museum. Tightly intertwining staircases not only occupy less space in the building than would two normal staircases, but also, if one set of steps is for ascending and the other for descending, then users would not interfere with one another and would also be entertained by the sight of the passing walkers going the other way, so close and yet so inaccessible. In science the double helix of DNA is a microscopic version of a Möbius prism. With his double-helix staircase at Chambord, yet again da Vinci was far ahead of his time.

In the area of architecture, several designers in the past few decades have worked to capture the fascination of the Möbius torus or the Möbius prism. Ben van Berkel and Caroline Bos of UN-Studio designed the Möbius House Het Gooi (see fig. 3.23), and Stephen Perrella of Archtectonics designed another Möbius House—both of these structures incorporate aspects of a Möbius torus. One of the most famous and most innovative architects of our time, Peter Eisenman, has designed the Max Reinhardt Haus (see fig. 3.24) to be located in Berlin; it unfortunately has not and may not reach construction stage. If it ever is built, perhaps the mayor should cut a Möbius ribbon (lengthwise) at the opening.

Fig. 3.23. Möbius House Het Gooi, a building incorporating a Möbius torus

Architecture has classically been tied to the Platonic solids in obvious ways, and so to incorporate Möbius forms represents a dramatic break. As explained by Vesna Petresin and Laurent Paul Robert (2002, p.3), "The Möbius strip is used as a diagram for post-Cartesian dwelling and is neither an interior space nor an exterior form; it is a transversal membrane reconfiguring these binary notions into a continuous, non-linear form, thus creating a basis of a complex, temporal experience ... the dualities in contemporary dwelling; e.g., the concepts of inside/outside, below/above, physical/virtual, private/nonprivate, material/media can be represented as such attractors."

Photograph courtesy of Eisenmann Architects

Fig. 3.24. Plans for the Max Reinhardt Haus, a building incorporating a Möbius torus

A different way of incorporating a Möbius torus into architectural structures was put forward by Jolly Thulaseedas and Robert Krawczyk (2003) of the College of Architecture at the Illinois Institute of Technology. Their intriguing Möbius Museum plan is shown in figure 3.25. The architect Santiago Calatrava has incorporated a Möbius prism in his Turning Torso in Malmo, Sweden (see fig. 3.26).

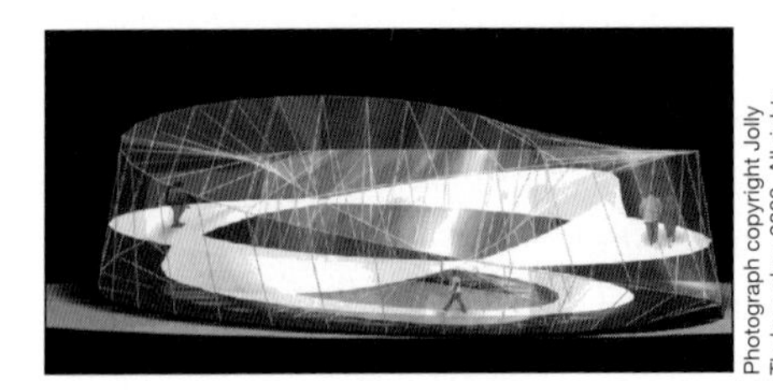

Fig. 3.25. Design for a Möbius Museum: the shell is a Möbius torus.

Fig. 3.26. Turning Torso, completed 2005, a Möbius-prism-shaped building

Further Exploration

Students are asked to sketch a set of plans for an innovative structure of their choice based on a Möbius concept. Some suggested ideas are (1) a building based on a Möbius strip, Möbius torus, or Möbius prism; (2) an amusement park ride based on a Möbius strip or Möbius torus; or (3) a bridge based on a Möbius torus. They then reflect on and describe some good points of their design as well as some of its disadvantages.

Guide for Students' Further Research at the High School Level

The properties of Möbius-type objects are fascinating to students and teachers alike, and they make excellent projects and presentations for students. The Möbius strip plays a role not only in mathematics but also in science, engineering, literature, art, music, dance, crafts, puzzles, magic tricks, scarves, and jewelry. An entire book, *The Möbius Strip* by Clifford Pickover (2006), is devoted to its properties and applications. A general World Wide Web search under "Möbius," "Mobius," or "Moebius" provides a wealth of information. Recent Möbius developments appear in articles by Ivars Peterson at the Web site for *Science News* online. Articles in Wikipedia under Möbius and Möbius Strip list interesting references to guide students to further sources. Martin Gardner's books, particularly *Mathematical Magic Show* (1977), are also rich and entertaining sources. Information and illustration of the works of each of the artists and architects mentioned is accessible through the Web. Many interesting links between art and mathematics, including some Möbius references, appear in *The Visual Mind,* edited by Michele Emmer (1993). Finally, the answer to Exploration 4 can be found in a *Mathematics Teacher* article by Charles J. Matthews (1972).

REFERENCES

Adams, Colin. *Why Knot? An Introduction to the Mathematical Theory of Knots.* Emeryville, Calif.: Key College Publishing, 1999.

———. *The Knot Book: An Elementary Introduction to the Mathematical Theory of Knots.* Providence, R.I.: American Mathematical Society, 2004.

Ball, Walter W. Rouse, and Harold S. M. Coxeter. *Mathematical Recreations and Essays.* 13th ed. New York: Dover Publications, 1987.

Emmer, Michele, ed. *The Visual Mind: Art and Mathematics.* Cambridge, Mass.: MIT Press, 1993.

Gardner, Martin. *Mathematics Magic Show.* New York: Knopf Publishing, 1977.

International Conference on the Arts in Society. "Art and the Mathematics of the National Museum of Australia." a06.cgpublisher.com/proposals/571/.

Mathematics and Knots/Edition Limitee. "Eternity." www.popmath.org.uk/sculpture/pages/1eternit.html.

Matthews, Charles Joseph. "Some Novel Möbius Strips." *Mathematics Teacher* 65 (February 1972): 123–25.

National Council of Teachers of Mathematics (NCTM). *Principles and Standards for School Mathematics.* Reston, Va.: NCTM, 2000.

Petresin, Vesna, and Laurent Paul Robert. "The Double Möbius Strip Studies." *Nexus Network Journal* 4 (2002): 38–53.

Pickover, Clifford. *The Möbius Strip.* New York: Thunder Mouth Press, 2006.

Thulaseedas, Jolly, and Robert Krawczyk. "Möbius Concepts in Architecture." *Proceedings of the ISAMA/Bridges 2003 Conference*, Granada, Spain, 2003.

Weeks, Jeffrey. *Exploring the Shape of Space.* Emeryville, Calif.: Key Curriculum Press, 2000.

———. *The Shape of Space.* 2nd ed. New York: Marcel Dekker, 2001.

Exploring Curvature with Paper Models

Howard T. Iseri

CURVATURE is a fundamental concept in modern geometry, and yet the topic is often neglected in the high school and undergraduate curriculum. A concept of curvature dating to Carl Friedrich Gauss (1777–1855) is used by differential geometers to study both Euclidean geometry and the non-Euclidean geometries discovered by Janos Bolyai (1802–1860), Nikolai Lobachevsky (1792–1856), and Bernhard Riemann (1826–1866) as curved surfaces. Although a formal investigation of the curvature of surfaces requires advanced multivariate calculus, the underlying concept can be explored using only basic plane geometry. The origin of this idea can be traced to Rene Descartes (1596–1650.) By measuring angles on polyhedral surfaces, Descartes measured the same quantities that Gauss would measure two centuries later and, at the same time, gave us a simple way to explore non-Euclidean geometry in terms of curvature.

Although curvature originates in mathematics, popular interest in the topic is due mostly to scientists. Einstein's discovery of the general theory of relativity (1915) introduced the notion that the behavior of gravity can be explained by equating the presence of matter in space with curvature. This curvature of space would allow a photon (with no mass, and hence no gravitational attraction) to travel in a straight line but also to appear to change direction when passing a massive object. This phenomenon was observed in the famous experiment that brought general relativity to the public's attention, when in 1919 Arthur Eddington, during a solar eclipse, observed that a distant star appeared to have changed position because the light coming from the star had traveled close to the sun. Today, the curvature and geometry of our universe are at the center of the debate concerning whether the universe contains enough matter to keep it from expanding forever.

The first part of this article examines how the curvature of curves and surfaces is related to the angles of polygons or polygonal curves and "solid" angles at the vertices of polyhedra. We examine the theorem at the center of Descartes' work in this area to give us a simple measure of the sharpness of a solid angle and then show how Descartes' theorem can be proved from the seemingly unrelated Euler theorem. The second part of the article presents examples of paper models that use Descartes' version of curvature to explore hyperbolic and elliptic figures in a precise and accessible way.

The Curvature of Curves

Figure 4.1 shows two paths, one polygonal and one smoothly curved. Following each from bottom to top, we see that the beginning direction and the ending direction are the same for both curves. The change in direction, therefore, is measured in both instances by the same angle, θ. For the polygonal curve, we call this angle the *exterior angle*. The *interior angle,* the one we typically measure, is supplementary to the exterior angle, so they sum to 180 degrees, or π radians, and we can easily find one from the other on a polygonal curve.

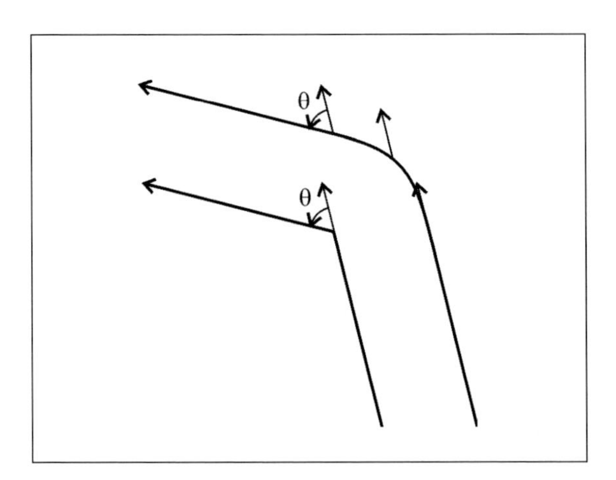

Fig. 4.1. These two curves change direction by the same amount.

If we interpret a curve as a path that we travel along, an external angle on a polygonal path measures an abrupt change in direction. On a smooth curve, this change in direction occurs gradually, and we have, in some sense, infinitely many, infinitesimally small external angles. So although we can measure the change in direction between two points on the same curve, another concept is needed to adequately describe how this change in direction takes place. This concept, called *curvature,* is defined to be the rate at which the direction changes with respect to arc length. From point A to point B on a curve, for example, the direction will change by some angle $\Delta\theta$, and the arc length along the curve between these two points will be a distance Δs. This notation gives us an *average* curvature of $\Delta\theta/\Delta s$. Taking shorter sections of the curve and a limit, we arrive at the curvature at each point, which is an instantaneous rate of rotation.

In figure 4.2, we have a circle of radius r. The arc between points A and B is a quarter of the circle and has length $2\pi r/4 = \pi r/2$. The change in direction between points A and B is $\pi/2$ radians. Dividing the change in direction by the arc length gives an average rate of change in direction (curvature) of $1/r$ radians per unit distance. The average curvature, not surprisingly, is the same for any arc of the circle, so in the limit, the curvature (denoted by the Greek letter κ; "kappa") at every point on a circle of radius r is $\kappa = 1/r$ radians per unit distance. Note that small circles will have more curvature, which should make sense. In general, if a curve has curvature κ at some point, then the part of the curve near that point will look like part of a circle of radius $1/\kappa$.

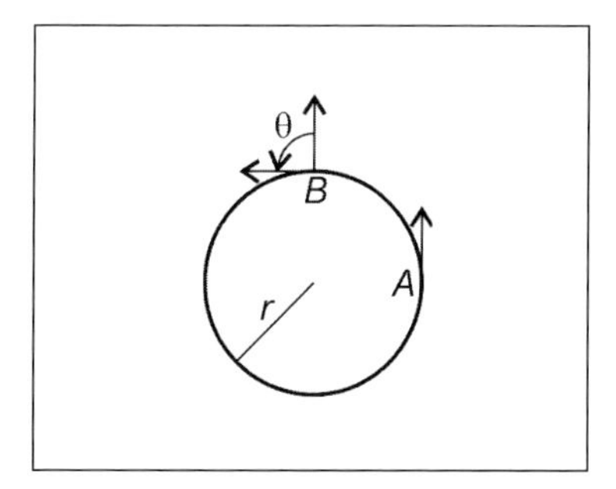

Fig. 4.2. The change in direction from point A to point B is $\pi/2$ radians over a distance of $\pi r/2$.

The significance of curvature to geometry becomes clearer in higher dimensions, and Descartes' theorem allows us to explore that significance. Before discussing surfaces, however, we will be helped by considering a lower-dimensional version of Descartes' theorem.

Imagine an ant walking around each shape in figure 4.3 in a counterclockwise direction. To do so it must turn as indicated by the shapes' exterior angles. The first figure is a square, and the four exterior angles each measure $\pi/2$ radians. The sum of the exterior angles in this instance is 2π radians. The sum is the same for the triangle. The figure on the right has one turn to the right, which we define to be negative, and with that, the exterior angles again sum to 2π radians.

> ***One-dimensional Descartes' theorem:*** Given any simple, closed polygonal curve in the plane, the exterior angles will sum to 2π radians.

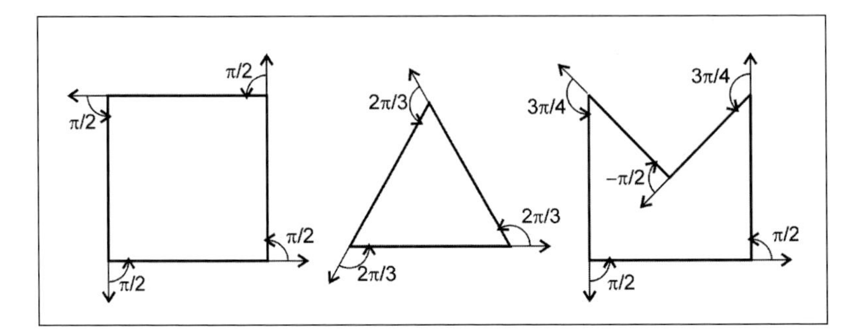

Fig. 4.3. The external angles on a polygon measure the rotations necessary to walk around it.

This theorem holds true for smooth curves. Imagine a circle painted on the floor. If you walk around it once, your body will make one complete rotation. If you face north initially, you will turn gradually until you face west, then south, and then north again. That is, you will make a 2π radian rotation. On a polygonal curve, the only difference is that the turns happen abruptly, but the net rotation is still 2π radians. For smooth curves, because curvature is essentially a derivative of the direction function with respect to the arc length, integrating the curvature around a simple, closed curve will give the net rotation, 2π radians. For this reason, an accumulated amount of curvature is called *total curvature,* and so total curvature is the same as a net change in direction. That is, the sum of the external angles measures total curvature.

In Euclidean (plane) geometry this sum of angle measures around a closed figure is 2π radians, and we will see that deviations from the number 2π serve as measures of how much a non-Euclidean geometry differs from the Euclidean model.

The Curvature of Surfaces

We have seen that the total curvature around a simple closed curve in the plane is always the same. As we turn to surfaces, the situation is similar, as long as we choose a suitable measure of curvature. *Gaussian curvature* is the standard measure of curvature for a smoothly curving surface, and it nicely ties the standard non-Euclidean geometries together as surfaces with constant Gaussian curvature. In particular, Gaussian curvature is 0 for the Euclidean plane, −1 for the hyperbolic "plane," and 1 for the elliptic and spherical "planes." Furthermore, we can show that integrating Gaussian curvature over any surface that is a smooth deformation of a sphere always yields the number 4π, which should remind us of the fact that the net rotation for a simple closed curve in the Euclidean plane is always 2π. Here, as with curves, an accumulated amount of Gaussian curvature is called *total curvature.* We can say, therefore, that closed curves (which are deformations of a circle) in the Euclidean plane always have total curvature 2π, and closed surfaces in Euclidean space that are deformations of a sphere always have total (Gaussian) curvature 4π. Being a deformation of a sphere here is important, because deformations of a torus, for example, always have total curvature 0.

Unfortunately, the definition of Gaussian curvature is somewhat involved, and calculating it for a typical surface is impractical. As with the curvature of curves, however, Gaussian curvature has an angular analog, which Descartes described in a manuscript with a tenuous existence. This manuscript was apparently not published during Descartes' lifetime. After Descartes died in 1650, when his papers were collected, it was accidentally dropped in a river, fished out, and hung to dry. Gottfried Leibniz (1646–1716), the coinventor of calculus, was given the opportunity to copy some of these papers including the manuscript containing our theorem. Leibniz' copy was then lost, found, and published in 1860 (Federico 1982; Phillips 1999).

Descartes' work focused on the angles of polyhedra. The *external angle* at a vertex of a polyhedral surface measures how much that vertex is "not flat," just as the external angle on a polygonal curve measures how much the curve is not straight. To see how this measure can be accomplished, consider a point in the plane. If we were to fit a collection of angles around this point, their measures would sum to 2π radians. Compare this result with the surface of a cube, a polyhedral surface with eight vertices. Each point on a face of the cube is surrounded by 2π radians. For example, on the one hand the point F shown in figure 4.4 has four right angles around it. Each point on the edge of the cube also is surrounded by 2π radians, as illustrated by the point E in figure 4.4. (This template can be enlarged with a photocopier or printed from a file on the CD accompanying this yearbook.[1]) On the other hand, each vertex point, such as the point V, is surrounded by only three right angles, which total $3\pi/2$ radians.

Compared with a point in the plane, therefore, each vertex comes up short by $\pi/2$ radians. Descartes called this deviation from 2π the *angulum externum,* or *external angle.* Since the eight vertices of the cube each have an external angle of $\pi/2$, the total external angle is 4π, or as Descartes says, "eight right angles"

1. This template, and those in the following figures, can be enlarged with a photocopier or printed from a file on the CD accompanying this yearbook.

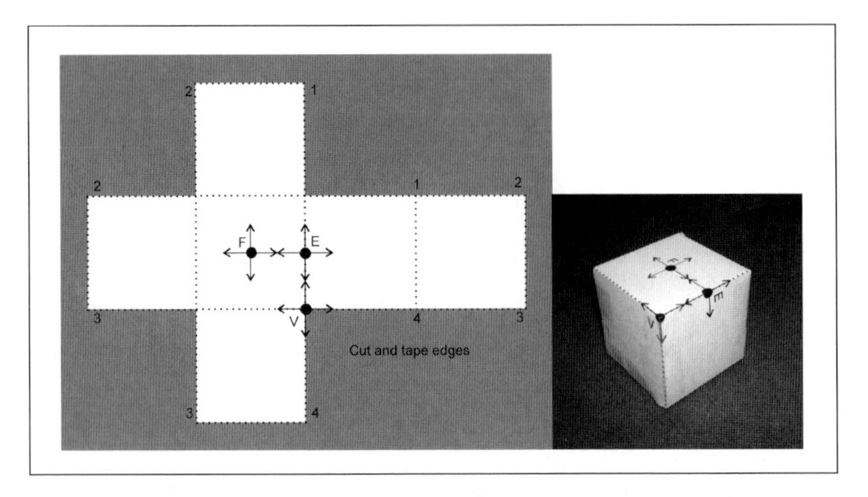

Fig. 4.4. The angles around points E and F sum to 2π. The angles around vertex V sum to $3\pi/2$. To construct the model, remove shaded regions and tape them together so that the vertices labeled with 1 come together, the vertices labeled 2 come together, and so on.

(Federico 1982). This is the same number that differential geometers find for the total amount of Gaussian curvature (i.e., the total curvature) on a sphere or any figure that can be deformed to a sphere.

Let us try this same computation with another simple polyhedron, a tetrahedron with four equilateral triangular faces. Each vertex is surrounded by three $\pi/3$-radian angles for a total of π radians. We are left with an external angle of $2\pi - \pi = \pi$ radians. Since four vertices are involved, we have an external angle sum of 4π. The result will be the same for any convex polyhedron, as we will prove. This outcome is Descartes' theorem.

> ***(Two-dimensional) Descartes' theorem:*** For any polyhedron the sum of the external angles is 4π.

This theorem is relatively easy to prove from Euler's formula,

$$f - e + v = 2, \tag{1}$$

where, for a given polyhedron, f is the number of faces, e the number of edges, and v the number of vertices. For simplicity, each of the faces can be subdivided into triangles by inserting diagonals, so we can assume that all the faces are triangles, although not necessarily on distinct planes. Each face now contributes three edges, each of which are shared by two faces. In other words, $3f = 2e$. Multiplying equation (1) on both sides by 2 and substituting $3f$ for $2e$ gives us

$$-f + 2v = 4. \tag{2}$$

Multiplying by π on both sides gives us

(3) $$2\pi v - \pi f = 4\pi.$$

We can now interpret equation (3). The (interior) angle sum of each of the f triangular faces is π, so the sum of all the angles of all the faces must be πf, which is also the sum of the angles at all the vertices. At each vertex the external angle is equal to 2π minus the angle sum. For all v vertices the sum of the external angles must be $2\pi v$ minus the sum of all the angles, which we already know is πf. Therefore, if we let ΣC be the sum of all the external angles, we get

(4) $$\Sigma C = 2\pi v - \pi f = 4\pi.$$

Note that if we were to allow an angle *excess* (i.e., angles surrounding a vertex that sum to *more* than 2π radians) to be a *negative* external angle, no modification to these computations is necessary. As a result, equation (4) applies to nonconvex as well as to convex polyhedra.

As an aside, we should note that Descartes considered only those polyhedra that were deformable to a sphere. However, both Euler's formula and Descartes' theorem can be extended to other surfaces. The external angles of a polyhedral torus, for example, sum to 0.

For curves in the Euclidean plane, we saw that the change in direction measured by the external angle occurs gradually for a smoothly curving curve, and the amount of change per unit arc length is called *curvature.* The external angle on a surface measures the "sharpness" at a vertex where the inclination of the faces abruptly changes. But again, on a smooth surface, this change occurs gradually. Therefore, to define Gaussian *curvature* as the amount of change per unit surface area makes sense. In fact, as noted earlier, integrating Gaussian curvature over a region on a surface yields a quantity called *total curvature,* and this phenomenon is what the sum of the external angles measures on a polyhedral surface. For both curves and surfaces, therefore, external angles provide a way to measure total curvature. Virtually any theorem about a smooth surface involving total curvature can be transformed into a theorem about polyhedral surfaces by replacing the words *total curvature* with the words *sum of the measures of the external angles*. The paper models discussed subsequently, therefore, are legitimate approximations for the geometry of curved surfaces.

The Geometry of Geodesics

Lines in the plane minimize distance in the sense that given two points, the shortest distance between them is measured along a line. For measurements taken along a surface, those curves that minimize distance are called *geodesics*. The lines on a plane are nothing more than the geodesics of a flat surface, and Euclidean geometry is just a particular example of the geometry of geodesics on surfaces.

We have already seen that the sum of the external angles is 4π for a cube, and coincidentally, the total (Gaussian) curvature for a sphere is also 4π. The geodesics on these two surfaces, as we shall see, also behave in a similar way.

On the sphere, the geodesics are the great circles, and on the sphere, a triangle with geodesic sides has an angle sum greater than 180 degrees, or π radians. The particular angle sum turns out to be directly related to the total curvature contained in the triangle.

For example, on a sphere a triangle can have three right angles. We can put one vertex at the north pole and two on the equator as in figure 4.5. This triangle contains one-eighth of the surface area of the sphere, and so it also must contain one-eighth of the total curvature, in particular, $C = 4\pi/8 = \pi/2$. The angle sum of the triangle is $3\pi/2$, which is $\pi/2$ greater than the Euclidean angle sum of π radians. The agreement between the angle sum excess and the curvature contained within the triangle is not a coincidence. In fact, the angle sum of a triangle is always $\pi + C$, where C is the curvature contained in the triangle. This outcome is the spherical version of the fundamental theorem of surface geometry, the Gauss-Bonnet theorem.

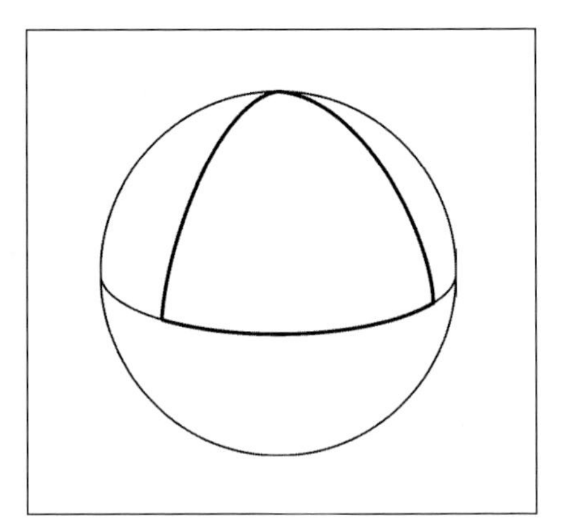

Fig. 4.5. A triangle on a sphere can have three right angles.

On the cube, a geodesic must be as straight as possible and, equivalently, must minimize distance. Plans for a paper model of a cube are shown in figure 4.6. Four straight-line segments are also drawn. Since folding along the edges will not change the length of the segments, nor will it change any small measurements taken along the surface, these segments will become geodesics on the cube after being folded and taped. Note also that after the edges are taped together, the segments AD and DC come together just as AE and EB do. The fact that AB, BC, and CA are each geodesic segments follows, and we have a triangle with three right angles. We again have a triangle with an angle sum $\pi/2$ more than the

Euclidean angle sum of π radians. Furthermore, this triangle contains one vertex with an external angle of $C = 2\pi - 3\pi/2 = \pi/2$. In other words, the excess angle sum equals the curvature contained within.

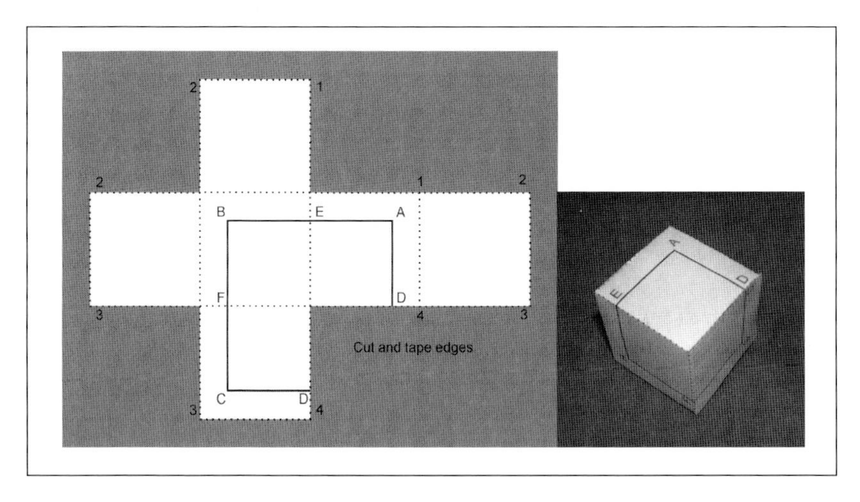

Fig. 4.6. A triangle on a cube. The triangle has three right angles, an angle sum of 270 degrees, and an angle-sum excess of 90 degrees, or $\pi/2$ radians.

Note, however, that for triangles on a sphere or a cube, the notion of "interior" is somewhat ambiguous. The three edges, in fact, separate the cube's surface into two regions. If we consider the larger region as the "interior" of the triangle, then the triangle's three "interior" angles each measure $3\pi/2$ radians for an angle sum of $9\pi/2$ radians. The angle sum excess, therefore, is $7\pi/2$, which is equal to the sum of the seven external angles of the cube, each measuring $\pi/2$ radians, that lie in the larger region. This result again matches the total curvature contained within this interpretation of the triangle, and so both interpretations of the triangle have this property. In general, for any triangle, we have

$$\Sigma = \pi + C, \tag{5}$$

where Σ is the angle sum of a triangle and C is the total curvature contained in the triangle.

The sphere is an example of an *elliptic* geometry, in which we see triangles with angle sums greater than π radians. A differential geometer would equate elliptic geometry with *positive* Gaussian curvature. In *hyperbolic* geometry, we see triangle angle sums that are less than π radians, and *negative* Gaussian curvatures characterize hyperbolic geometries. As we shall see, the corresponding solid angles have *more* than 2π radians surrounding them, and so extra radian measure beyond 2π can naturally be viewed as a *negative* external angle.

In his book *The Shape of Space*, Jeffrey Weeks (1985) presents a visualization of the hyperbolic plane, and therefore negative curvature, with something he calls *hyperbolic paper*, which he attributes to the well-known geometer William Thurston. The idea is based on the fact that equilateral triangles tile the plane with six triangles coming together at each vertex and each contributing an angle measuring $\pi/3$ radians around that vertex. These six angles total 2π radians around each of these vertex points. Hyperbolic paper, in contrast, squeezes *seven* equilateral triangles around each vertex, creating a very wavy surface and an approximation to the hyperbolic plane.

In figure 4.7 the seven equilateral triangles, when taped together, create a small piece of hyperbolic paper. As in the cube example, several line segments are drawn in this figure, and after the seven triangles are taped together, the lines come together to form geodesic ΔABC with angle sum $2\pi/3$ radians. In constructing geodesics across "cuts," one must make sure that vertical angles formed by the cut and the geodesic are equal. The photograph shows what the paper looks like after being taped together. Note that the paper takes a saddle shape, which is characteristic of a hyperbolic surface with negative curvature. The triangle in this figure has an angle sum that is less than π radians; in particular, the angle sum "excess" is $-\pi/3$ radians. This result agrees with the fact that the external angle at the central vertex of the surface is $-\pi/3$ radians. Again the deviation of the angle sum of a triangle from π radians equals the total curvature contained in the triangle.

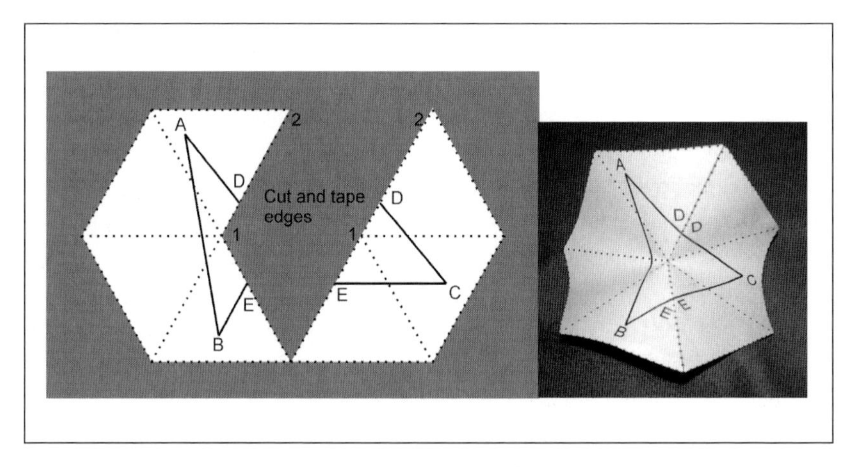

Fig. 4.7. A triangle on hyperbolic paper. The angle at the top is 30 degrees, at the bottom is 40 degrees, and on the right is 50 degrees, for an angle sum of 120 degrees and an angle-sum deficit of 60 degrees, or $\pi/3$ radians.

Exploring Euclidean and Non-Euclidean Geometry

Virtually all the phenomena observed in Euclidean and non-Euclidean geometry can be explored concretely on our paper models. For example, Euclid's fifth postulate states that given a figure like that shown in figure 4.8 with $\alpha < \pi/2$, the lines l and m must intersect. In hyperbolic geometry, however, the two lines might not intersect. In Euclidean geometry the distance between the lines decreases linearly, but in hyperbolic geometry this rate of decrease may get smaller. Negative curvature between the lines lessens the rate of decrease in the distance between l and m. In fact, total curvature equal to $-\alpha$ between the lines stops this decrease completely. The ability to contain this amount of curvature in a small region helps make this phenomenon easier to see.

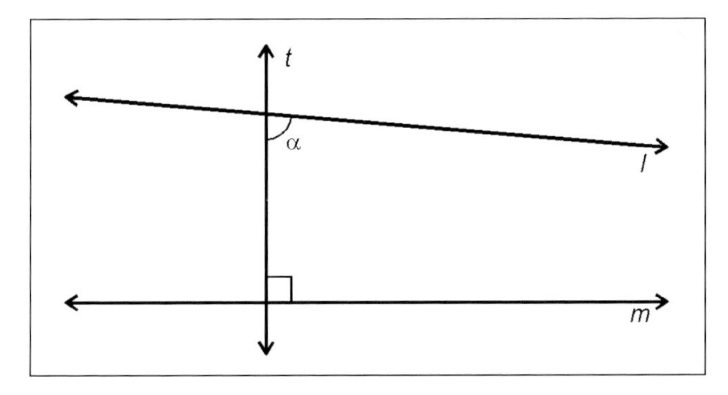

Fig. 4.8. Two lines cut by a transversal. In Euclidean geometry, if $\alpha < 90$ degrees, then l and m intersect.

If $\alpha = \pi/4$, then curvature totaling $-\pi/4$ can prevent these lines from intersecting, and we can build a model that illustrates this outcome. Adding a $\pi/4$-radian wedge gives us the necessary Cartesian curvature, and we can construct a model as in figure 4.9. Note that since the edges on a polyhedral surface do not affect the geometry, we have no need to crease edges in the paper models. With the introduction of negative curvature, the lines l and m diverge in one part of the model and remain equidistant in another. Also notice that $ABCD$ has three right angles and one acute angle. A quadrilateral with three right angles is called a *Lambert quadrilateral* after one of the mathematicians whose work paved the way for the development of non-Euclidean geometry. In Euclidean geometry the fourth angle of a Lambert quadrilateral is necessarily a right angle, but in hyperbolic geometry it is always acute. This outcome illustrates how the history of non-Euclidean geometry can be investigated with these paper models by building counterexamples to Euclidean objects.

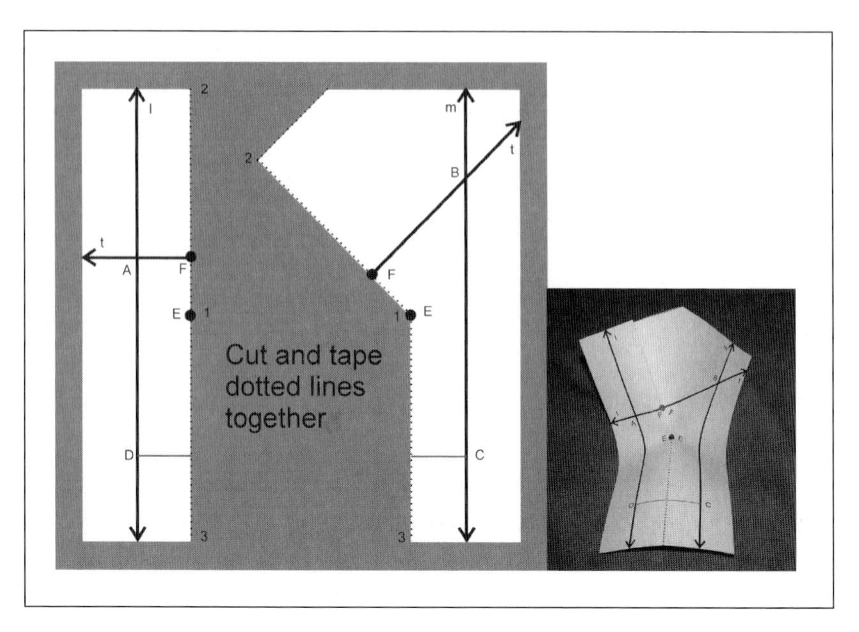

Fig. 4.9. Two lines cut by a transversal near a point with negative curvature. The interior angles below the transversal measure 90 degrees and 45 degrees, but the two lines do not intersect.

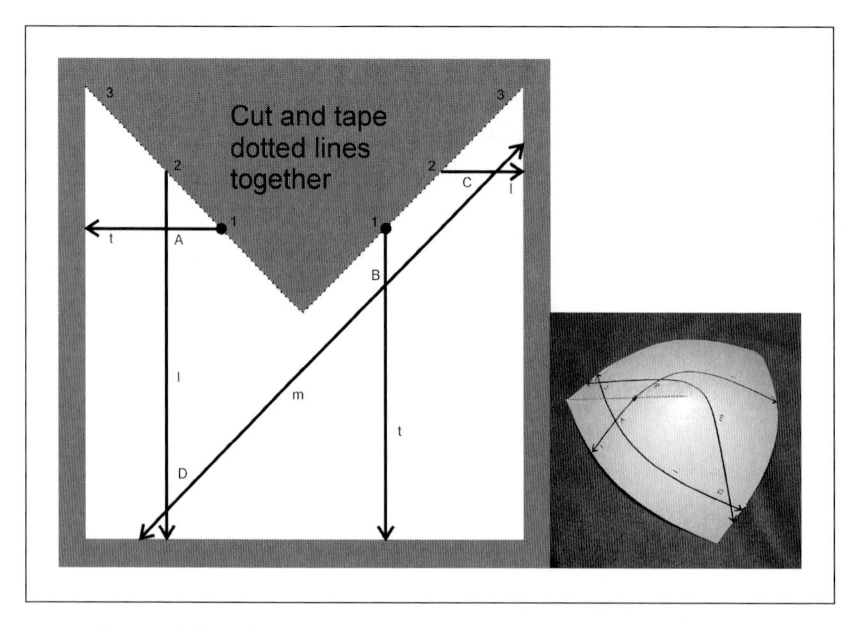

Fig. 4.10. Two lines cut by a transversal near an "elliptic" vertex. The interior angles, $\angle A$ and $\angle B$, measure 90 degrees and 135 degrees, respectively, but the two lines still intersect on that side of the transversal.

Elliptic comparisons with Euclid's fifth postulate can be made just as easily. In figure 4.8, if $\alpha > \pi/2$, then the lines l and m can still intersect on the right if enough positive curvature occurs between the lines. For example, when $\alpha = 3\pi/4$ radians, in a Euclidean plane l and m diverge on that side of the transversal. Positive curvature of $C = \pi/4$ is sufficient to stop that divergence, and more curvature forces the lines to intersect. In figure 4.10, positive curvature of $C = \pi/2$ is introduced between l and m by removing a $\pi/2$-radian wedge. This removal results in the two lines' intersecting on the side of the transversal with interior angles summing to more than π radians. In this instance, the lines l and m have two points of intersection. Here we can see why in elliptic geometry we can have "2-gons," which are impossible in Euclidean geometry.

Summary

Through all these examples, we see that a direct relationship exists between the curvature of a surface and such properties as the angle sum of a triangle. We have seen how total curvature on paper models can be used to complement and contrast with the phenomena encountered in the usual Euclidean and non-Euclidean geometries. These polyhedral models make possible the construction of explicit, concrete models with precise specifications, and with these come the ability to explore isolated properties by comparing and contrasting spaces with and without these properties. This ability to play with and manipulate geometric spaces is crucial to reaching a deep understanding of geometry.

REFERENCES

Federico, Pasquales J. *Descartes on Polyhedra.* New York: Springer Verlag, 1982.

Phillips, Tony. "Descartes's Lost Theorem." American Mathematical Society Web Site Feature Column, September 1999. www.ams.org/featurecolumn/archive/descartes1.html.

Weeks, Jeffrey. *The Shape of Space.* New York: Marcel Dekker, 1985.

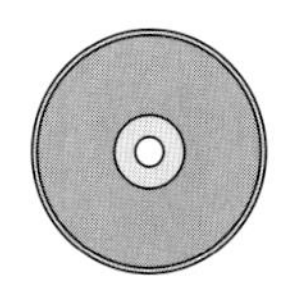

Blackline masters of the curvature templates for figures 4.4, 4.6, 4.7, 4.9, and 4.10 are found on the CD-ROM disk accompanying this Yearbook.

Prairie Plants: Exploring Fractal Forms in Nature

Dane Camp
Erich Hauenstein

NATURE is the ultimate geometry laboratory. Much of the geometry taught in school has roots in efforts of ancient civilizations to understand the universe. However, the jagged and self-similar shapes that are inherent in many forms found in nature are not easily replicated by using traditional Euclidean geometry, but rather by using other tools, such as transformations of coordinates, image compression, and fractal geometry. The two trees shown in figures 5.1 and 5.2 illustrate the contrast between Euclidean and fractal forms. Investigating fractal geometry enables students to explore and be creative with mathematics by discovering shapes and figures that are reminiscent of what they observe in nature. By examining these relationships through experimentation, students can not only see applications to the world outside but also better understand geometric relationships and develop mathematical habits of mind.

The discussion that follows consists of two parts. The first demonstrates how to generate realistic plantlike structures by hand using geometric transformations. The second formalizes the process and illustrates how to generate similar diagrams using technology by way of iterated function systems. We have used the hand-drawing techniques with students starting as early as sixth grade. The formal procedure has been used in varying degrees of complexity with high school classes of sophomores and above. The more sophisticated the students' knowledge of matrix algebra and transformations of coordinates, the better their understanding of the mechanics of the procedure. Such an understanding, however, is not required for students to appreciate the beauty of the process. Although none of the ideas are new, the interplay between the formal and informal perspectives never ceases to fascinate both novices and geometric veterans alike.

Fig. 5.1. Rudimentary Euclidean tree

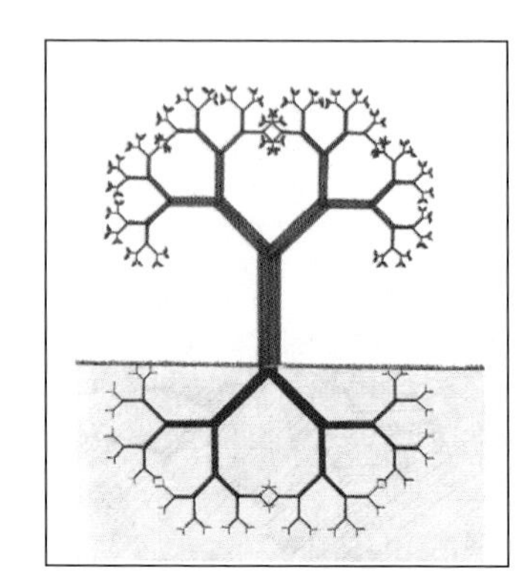

Fig. 5.2. Rudimentary fractal tree

Hand Drawings of Prairie Plants

The four "prairie plant" illustrations in figures 5.3, 5.4, 5.5, and 5.6 were formed by combining an iterative mathematical procedure with an artistic perspective to emulate a living plant that one might find in a field or prairie. Probably the most striking thing to note about these illustrations is how simply they can be generated. Yet when combined with a little color and artistic imagination, the iterative process produces pictures of very complicated and aesthetically beautiful plants similar to those found during a leisurely walk through a wood or glen.

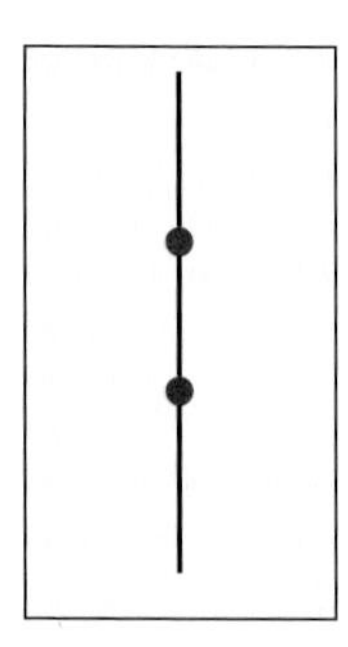

Fig. 5.3. Stage 0

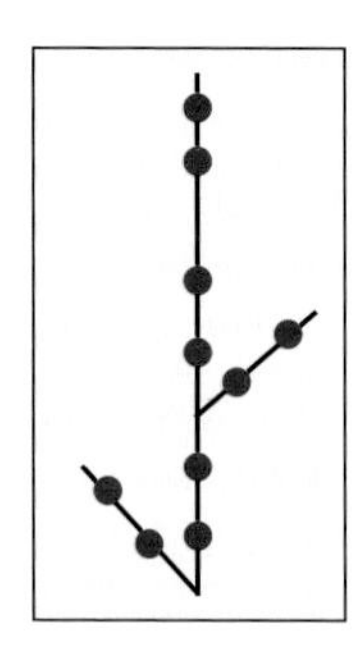

Fig. 5.4. Stage 1

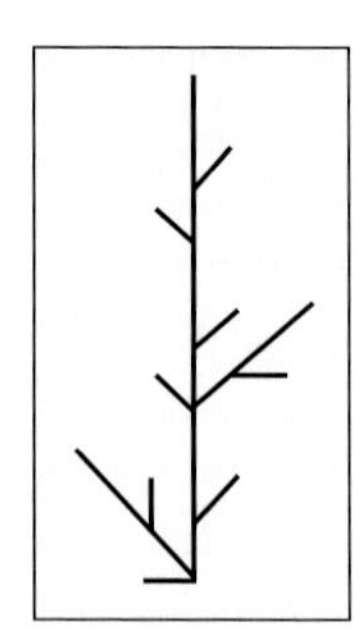

Fig. 5.5. Stage 2

The Prairie Flower

Draw a vertical line (fig. 5.3), and divide it into three equal segments. We refer to this initial figure as stage 0. From the base of both the *bottom* and *middle* segments, add a segment of equal length at a 45-degree angle. Put the new segment facing to the left from the bottom segment and to the right from the middle segment, as shown in figure 5.4, referred to as stage 1.

Perform the same operation on each of the five "subsegments" of this first stage to derive stage 2, as in figure 5.5. Continue the process on each stage until successive iterations produce no apparent changes in the overall figure. Accent the final diagram: for example, place orange flowers at the end of stage 1 and yellowish orange blossoms at the ends of stage 2, as shown in figure 5.6.

Fig. 5.6.
Prairie flower

The Prairie Grass

Draw stage 0 as a vertical segment (fig. 5.7), and divide it into three equal parts, as before. This time, however, add segments of equal length at 30-degree angles, putting one segment on the base of the *middle* segment angled to the left and the other on the base of the *top* segment angled to the right, as shown in figure 5.8, to get stage 1. Again, continue this process to get stage 2 by treating each subsegment with the same two operations, as in figure 5.9. Continue at each stage until successive iterations produce no apparent change to the overall figure. Use some creativity to touch up the final diagram: for example, color the entire figure brown, but at the extremities of the last stage place small yellow buds as illustrated in figure 5.10.

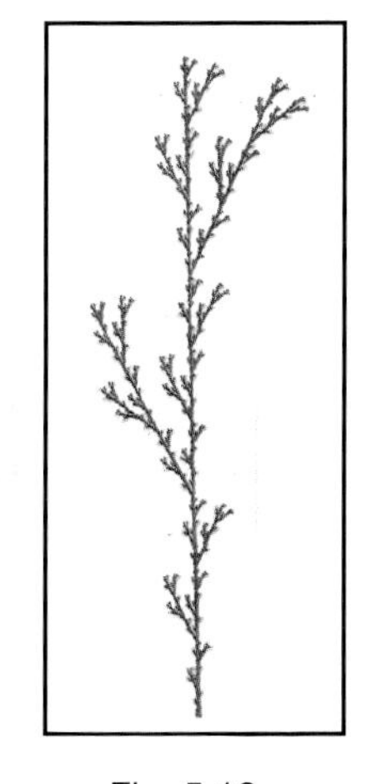

Fig. 5.10.
Prairie grass

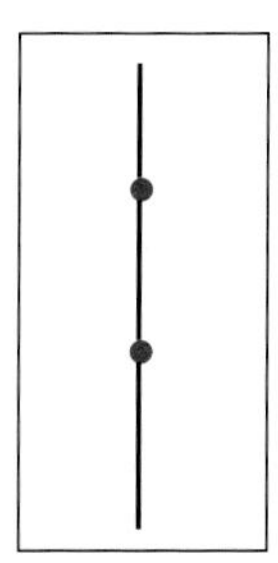

Fig. 5.7. Stage 0

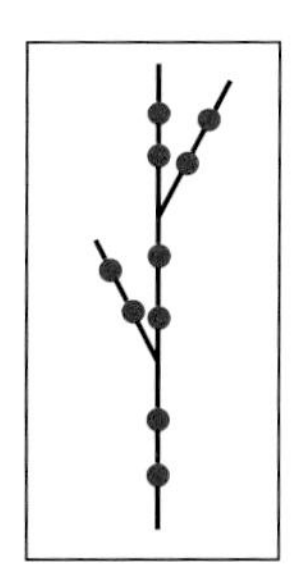

Fig. 5.8. Stage 1

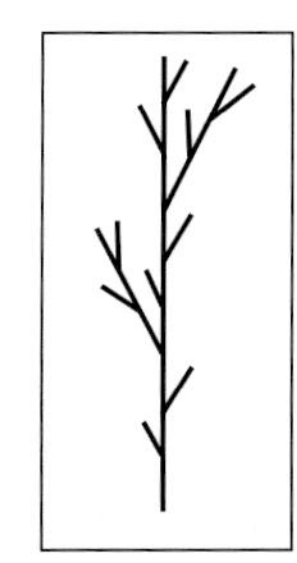

Fig. 5.9. Stage 2

The Prairie Thistle

This time divide the original vertical stage-0 segment (fig. 5.11) into *four* equal parts. Where the bottom and second segments meet, insert a segment of equal length at a 90-degree angle to the right. Also, where the third and top segments meet, insert a segment of equal length at a 90-degree angle to the left, as shown in figure 5.12. Repeat the addition of segments with one-fourth the length to each of the subsegments using the same rules on stage 1 to get stage 2, as illustrated in figure 5.13. Iterate and garnish as desired. We have colored the entire figure green but have left the extremities of the final stage colored black to signify thorns. For effect, a purple thistle flower is placed on top in figure 5.14.

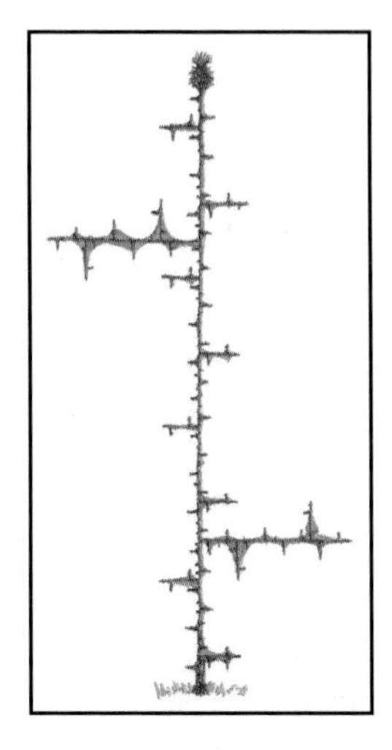

Fig. 5.14.
Prairie thistle

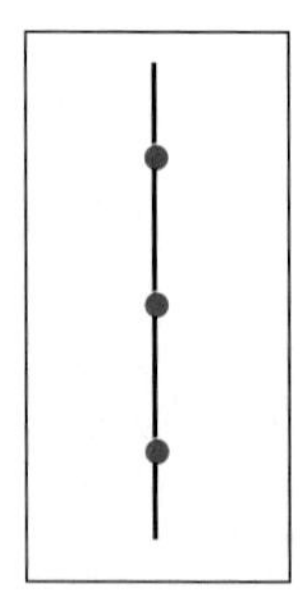

Fig. 5.11. Stage 0

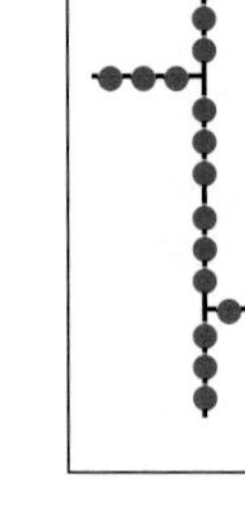

Fig. 5.12. Stage 1

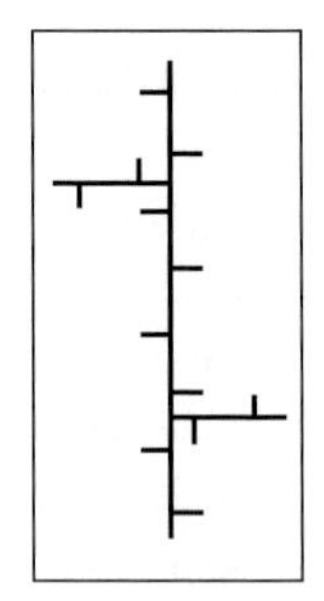

Fig. 5.13. Stage 2

Queen Anne's Lace

For the final example, we try something a little different. Draw the vertical stage-0 segment (fig. 5.15) but divide it into only *two* equal parts. This time, however, we derive stage 1 by *removing* the upper half and *replacing* it with two segments of equal length symmetrically placed to include a 60-degree angle between them, as shown in figure 5.16. Naturally, we continue by dividing each of the three segments into two congruent segments. In the next iteration, however, we do not remove the upper half of the *base* segment; instead, we leave it so as to preserve a connected stem for stage 1, as illustrated in figure 5.17. Notice

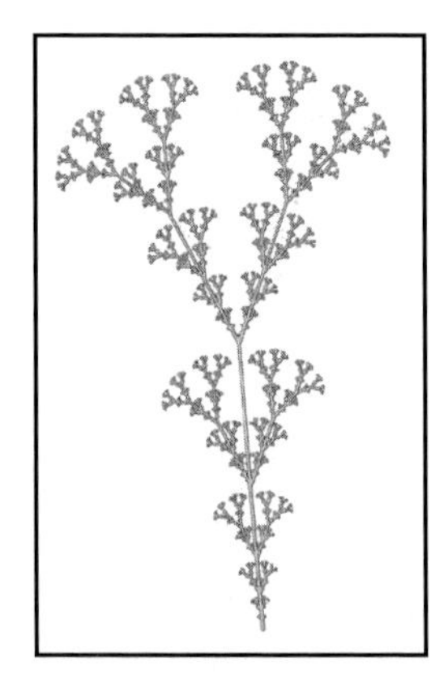

Fig. 5.18.
Queen Anne's Lace

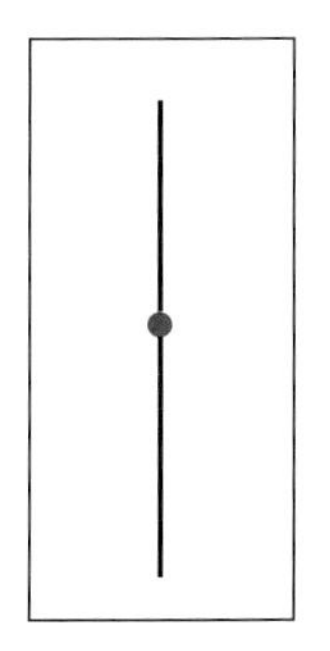

Fig. 5.15. Stage 0

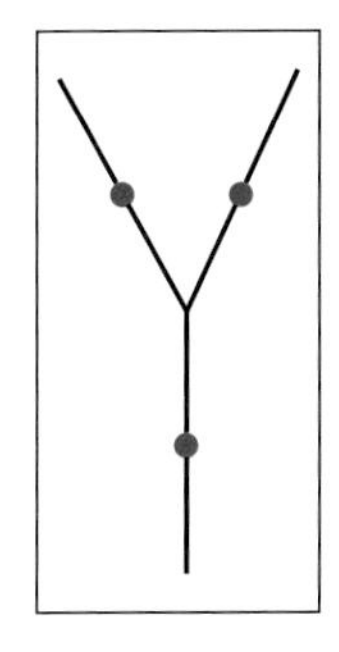

Fig. 5.16. Stage 1

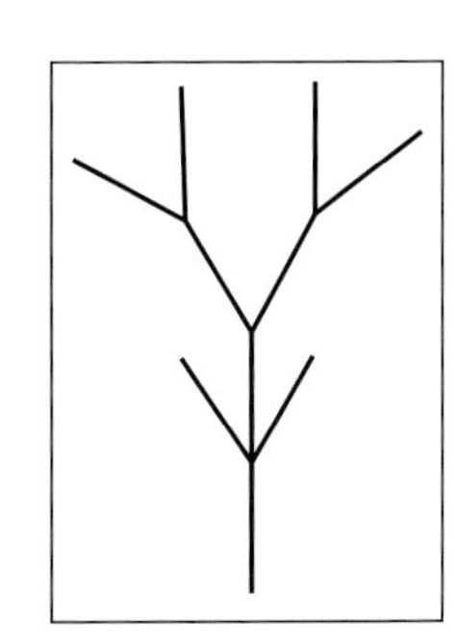

Fig. 5.17. Stage 2

that the upper half *was* removed for the figures added to the two branches. As we continue the iterations, we appreciate that this slight modification keeps the plant connected. For our final figure, we have colored the entire plant green, except for the extremities, which are colored white (fig. 5.18).

Simulating Prairie Plants Using Iterated Function Systems

All the figures that we have generated can be drawn using technology. Computers can quickly and accurately provide as much detail as we require. Here, however, we use the rather primitive computing and graphing capability of the calculator to illustrate the simplicity of the procedure. The transformations that we use are familiar to students of analytic geometry and are introduced as a culminating activity that assists in making the connections between the matrices and the transformations they represent. Let us review these briefly, introduce an algorithm for applying iterated function systems, and then illustrate how to use technology to simulate the plants we constructed by hand.

Transformations of Coordinates

In what follows, we start with a square in the coordinate plane with vertices at (–1, 0), (1, 0), (1, 2), and (–1, 2). Points lying within this square can be transformed through scaling (stretching or shrinking the point's distance from the x- or y-axis), rotation (turning the point counterclockwise about the origin), and translation (shifting the point in the x or y direction a specified distance).

Here are the transformations and the order in which they will be performed. Note that the output of one transformation becomes the input for the next one.

1. Scaling a point with original coordinates of (x, y), with respect to the origin, first by a horizontal factor of s_x transforms it to coordinates (x', y) and then by a vertical scale factor of s_y transforms it to coordinates (x', y'), as shown in figure 5.19.

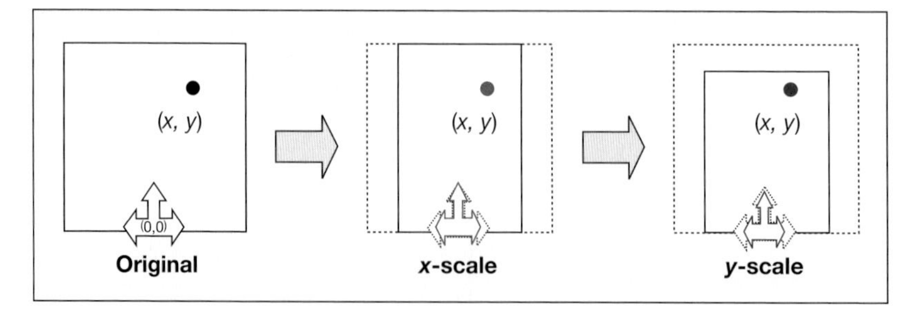

Fig. 5.19. Scaling a point (x, y) first by a horizontal factor s_x and then by a vertical factor s_y

This transformation has the matrix representation

$$\begin{bmatrix} x' \\ y' \end{bmatrix} = \begin{bmatrix} s_x & 0 \\ 0 & s_y \end{bmatrix} \bullet \begin{bmatrix} x \\ y \end{bmatrix} = \begin{bmatrix} s_x x \\ s_y y \end{bmatrix}. \quad (1)$$

2. Rotation of the resulting point (x', y') about the origin through an angle of r degrees yields (x'', y''); note that x' and y' act as hypotenuses of the right triangles in the diagram (fig. 5.20).

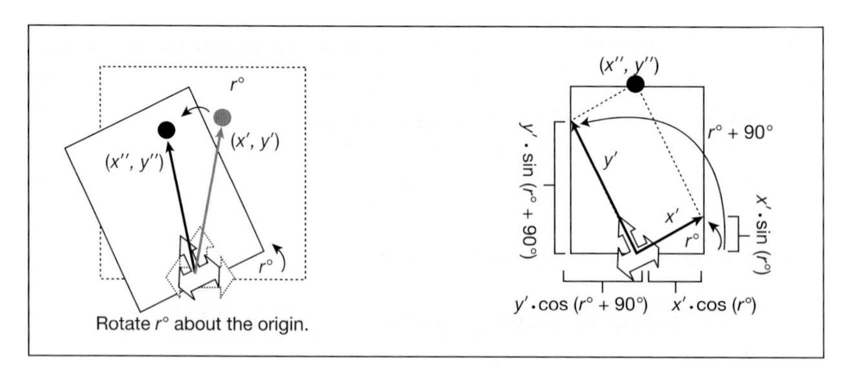

Fig. 5.20. Rotating point (x', y') around the origin through an angle of r degrees

So the coordinates for (x'', y'') after rotation are

$$x'' = x'\cos(r^\circ) + y' \cos(r + 90^\circ) = s_x x\cos(r^\circ) - s_y y\sin(r^\circ),$$
$$y'' = x'\sin(r^\circ) + y' \sin(r + 90^\circ) = s_x x\sin(r^\circ) + s_y y\cos(r^\circ).$$

The matrix representation of the transformations so far is

$$\begin{bmatrix} x'' \\ y'' \end{bmatrix} = \begin{bmatrix} \cos(r^\circ) & -\sin(r^\circ) \\ \sin(r^\circ) & \cos(r^\circ) \end{bmatrix} \bullet \left(\begin{bmatrix} s_x & 0 \\ 0 & s_y \end{bmatrix} \bullet \begin{bmatrix} x \\ y \end{bmatrix} \right)$$

(2)

$$= \begin{bmatrix} s_x x\cos(r^\circ) - s_y y\sin(r^\circ) \\ s_x x\sin(r^\circ) + s_y y\cos(r^\circ) \end{bmatrix}.$$

3. Finally, translating the point (x'', y'') shifting by t_x horizontally and t_y vertically yields (x''', y''') (fig. 5.21).

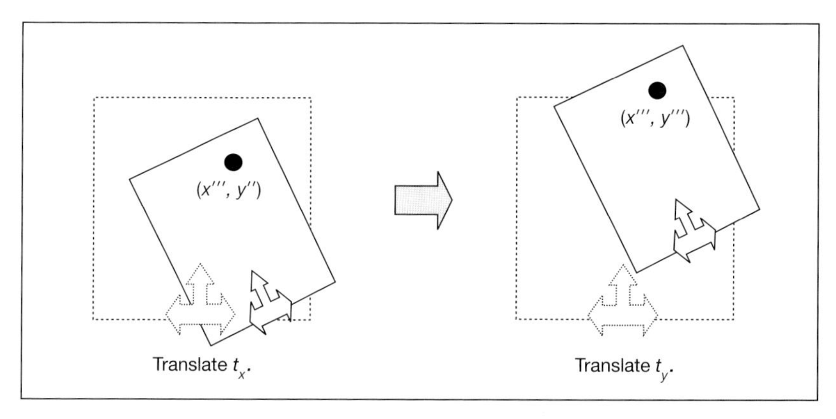

Fig. 5.21. Translating point (x'', y''), shifting by t_x horizontally and t_y vertically yields (x''', y''').

This shift gives the overall transformation

$$\begin{bmatrix} x''' \\ y''' \end{bmatrix} = \begin{bmatrix} \cos(r^\circ) & -\sin(r^\circ) \\ \sin(r^\circ) & \cos(r^\circ) \end{bmatrix} \bullet \left(\begin{bmatrix} s_x & 0 \\ 0 & s_y \end{bmatrix} \bullet \begin{bmatrix} x \\ y \end{bmatrix} \right) + \begin{bmatrix} t_x \\ t_y \end{bmatrix}$$

(3)

$$= \begin{bmatrix} s_x x\cos(r^\circ) - s_y y\sin(r^\circ) + t_x \\ s_x x\sin(r^\circ) + s_y y\cos(r^\circ) + t_y \end{bmatrix}.$$

Note that it is possible for the transformed coordinates to end up outside the original window.

Transformations and Prairie Plants

Next consider the construction of a prairie plant from a transformational point of view. For example, in the first drawing of the prairie flower, we perform five actions going from stage 0 to stage 1, each action corresponding to one of the five smaller segments. Each action is the result of three transformations—scaling, rotating, and translating. Let us examine each from an original 2-by-2 window. The first action, or *lens,* represents the bottom segment. The horizontal and vertical scales are both 1/3 because each segment is 1/3 the original, and we did not rotate or translate (fig 5.22). The term *lens* comes from an analogy with a copy machine that can not only scale but also translate and rotate an image (see Peitgen, Jurgens, and Saupe [1992, pp. 23–26]). These "lens 1" transformations are summarized by the following table:

Lens	s_x	s_y	r°	t_x	t_y
1	1/3	1/3	0°	0	0

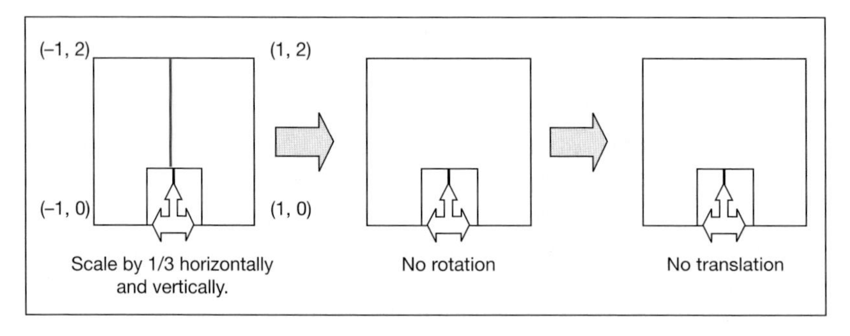

Fig. 5.22. Lens 1 transformations

If we let the second lens represent the "middle" segment, again the scaling factors are both 1/3, and no rotation occurs. However, the figure is shifted up to the middle, so the vertical translation is 2/3—since the height of the window is 2 and the scale factor is 1/3 (fig. 5.23). These lens 2 transformations are summarized by the following table:

Lens	s_x	s_y	r°	t_x	t_y
2	1/3	1/3	0°	0	2/3

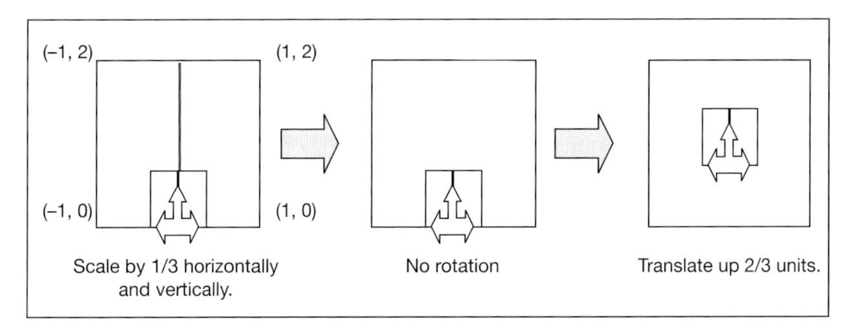

Fig. 5.23. Lens 2 transformations

Allowing the third lens to represent the "top" segment, again we have a horizontal and vertical scale factor of 1/3 and no rotation; however, we must translate up 4/3 units—two-thirds of the height of 2 (fig. 5.24). These lens 3 transformations are summarized by the following table:

Lens	s_x	s_y	$r°$	t_x	t_y
3	1/3	1/3	0°	0	4/3

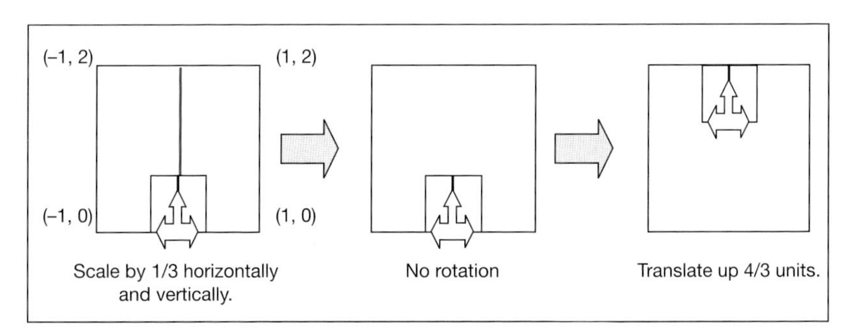

Fig. 5.24. Lens 3 transformations

For the fourth lens, we encode the segment that is tilted to the left. Although the scale factor is still 1/3, the segment is rotated 45 degrees counterclockwise and then translated up as the second lens was (fig. 5.25). These lens 4 transformations are summarized by the following table:

Lens	s_x	s_y	$r°$	t_x	t_y
4	1/3	1/3	45°	0	2/3

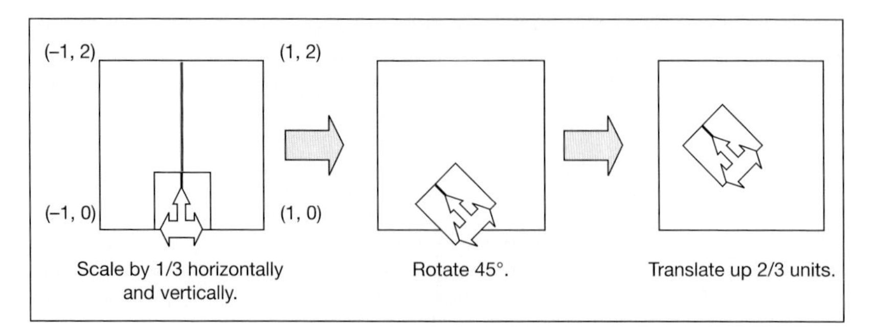

Fig. 5.25. Lens 4 transformations

Finally, the last lens represents the segment that is tilted to the right. The scale factors are the same, but since it is rotated clockwise 45 degrees, the rotation factor is –45 degrees. No translation occurs this time (fig. 5.26). These lens 5 transformations are summarized by the following table:

Lens	s_x	s_y	$r°$	t_x	t_y
5	1/3	1/3	–45°	0	0

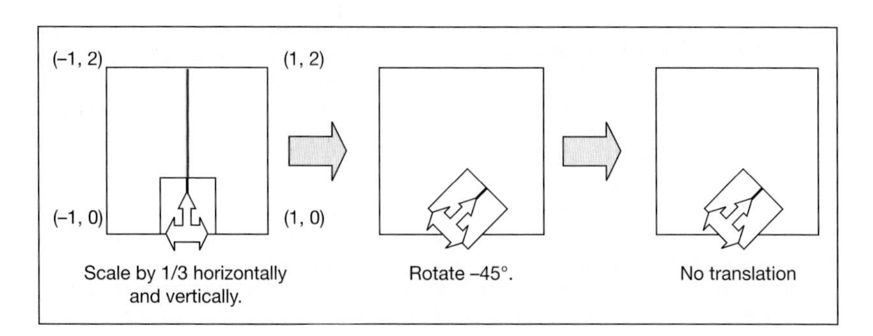

Fig. 5.26. Lens 5 transformations

Iterated Function Systems

We are now ready to use the power of an *iterated function system* (IFS) to produce a fractal image of the prairie flower. We accomplish this task by taking an arbitrary point in the plane, choosing one of the five lenses at random, and performing the transformations prescribed by that lens on that point. We plot the output from this first iteration. Then we repeat the process by again choosing a lens at random and performing its transformations on the output from the previous stage. As hundreds of points are generated and plotted in this way, they rapidly become very close to the desired image. In technical terms, the fractal is an *attractor* for the IFS. More details about this process can be found in Peitgen, Jurgens, and Saupe (1992).

Figure 5.27 presents a diagram and a table that summarizes all the lenses for the prairie flower (fig. 5.27).

Drawing 1 (prairie flower)					
Lens	s_x	s_y	$r°$	t_x	t_y
1	1/3	1/3	0°	0	0
2	1/3	1/3	0°	0	2/3
3	1/3	1/3	0°	0	4/3
4	1/3	1/3	45°	0	2/3
5	1/3	1/3	–45°	0	0

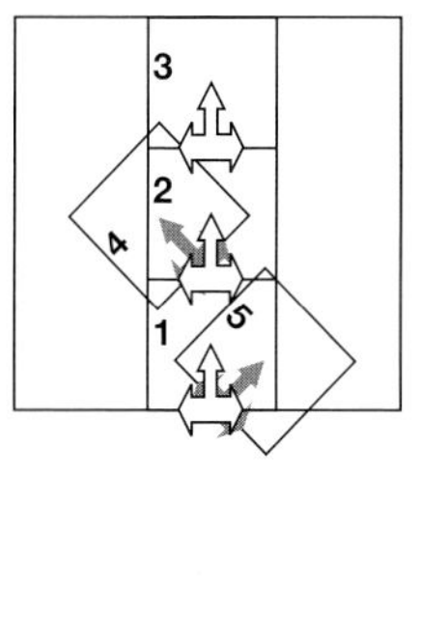

Fig. 5.27. Summary of all lenses for the prairie flower

Now that we have a description of the iterated function system for this image, all we need to do is write a program. We can do so for a computer or even a graphing calculator. First we use the matrix feature of the calculator to store data from the table. To simplify the process, we eliminate the horizontal translation in the table because it is always zero. To make computations go quickly, we enter two values for the angle of rotation, one for the sine and one for the cosine. So for the first drawing we enter the following table into a 5-by-5 matrix called "lens."

x scale	y scale	$\cos(r°)$	$\sin(r°)$	y trans
1/3	1/3	1	0	0
1/3	1/3	1	0	2/3
1/3	1/3	1	0	4/3
1/3	1/3	.707	.707	2/3
1/3	1/3	.707	–.707	0

We start with a single point at the origin (0, 0), although the choice of starting point is arbitrary. The program randomly selects one row of the table (corresponding to one of the five lenses) to copy the original point. Then the program randomly selects another row and copies the output of the first iteration.

This process continues for 2000 iterations, after which an image has emerged. The program that follows was designed for a TI-89 calculator but can be easily adapted to other technology and modified to enhance performance. Notice that the matrix operations represented by equation (3) are embedded in the thirteenth and fourteenth lines of the program.

```
ClrDraw              [This command clears the drawing window.]
dim(lens)[1]→d       [d will represent the number of rows, hence lenses.]
0→n                  [n will be the number of copies the machine makes.]
0→x                  [Set the original to the origin where x = 0]
0→y                  [and y = 0.]
While n<2000         [We will make 2000 copies or points.]
rand( )→p            [Select a random number between 0 and 1.]
1→c                  [c is the counter in a loop; initially c is the first lens.]
While p>c/d          [Selects which lens will be used with equal probability.]
c+1→c                [Increments c.]
EndWhile             [Ends loop starting at While p>c/d.]
x→w                  [w is a temporary variable because x changes.]

x*lens[c,1]*lens[c,3]-y*lens[c,2]*lens[c,4]→x                 [Calculates the x value.]
w*lens[c,1]*lens[c,4]+y*lens[c,2]*lens[c,3]+lens[c,5]→y       [Calculates the y value.]

PtOn x,y             [Plots the point.]
n+1→n                [Increments n.]
EndWhile             [Ends loop starting with While n < 2000.]
```

Before running the program, the calculator has to be set up. We could do so within the program itself; however, to simplify things we do not clutter the program any further. We just make sure to set a window that allows x to range from -1 to 1 and y from 0 to 2. Because the aspect ratio on the TI-89 distorts things a bit, we use window with -2 to 2 for x and 0 to 2 for y. We also need to make sure all functions in "Y = " are off and that the axes are turned off. If the program is run with the matrix for the first drawing, the output should look like figure 5.28.

The outputs and the matrix "lenses" for the other three drawings are shown in figures 5.29, 5.30, and 5.31.

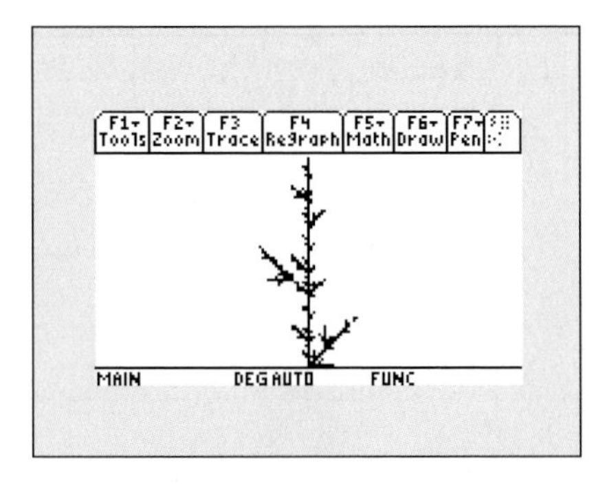

Fig. 5.28. The output of the program run with the matrix for the first drawing (prairie grass)

x scale	*y* scale	cos($r°$)	sin($r°$)	*y* trans
1/3	1/3	1	0	0
1/3	1/3	1	0	2/3
1/3	1/3	1	0	4/3
1/3	1/3	.866	.5	4/3
1/3	1/3	.866	–.5	2/3

Fig. 5.29. The output of the program run with the matrix for the prairie flower

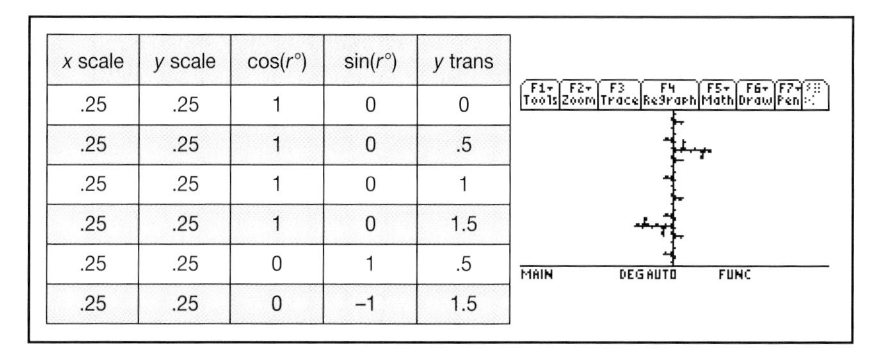

x scale	*y* scale	cos($r°$)	sin($r°$)	*y* trans
.25	.25	1	0	0
.25	.25	1	0	.5
.25	.25	1	0	1
.25	.25	1	0	1.5
.25	.25	0	1	.5
.25	.25	0	–1	1.5

Fig. 5.30. The output of the program run with the matrix for the prairie thistle

x scale	*y* scale	cos($r°$)	sin($r°$)	*y* trans
.5	.5	1	0	0
.5	.5	.866	.5	1
.5	.5	.866	–.5	1
0	.25	1	0	.5

Fig. 5.31. The output of the program run with the matrix for the Queen Anne's Lace

When we look at the figures we have generated, their close resemblance to real plants that one might find in the prairie is surprising. Also amazing is the small amount of information that is needed to generate these figures with all their intricate detail. As a matter of fact, if we could approximate the lenses used for a real prairie plant by employing a graphical device with color and better resolution, we would be able to create a reasonable counterfeit of the original. But how can we approximate these lenses? Mathematicians use something called the *collage theorem* to do just that (Barnsley 1988). It allows them to make beautiful "fractal forgeries." This technique can be used to compress data, a field that is of increasing importance in an age in which information is exploding (Barnsley and Hurd 1993). That such simple tools can be combined to give us a glimpse of something so complex is fascinating. We should not be surprised if we remember that the roots of geometry go back to humankind's attempt to understand the universe.

Introducing our students to this simple application of iteration has taught them, and us, a number of invaluable lessons. It allows for learners across a wide range of experience and abilities to become actively engaged in the discovery of meaningful geometric concepts and connections. They get an immediate, concrete, visual example of how mathematics is relevant to their daily lives. In addition, they also develop an appreciation for geometry as more than just a formal exercise—it can be an artistic and creative experience as well. As mathematics teachers, we appreciate the fact that exploring the replication of natural objects using fractal geometry allows us to describe figures we could not handle using more traditional methods, and we naturally enjoy the fact that this process is a wonderful synthesis of many simple, yet powerful, intellectual tools. Playing with such tools and sharing the joy of discovery with others keeps the adventure of learning and teaching mathematics exciting and new.

REFERENCES

Barnsley, Michael. *Fractals Everywhere.* New York: Academic Press, 1988.

Barnsley, Michael, and Lyman P. Hurd. *Fractal Image Compression*. Wellesley, Mass.: A. K. Peters, 1993.

Peitgen, Heinz-Otto, Hartmut Jürgens, and Dietmar Saupe. *Chaos and Fractals: New Frontiers of Science*. New York: Springer-Verlag, 1992.

Folding Polygons to Convex Polyhedra

Joseph O'Rourke

A LINE of research investigation has opened in the last decade that spans two millennia of geometry: from Greek explorations of convex polyhedra to cutting-edge geometrical research today. And yet the topic can be understood by high school and middle school students and lends itself to hands-on experimentation.

The main question driving this research is simply stated:

Question 1: Which polygons can fold to a convex polyhedron?

Unpacking this question requires defining its four technical terms. A *polygon P* is a planar shape whose boundary is composed of straight segments. It may be considered a shape that could be cut as a single piece from a sheet of paper by straight scissors cuts. A *polyhedron Q* is the three-dimensional analog of a two-dimensional polygon. It is a solid in space whose boundary is composed of polygonal *faces*. Because we are concerned mainly with this surface boundary, we use Q to refer to the surface rather than the solid. A *convex* polyhedron is one without dents or indentations. Examples include the five "regular" Platonic solids (tetrahedron, octahedron, cube, dodecahedron, icosahedron), the thirteen "semi-regular" Archimedean solids (e.g., the truncated icosahedron, which is modeled by a soccer ball), or any of an infinite variety of irregular convex polyhedra. Because we discuss only convex polyhedra in this article, the modifier "convex" will often be left implicit. Finally, to *fold* a polygon to a polyhedron means to make creases that allow the polygon be folded to form the surface of a polyhedron, without any wrapping overlap and without leaving any gaps. Another way to view this process is in reverse: a polygon P can fold to a polyhedron Q if Q could be cut open and unfolded flat to P.

Two examples are shown in figure 6.1. Note from figure 6.1(a) that creases of P, which become edges of Q, do not necessarily begin or end at vertices (corners) of P. Note from figure 6.1(b) that a nonconvex polygon might fold to a convex polyhedron. If we view the process in reverse as unfolding, the cuts in both these examples are along polyhedron edges. We see that in general the cuts are arbitrary surface segments.

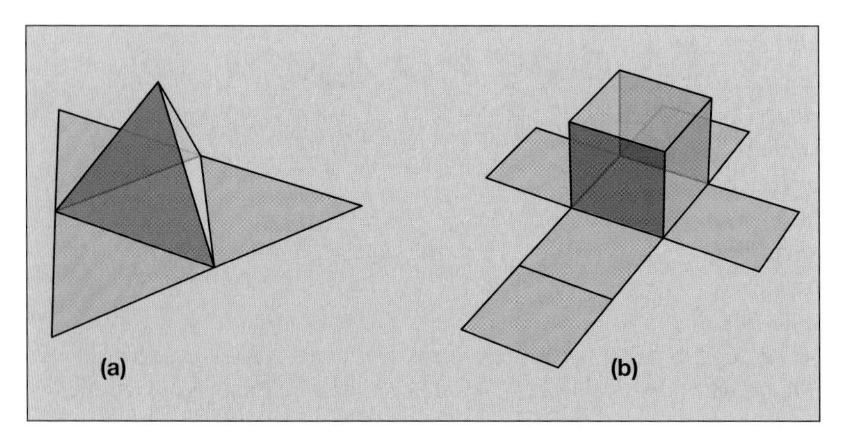

Fig. 6.1. (a) Folding an equilateral triangle to a regular tetrahedron; (b) folding the Latin cross to a cube.

These examples already show that at least some polygons can fold to some polyhedra. This outcome naturally raises this question:

Question 2. Does every polygon fold to some convex polyhedron?

The answer is no; a counterexample is the polygon shown in figure 6.2.

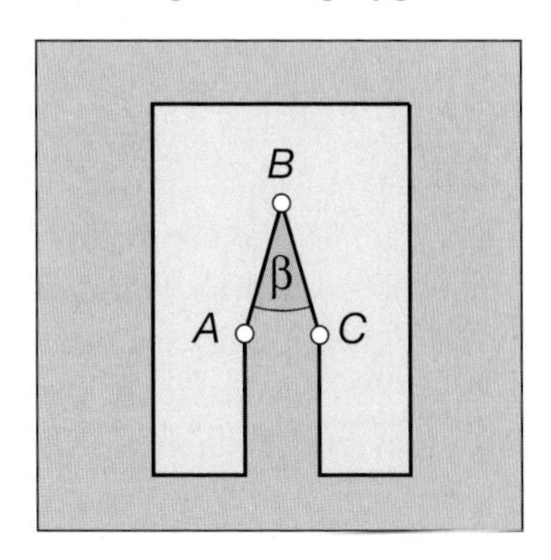

Fig. 6.2. An unfoldable polygon (fig. 25.1 in Demaine and O'Rourke [2007])

We offer an elementary proof that the polygon in figure 6.2 cannot fold to a convex polyhedron. The primary observation is that the sum of the face angles surrounding any true vertex v (corner) of a convex polyhedron is less than 360 degrees. In technical language, the sum of the face angles *incident to* v is less than 360 degrees. This result is one reason that a sixth Platonic solid does not exist: gluing together three equilateral triangles at a vertex produces the tetrahedron, gluing four yields the octahedron, and gluing five gives the icosahedron. But six times 60 degrees is 360 degrees, so six equilateral triangles lie flat and do not form a true vertex. Another way to phrase this constraint, more useful for our purposes, is that the total angle surrounding any point p on a convex polyhedron (vertex or not) is less than or equal to 360 degrees. When the angle is exactly 360 degrees, p lies in the interior of a face or on the interior of an edge; when the total angle is less than 360 degrees, the point is a vertex. This constraint does not hold, incidentally, for nonconvex polyhedra for which no *a priori* bound exists on the total face angle surrounding surface points.

The consequence of the angle constraint is that when we *glue* the perimeter of any polygon P to itself to form the folding, we can never glue more than 360 degrees around any one point, for otherwise, the resulting polyhedron Q would not be convex. Next classify the vertices of a polygon P as either *convex*, having interior angle less than 180 degrees, or *reflex*, having interior angle greater than 180 degrees. The polygon in figure 6.2 has three consecutive reflex vertices (A, B, C), with a very small, complementary exterior angle β at B. All other vertices are convex, with interior angles strictly larger than β.

Next we imagine how we might glue up the perimeter of the polygon in figure 6.2 in the vicinity of the problematic vertex B. We have only two options. Either we "zip" together edges BA and BC, or we glue some other point or points of the perimeter into B. The first possibility forces A to glue to C, exceeding 360 degrees there and violating the angle constraint. So this option is ruled out. The second possibility cannot occur with this polygon because no point on the perimeter has a sufficiently small interior angle to fit inside β at B. This constraint rules out the second possibility, and shows that this polygon cannot fold to any convex polyhedron.

So now we know that sometimes polygons can, and sometimes they cannot, fold to a polyhedron, thus justifying the wording of question 1: *Which* polygons can fold to a polyhedron? Before pursuing this question further, we naturally wonder how common *foldability* in this sense is. This pursuit would lead into a thicket of questions of how to define a "random polygon." Suffice it to say that under reasonable assumptions, foldability turns out to be relatively rare: if we cut out a random polygon of n sides from a sheet of paper, the probability that it will fold to a convex polyhedron approaches zero as n gets large.

Having explained question 1 and explored a few basic issues, we admit that as yet we have no satisfactory answer to it. In particular, we have found no characterization of which polygons fold and which do not, except in certain special cases, explored below. Nevertheless, we now have available an algorithm, implemented in publicly available software, that will take any specific polygon P and tell us whether it can fold, and if so, give some information about the polyhedron Q to which it can fold. Before we can explain this somewhat mysterious statement, we turn to the powerful theorem that sits at the heart of this research.

Alexandrov's Theorem

Alexandrov's theorem is both beautiful and elementary, that is, elementary in statement but not in proof. However, it is not taught in any Western secondary school or undergraduate curriculum as far as I know. Part of the reason is language: the theorem was published in Russian in 1941 (Alexandrov 1941), but it has only recently appeared in English (Alexandrov 2005). Fortunately, the beauty and utility of this theorem are now more widely recognized; it is essential for our topic.

Here I simplify and specialize his theorem to our needs. First, let us define an *Alexandrov gluing* of a polygon to be just what we need for a folding to a convex polyhedron. Three conditions must be satisfied for a gluing to be an Alexandrov gluing:

> *Condition 1:* The gluing must entirely consume the perimeter of the polygon with matches: every point p of the perimeter must be matched with one or more points of the perimeter. Here we allow isolated points to be mateless (or to match with themselves), as we did in figure 6.2 when considering "zipping" in the neighborhood of B.
>
> *Condition 2:* The gluing creates angles totaling no more than 360 degrees at any point. This constraint is our angle condition for convex polyhedra.
>
> *Condition 3:* The gluing should result in a *topological sphere*, that is, a surface that could be deformed to a sphere, not a torus (donut) or a fundamentally twisted shape, and so on, but rather what amounts to a lumpy, closed bag.

This third condition is difficult to state precisely without introducing technical language from topology. Regardless, I hope it is clear that if a gluing has any hope of producing a convex polyhedron, it must be an Alexandrov gluing, for the reason that the three conditions just specify what is obviously necessary—no gaps, the 360-degrees condition, and producing a spherical shape. Here is Alexandrov's celebrated theorem.

> *Theorem (Alexandrov).* Any Alexandrov gluing corresponds to a

unique convex polyhedron (where a doubly covered polygon is considered a polyhedron).

Let us ignore the parenthetical caveat for a moment to emphasize what this theorem is saying: the obvious necessary conditions for a polygon to fold to a polyhedron are also sufficient. Not only that, the resulting polyhedron is unique, meaning that any time we find an Alexandrov gluing, we have created a convex polyhedron. The one catch is that Alexandrov's proof guarantees the existence of the polyhedron but gives no indication of what it looks like!

We have already seen two Alexandrov gluings in figure 6.1, but the foldings were obvious, both owing to their regularity and because the crease lines are self-evident. But let us consider the unusual folding of an equilateral triangle in figure 6.3(a) that, as we shall see, produces the irregular tetrahedron in figure 6.3(c). We can easily check that this folding is an Alexandrov gluing. Condition 1 is satisfied because no perimeter sections are left unmatched. Condition 2 is met: at the gluing together of $\{X, A, B, C\}$, the three interior angles of the triangle sum to 180 degrees and together with the straight angle at X give a total of 360 degrees. Furthermore, the four "pinch" points $\{P, Q, R, S\}$ are glued to themselves, with each total angle equal to 180 degrees, because each of these points lies on the interior of a side of the triangle. All other glued points bring two sides of the triangle together for a total of 180 degrees + 180 degrees, or 360 degrees. That Condition 3 is satisfied is best seen by folding the triangle, which we encourage the reader to attempt by following these instructions.

Start by selecting a point X about one-third along side AB from right to left. Locate P, Q, R, and S, the midpoints of AX, XB, BC, and CA, respectively. Cut out the triangle ABC. Draw segments PR, SR, SQ, and RQ, and crease along them. Fold CR to meet AR, and tape these two segments together. Fold CS to meet BS, and tape these two edges together. You should now be able to tape the edge containing the three vertices of the original triangle (A, B, and C) to edge PQ. The result is something that looks like a bag, which can be nudged to form the irregular tetrahedron shown in figure 6.3(c), the unique convex polyhedron predicted by Alexandrov's theorem. (See the blackline master on the accompanying CD for more detailed instructions.) The folding in figure 6.3(b) is also an Alexandrov gluing, but it produces simply a doubly covered, flat 30-60-90-degree triangle. This result is the reason for the exception clause in Alexandrov's theorem: some gluings produce zero-volume, flat "polyhedra."

Folding Convex Polygons

Although foldability in general is rare, every convex polygon folds to a polyhedron. A *convex polygon* is one without dents: every vertex is convex. Not only do all these polygons fold, they all fold to an infinite variety of polyhedra:

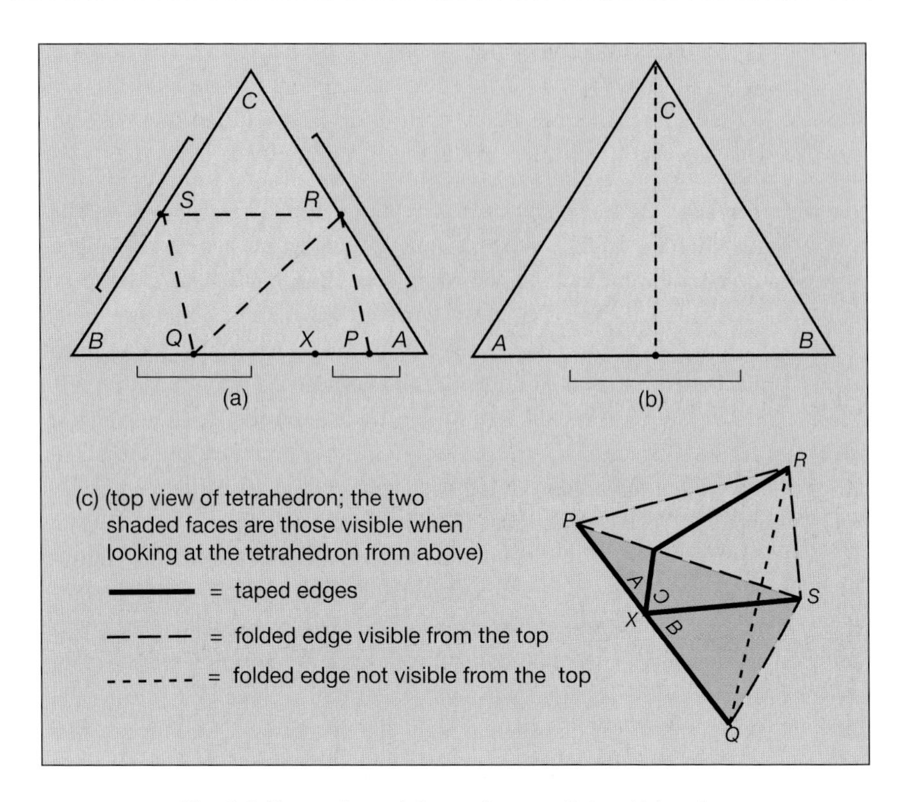

Fig. 6.3 Alexandrov gluings of an equilateral triangle.

a) The three corners (*A, B, C*) all glue to point *X*. The four fold points (*P, Q, R,* S) become the four vertices of the resulting tetrahedron in (c), two of whose faces are *PQS* and *QRS*.

(b) A folding that creases the triangle down an altitude, gluing *A* and *B* together and edge *AC* to *BC*

(c) The tetrahedron that results from the gluing in (a), folding toward the viewer. Note the angles at the three vertices (*A, B, C*) "disappear," forming 180 degree at *X* on tetrahedron edge *PQ*.

> *Theorem.* Every convex polygon folds to an infinite number (a continuum) of noncongruent convex polyhedra.

The essence of this claim follows from Alexandrov's theorem. Take any convex polygon *P*, and mark a point *X* on its boundary. Walk around half the perimeter, and mark an opposite point *Y*. Next glue the perimeter half from *X* to *Y* to the half from *Y* to *X;* see figure 6.4(a). We argue that this gluing is an Alexandrov gluing. Certainly it consumes the entire perimeter (Condition 1). The crucial requirement is Condition 2: no more than 360 degrees of angle is glued at any one point. At the fold points *X* and *Y*, the amount of angle is 180 degrees. Any other two points glued together either sum to exactly 360 degrees if the points are interior to a side of P or to less than 360 degrees if one or the other is a vertex of P. Here convexity of P is used: any convex vertex has an interior angle less than 180 degrees. That the gluing is a topological sphere (Condition 3) can be seen if one views the perimeter-halving gluing as zipping up a pocketbook.

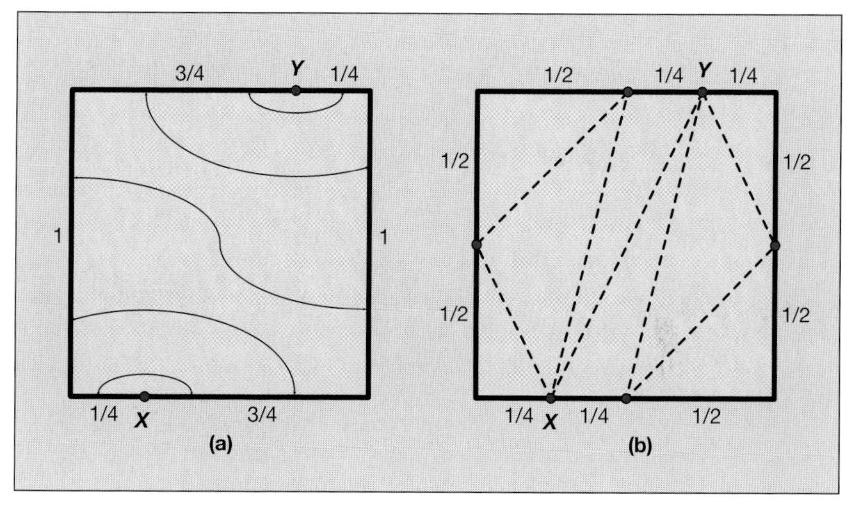

Fig. 6.4. (a) A perimeter-halving folding of a unit square. X is 1/4 from the lower left corner, and Y is 1/4 from the upper right corner. The perimeter half (of length 2) is glued symmetrically as indicated. The folding produces, not obviously, an octahedron.
(b) Crease pattern of edges, and vertices of octahedron. As in figure 6.3, the corners of the polygon "disappear" in the folding.

So Alexandrov's theorem says that every perimeter-halving folds to a convex polyhedron. Sliding X around the boundary, and Y correspondingly, leads to a continuum of foldings and produces, amazingly, an infinite number of noncongruent polyhedra.

Figure 6.5 shows the continuum achieved by perimeter-halving foldings of a square (Alexander, Dyson, and O'Rourke 2003). Starting at the 3-o'clock position with the doubly covered $1 \times 1/2$ rectangle achieved by creasing down a midline, the continuum continues clockwise to the doubly covered right triangle achieved by creasing down a diagonal at the 9-o'clock position. This creasing corresponds to sliding X from the midpoint of a side of the square to an adjacent vertex. Sliding X beyond the vertex repeats the shapes in mirror image, clockwise from 9 o'clock to 3 o'clock. Incidentally, figure 6.5 shows only a portion of the polyhedra foldable from a square. See Demaine and O'Rourke (2007, p. 416) for the full variety.

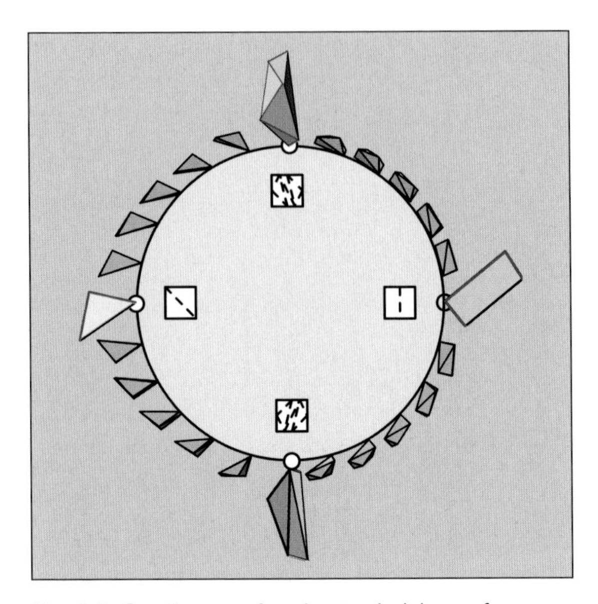

Fig. 6.5. Continuum of perimeter-halvings of square. Four crease patterns are shown. The octahedron at the 6-o'clock and 12-o'clock positions corresponds to figure 6.4(b).

As a practical experiment, we can cut out of paper any convex polygon, start creasing it at an arbitrary *X*, and "zip" up the boundary from there with tape, and eventually arrive at *Y;* we need not make a measurement of the perimeter. The result will be a handbaglike or pitalike shape, which, by Alexandrov's theorem, we can coax (with patience!) to reveal the creases that fold it into its unique polyhedral form.

To date the "space" of all foldings of regular polygons has been explored, but there is as yet little general understanding of the phenomenon.

The Foldings of the Latin Cross

When we consider foldings of nonconvex polygons, we enter largely unknown territory. My colleagues, including five college students, and I decided to explore the foldings of the Latin cross as a test case (Demaine et al. 1999; Demaine and O'Rourke 2007, section 25.6). We found, to our surprise, that the Latin cross folds not only to the cube (fig. 6.1(b)) but to 22 other distinct convex polyhedra—two flat quadrilaterals, seven tetrahedra, three pentahedra, four hexahedra, and six octahedra (see fig. 6.6). Here no continuum exists—the nonconvexities prevent the continuous sliding possible with convex polygons.

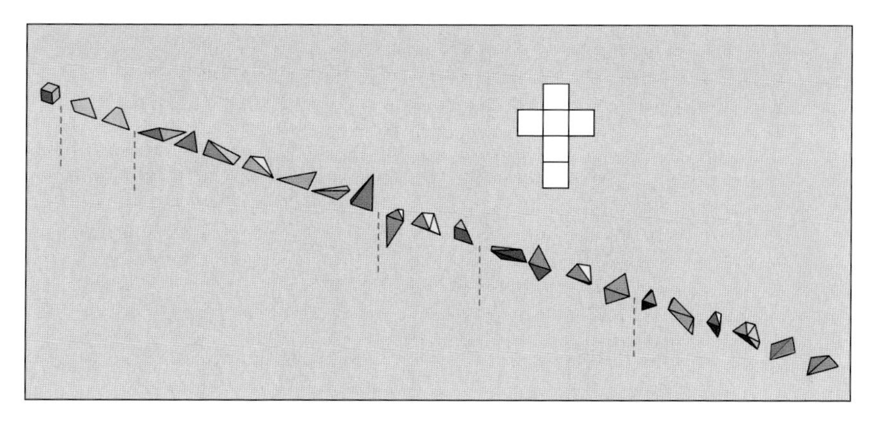

Fig. 6.6. The 23 polyhedra foldable from the Latin cross. (fig. 25.30 in Demaine and O'Rourke [2007])

How these foldings are achieved is by no means obvious. Figure 6.7 illustrates just one of the 23 foldings in detail, a delicate folding to a tetrahedron. The other foldings are equally intricate.

Aside from this one detailed example, we are left largely without a general theory encompassing the foldings of nonconvex polygons. In particular, the polyhedra achievable from the other ten hexamino unfoldings of a cube, besides the Latin cross, remain to be explored.

Reconstruction of Three-Dimensional Polyhedra

As mentioned previously, an algorithm and software exist to take any given polygon *P* and list all the Alexandrov gluings of *P*. But to which unique polyhedra these gluings correspond is unknown. The polyhedra displayed in figures 6.5 and 6.6 were reconstructed by laborious ad hoc techniques that cannot extend much beyond octahedra. Quite recently a group of researchers (Bobenko and Izmestiev 2006; O'Rourke 2006) discovered a way to convert Alexandrov's existence proof into a constructive proof, with the implication that solving a particular differential equation will lead to the unique three-dimensional shape guaranteed by Alexandrov's theorem. Whether this advance will lead to a practical numerical method of producing these polyhedra remains to be seen.

Nonconvex Polyhedra

Here we reach the frontier of knowledge on this topic. Let me close with one outstanding unsolved ("open") problem, which involves nonconvexity:

Fig. 6.7. Folding the fourth polyhedron in figure 6.6 (fig. 25.10 in Demaine and O'Rourke [2007], from the video Demaine et al. [1999])

Question 3. Does every polygon fold to some (perhaps nonconvex) polyhedron?

Even restricting the fold to be perimeter-halving leaves the question unresolved. If the answer to Question 3 is no, which could be established by a single counterexample, then immediately we enter new, uncharted territory by deleting the qualifier "convex" from question 1: *Which* polygons can fold to a polyhedron? Rarely can an open problem at the frontier of mathematical research be stated so succinctly.

Acknowledgments. I am indebted to Erik Demaine, my coauthor on Demaine and O'Rourke (2007) and on most of my work in this area, and to Joseph Malkevitch for useful comments.

REFERENCES

Alexander, Rebecca, Heather Dyson, and Joseph O'Rourke. "The Convex Polyhedra Foldable from a Square." In *Proceedings of the Japan Conference on Discrete and Computational Geometry. Lecture Notes Computer Science,* vol. 2866 (2003): 38–50.

Alexandrov, Aleksandr D. "Existence of a Convex Polyhedron and a Convex Surface with a Given Metric." In *A. D. Alexandrov: Selected Works: Part I,* edited by Yuri G. Reshetyak and Semën S. Kutateladze, pp. 169–73. Hawthorne, Victoria, Australia: Gordon and Breach, 1996. Translation of *Doklady Akademii Nauk SSSR, Matematika*, vol. 30, no. 2 (1941): 103–6.

———. *Convex Polyhedra.* Monographs in Mathematics. Berlin: Springer-Verlag, 2005. Translation of *Vupyklue Mnogogranniki,* 1950.

Bobenko, Alexander I., and Ivan Izmestiev. "Alexandrov's Theorem, Weighted Delaunay Triangulations, and Mixed Volumes." *Annales de l'Institut Fourier* 58, no. 2 (2008): 447–505.

Demaine, Erik D., Martin L. Demaine, Anna Lubiw, Joseph O'Rourke, and Irena Pashchenko. "Metamorphosis of the Cube." Video and abstract. In *Proceedings of the 15th Annual ACM Symposium on Computational Geometry*, pp. 409–10, Miami Beach, Florida, 1999.

Demaine, Erik D., and Joseph O'Rourke. *Geometric Folding Algorithms: Linkages, Origami, Polyhedra.* New York: Cambridge University Press, 2007. www.gfalop.org.

O'Rourke, Joseph. "Computational geometry column 49." *SIGACT News* 38, issue 143 (2007): 51–55.

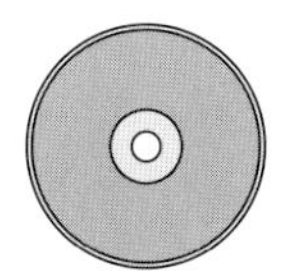

Blackline masters for three activities: Folding a Latin cross to make a tetrahedron, Folding a square into an octahedron, and Folding a triangle into a tetrahedron, are found on the CD-ROM disk accompanying this Yearbook.

Part II

Learning Geometry

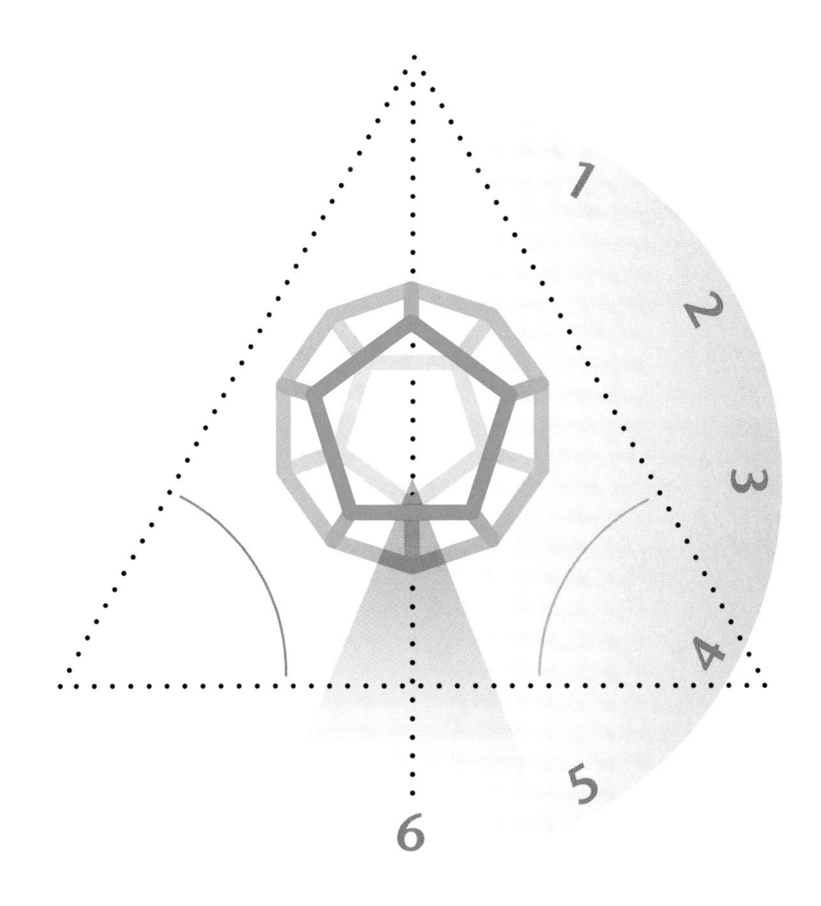

Highlights of Research on Learning School Geometry

Michael T. Battista

IN THIS article, I highlight ideas from research that foster insights on the learning and teaching of geometry in grades K–12. In particular, I focus on research that helps us understand students' geometric sense making and reasoning, including how they convince themselves of the validity of geometric ideas.

How Do Students Learn Geometry?

In this portion of the article I examine several theories that are important for understanding geometry learning.

Theory 1: The van Hiele Levels

In the late 1950s two Dutch educators, Pierre van Hiele and Dina van Hiele-Geldof, developed a model that describes how students' geometric reasoning develops. According to the van Hiele theory, students progress through discrete, qualitatively different *levels* of geometric thinking. As originally conceived, the levels are sequential and hierarchical, so for students to attain a level, they must have passed through the levels that precede it. Research generally indicates that

Time to prepare portions of this material was partially supported by the National Science Foundation under Grant numbers ESI 0099047 and 0352898. The opinions, findings, conclusions, or recommendations, however, are those of the author and do not necessarily reflect the views of the National Science Foundation. This article is derived in large part from the author's article in the *Second Handbook of Research on Mathematics Teaching and Learning* (Battista 2007).

the van Hiele theory is accurate in describing the development of students' geometric reasoning (Battista 2007; Burger and Shaughnessy 1986; Clements and Battista 1992; Fuys, Geddes, and Tischler 1988). However, some researchers have argued that because of differing experiences and instruction, students may have different van Hiele levels for different topic domains in geometry (e.g., quadrilaterals and triangles versus geometric motions), and that the types of thinking that characterize different levels may be developing simultaneously, but at different rates (Battista 2007). I next describe the levels as elaborated and refined by Clements and Battista (1992) and more recently by Battista (2007)[1].

Level 1: Visual-Holistic Reasoning

Students identify, describe, and reason about shapes and other geometric configurations according to their appearance as visual wholes. They often justify their shape identifications using vague, holistic judgments, such as saying that a square is not a rectangle because rectangles are "long" or claiming that two figures are the "same shape" because they "look the same." Students often refer to visual prototypes, saying, for example, that a figure is a rectangle because "it looks like a door." Or they use imagined visual transformations, saying, for instance, that a shape is a square because if it is turned, it looks like a square. The orientation of figures may strongly affect Level 1 reasoning. For instance, students using Level 1 reasoning often identify squares only when their sides are horizontal and vertical. In Level 1, some students are unable to identify many common shapes, whereas other, more advanced students correctly identify those shapes.

Level 2: Descriptive-Analytic[2] Reasoning

Students explicitly attend to, conceptualize, and specify shapes by describing their parts and spatial relationships among the parts. However, students' descriptions and concepts vary greatly in sophistication. At first, students describe parts and properties of shapes informally and imprecisely using strictly informal language typically learned in everyday experience. As students begin to acquire formal geometric concepts explicitly taught in mathematics curricula (such as angle measure and parallelism), they start to use a combination of informal and formal descriptions of shapes. However, the formal portions of students' shape descriptions are insufficient to specify shapes completely. Finally, students explicitly and exclusively use formal geometric concepts and language to describe and conceptualize shapes in a way that attends to a sufficient set of properties to specify the shapes. Students can use and formulate formal definitions for class-

---

1. Over the years, several numbering systems and verbal labels have been used to describe the van Hiele levels. I use numbers and labels that I believe best reflect modern research on the levels.

2. I use the term *analysis* to refer to the process of understanding objects by decomposing them into components.

es of shapes. However, their definitions are not minimal, because they do not interrelate properties or see that some subset of properties implies other properties. They simply think in terms of unconnected lists of formally described characteristics.

Two major interrelated factors contribute to the development of Level 2 reasoning. The first is an increasing ability and inclination to account for the spatial structure of shapes by analyzing shape parts. The second is an increasing ability to understand and apply formal geometric concepts to specify relationships among shape parts.

Illustrations of Levels 1 and 2

Students at the beginning of Level 1 might identify figures 7.1 and 7.5 as squares and figure 7.2 as a triangle. Beginning Level 1 reasoning about figure 7.5 might be "It's a square because if you turn it this way (45 degrees), it would be a square." Students at the end of Level 1 might reject figure 7.3 as a square for a visual-holistic reason, such as, "Usually a square is not stretched out and this has more added on to it." Students at the beginning of Level 2 might say that figure 7.3 is a rectangle because it has "two long sides and two short sides." Students at the end of Level 2 would identify figure 7.4 as a rectangle because, "It has opposite sides equal and four right angles." Before reaching the last phase of Level 2, most students would identify figure 7.4 as a square.

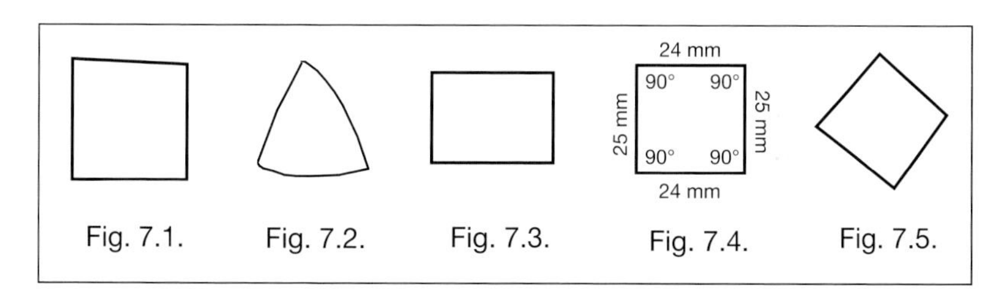

Fig. 7.1. Fig. 7.2. Fig. 7.3. Fig. 7.4. Fig. 7.5.

Level 3: Relational-Inferential Reasoning

Students infer relationships among geometric properties of shapes. For example, a student might say, "If a shape has property X, it also has property Y." However, the sophistication of students' inferences varies greatly. Students start with *empirical* inference, for instance, noticing that whenever they have seen property X occur, property Y occurs. Next, by analyzing how shapes can be built one part at a time, students conclude that when one property occurs, another property must occur. For instance, students might conclude that if a quadrilateral has four right angles, its opposite sides are equal because when they draw a rectangle by making a sequence of perpendiculars, they cannot make the opposite sides unequal.

Next, students make simple *logical* inferences about properties. For example, a student might reason that because a square has all sides equal, it has opposite sides equal. Such reasoning enables students to make the inferences needed for hierarchical classification. For instance, a student whose definition for a rectangle is "four right angles and opposite sides equal" might infer that a square is a rectangle because "a square has four right angles, which a rectangle has to have; and since a square has four equal sides, it has opposite sides equal, which a rectangle has to have." However, the ability to make such inferences does not necessarily lead students to adopt a hierarchical shape-classification system; they may still resist the notion that a square is a rectangle even though they can follow the logic justifying such a statement. In the final phase of Level 3, students use logical inference to *reorganize* their classification of shapes into a logical hierarchy. Not only does the reason a square is a rectangle become clear, but classifying a square as a rectangle becomes a necessary part of reasoning. Students give logical arguments to justify their hierarchical classifications. Finally, as students progress through Level 3, they are increasingly able to understand and appreciate minimal definitions for classes of shapes.

Level 4: Formal Deductive Proof

Students understand and can construct formal geometric proofs. That is, within an axiomatic system, they can produce a sequence of statements that logically justifies a conclusion as a consequence of the "givens." They recognize differences among undefined terms, definitions, axioms, and theorems. Only at this level are students capable of the kind of reasoning demanded in the axiomatic, proof-based geometry courses that have been traditionally taught in United States high schools.

Level 5: Rigor

Students understand, and can use and analyze, alternative axiomatic systems. This level generally corresponds to university studies of geometry.

Theory 2: Abstraction

Abstraction is the process by which the mind registers objects, actions, and ideas in consciousness and memory. Once objects, actions, and ideas have been abstracted at a sufficiently deep level, they themselves can be mentally operated on (e.g., compared, decomposed, analyzed).

Two special forms of abstraction are fundamental to geometry learning and reasoning (Battista 1999; Battista and Clements 1996). *Spatial structuring* is the mental act of organizing an object or set of objects by identifying its components and establishing interrelationships among them. *Mental models* are nonverbal, mental versions of situations that capture the structure of the situations they represent (Battista 1994; Greeno 1991; Johnson-Laird 1983, 1998). Individuals reason by activating mental models that enable them to imagine possible scenarios and solutions to problems. Learning occurs as individuals recursively cycle through phases of action (physical and mental), reflection, and abstraction in a way that enables them to develop ever more sophisticated mental models.

Here is an example of the use of mental models in geometric reasoning. A student was attempting to decide on the validity of the following statement: "If all the angles in a polygon are equal, then all the sides in the polygon are equal." The student said that in thinking about the problem, he first imagined a regular pentagon, which he attempted to transform so that it had unequal sides. But he could not imagine how to change some but not all the sides of the pentagon without changing the angle measures. The mental model this student activated to think about this problem made the problem impossible for him to solve. Had the student originally activated a mental model of a square or a regular hexagon, he could have much more easily implemented his transformational strategy successfully.

Theory 3: Concept Learning and the Objects of Geometric Analysis

Understanding students' geometric reasoning requires an analysis of the "objects" and mental entities that they reason about. Three major types of "objects" must be considered. *Physical objects* are things such as a door, box, ball, geoboard figure, picture, drawing, or dynamic, draggable computer figure. *Concepts* are the mental representations that individuals abstract for categories of like objects. These representations can include prototypical examples, images, mental models, and verbal descriptions. *Concept definitions* are formal, mathematical verbal or symbolic specifications of categories of objects or entities. An important thing to note is that students' concepts can be very different from concept definitions. For example, a young child's concept of triangle might consist of a mental image of an equilateral triangle, whereas a concept definition for triangle might be "a polygon with three sides." Because research on the formation and use of concepts fosters important insights into the development of geometric reasoning, especially at van Hiele Levels 1 through 3, I summarize some of the relevant findings here.

People form two fundamentally different types of concepts (Pinker 1997). *Natural concepts,* such as apples or dogs, are formed in everyday activity and are rarely accompanied by concept definitions. Generally, such concepts are induced from instances and are thought about through visual resemblance to prototypical examples. *Formal concepts,* such as rectangles, have definitions that explicitly specify a sufficient set of properties to identify instances. When forming natural concepts, the goal is to recognize category instances; when forming formal concepts, the goal is to specify precisely the defining features of category instances.

Learning geometry involves forming both natural and formal concepts. Students form natural shape concepts outside school as well as in primary-grade instruction when adults help them identify, but not define, shapes. Subsequent school instruction demands that students learn formal concepts for shapes as verbally stated, property-based definitions. Difficulties often arise as students attempt to reconcile natural and formal concepts for the same category. For instance, students who possess a natural concept of square might identify as squares all shapes that are approximately square. Because such students do not know any properties of a square, they have difficulty connecting their imprecise recognition procedure with a formal definition of square as a quadrilateral with four right angles and four congruent sides. One crucial component of students' move from van Hiele Level 1 to Level 2 involves progressing from natural to formal shape concepts.

A second crucial component in moving from van Hiele Level 1 to Level 2 is developing understanding of formal concepts that specify relationships between parts of shapes. The concepts of angle measure, length measure, congruence, and parallelism are used to specify the defining spatial properties of shapes. For instance, saying that a rectangle has four 90-degree angles describes precisely how a rectangle's sides are related spatially.

Relating Abstraction and Concept Learning

In an investigation of the development of the concept of angle, Mitchelmore and White (2000) describe several stages of abstraction, which we might conjecture occur for other geometric concepts. In the first stage students abstract isolated concepts from specific situations. For instance, students abstract one angle-related idea for hills and another angle-related idea for roofs. In the next stage, a somewhat more general abstraction of angle enables students to start seeing similarities between different situation-specific angle concepts. For instance, students might merge their hill and roof-angle concepts into a "tilted" angle-related concept. In the last stage, students integrate various context-specific, angle-related concepts into a generalized concept of angle that is encapsulated in the standard definition of two rays with a common endpoint. For instance, students might integrate the angle-related ideas in line intersections, rotations, and slope-like contexts, seeing how each can be viewed in terms of two rays with a common endpoint.

Theory 4: Diagrams and Representations

Diagrams: Representing or Represented?

Diagrams and physical objects play two major roles in geometry. On the one hand, physical objects can be thought of as the *input* for geometric conceptualization. Geometric concepts are generally derived through an analysis of such objects: the objects become *represented* or described by the concepts. On the other hand, physical objects, including diagrams, are often used to *represent* formal geometric concepts. Thus a triangle diagram can be thought of as representing the formal concept of triangles, or as a single object that we wish to represent or describe geometrically. Geometry instruction and curricula generally neglect the process of forming concepts from physical objects and instead focus on using diagrams and objects to represent formal shape concepts. Furthermore, students often get confused about the represented and representing role of diagrams. That is, often, when teachers use diagrams to represent formal concepts, students reason about the diagrams, not the formal concepts. Consequently, students often attribute irrelevant characteristics of a diagram to the geometric concept it is intended to represent (Clements and Battista 1992; Yerushalmy and Chazan 1993). For instance, students might not recognize right triangles in nonstandard orientations because they have unintentionally abstracted a particular orientation as an attribute of such triangles.

More Student Difficulties with Diagrams

Another difficulty occurs when students use diagrams in proofs (Clements and Battista 1992; Presmeg 1997; Yerushalmy and Chazan 1993). For example, students may assume that sides that look parallel in an accompanying diagram *are* parallel (adding unintended conditions). Or students might link a theorem too tightly to the example diagram given with the theorem statement. For instance, if a theorem is originally illustrated by a diagram of an acute triangle, students might believe that the theorem does not apply, or might not think to apply it, to obtuse triangles. Another difficulty occurs when drawings do not capture appropriate geometric relationships—for example, a tangent to a circle might be drawn freehand so that it is not perpendicular to the radius it intersects.[3] Finally, students might mistake geometric diagrams for pictures of objects, profoundly changing their interpretation. For instance, a bright fourth grader identified nonrectangular parallelograms as rectangles because he interpreted them as pictures of rectangles viewed "sideways" (Clements and Battista 1992).

Selected Studies of Learning and Teaching Geometry

Poor Geometry Achievement in the United States

An abundant amount of research shows that a great majority of students in the United States have inadequate understanding of geometric concepts and poorly developed skills in geometric reasoning, problem solving, and proof (Clements and Battista 1992; Beaton et al. 1996; Mullis et al. 1997; Mullis et

3. Of course, a sloppy drawing might not interfere with reasoning if the individual creating it is aware of the relationships that *should* exist in it.

al. 1998). Furthermore, students in the United States do far more poorly in geometry than students in many other countries (see, e.g., Clements [2003]). The primary cause for this poor performance is both *what* and *how* geometry is taught (Clements and Battista 1992). Most U.S. geometry curricula consist of a hodgepodge of superficially covered concepts with no systematic support for students' progression to higher levels of geometric thinking. Indeed, according to Senk (1989), more than 70 percent of U.S. students begin high school geometry below van Hiele Level 2 even though only students who enter at Level 2 or higher have a good chance of becoming competent with proof by the end of the course (Clements and Battista 1992; Senk 1989). Further, almost 40 percent of students end the year of high school geometry below Level 2, and only about 30 percent of students in full-year geometry courses that teach proof reach a 75-percent mastery level in proof writing (Clements and Battista 1992; Senk 1985). Clearly, preproof geometry instruction should be designed so that students fully attain Level 2 and start moving into Level 3 *before* they reach high school. Also, as long as preproof geometry instruction is inadequate, high school geometry instruction should be designed to accommodate students who enter below Level 2.

Interactive Geometry Software

Interactive geometry software has much potential for improving students' geometry learning, so here I discuss several research studies that illuminate and improve its use.

Interactive Geometry Software in Elementary School

Because interactive geometry software, such as Cabri (Baulac, Bellemain, and Laborde 1994) and The Geometer's Sketchpad (GSP) (Jackiw 1995), was originally designed for students at the secondary school level, the use of this software in elementary school classrooms requires software adaptations or additions. For instance, the Shape Makers computer microworld is a special add-on to GSP that provides students with geometric shape-making objects (Battista 1998). An example is the Parallelogram Maker, which can be used to make any desired parallelogram that fits on the computer screen, no matter what its shape, size, or orientation—but only parallelograms. The appearance of a Parallelogram Maker is changed by dragging its vertices with the mouse (see fig. 7.6).

The Shape Makers environment enables elementary school students to explore dynamically draggable shapes without having to learn any of the construction operations of the software. Furthermore, draggable shapes, such as the Parallelogram Maker, are interesting, manipulable, visual-mechanical objects with movement constraints that can be conceptualized and geometrically analyzed (Battista 2008). With proper instructional guidance, students' analyses of Shape Makers can promote their construction of meaningful concepts of geo-

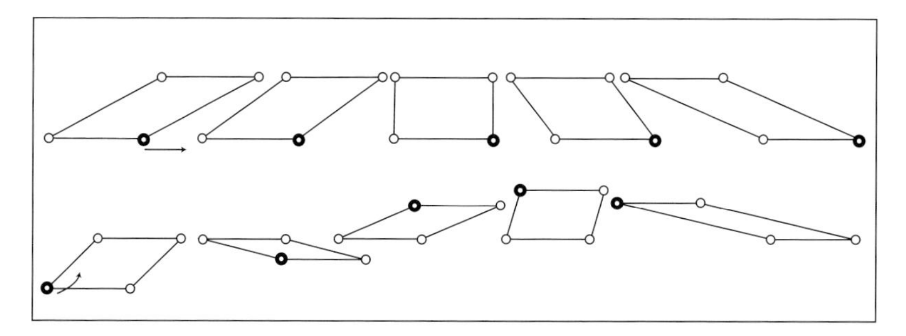

Fig. 7.6. Shape Makers computer microworld's Parallelogram Maker

metric properties that can be applied not only to draggable drawings but also to the category of geometric objects these drawings generate. Working with Shape Makers can also help students move from van Hiele Level 1 to Levels 2 and 3.

To illustrate the variety of geometric reasoning exhibited by elementary school students with interactive geometry software, I describe an episode with fifth graders at the beginning of their work with Shape Makers (Battista 1998, 2001, 2008). Students are investigating the Square Maker by dragging its parts in various ways. Consider their remarks:

M. T.: I think maybe you could have made a rectangle.

J. D.: No, because when you change one side, they all change.

E. R.: All the sides [in the Square Maker] are equal.

M.T., J. D., and E. R. abstracted different things from their Square Maker manipulations. M.T. noticed the visual similarity between squares and rectangles, causing him to conjecture that the Square Maker could make a rectangle that is not a square. J. D. abstracted a holistic movement regularity—when one side changes length, all sides change (thus, he could not get the sides of the Square Maker to be different lengths, which he thought was necessary for a rectangle). E. R. conceptualized the movement regularity noticed by J. D. with more precision by expressing it as a formal geometric property. From the van Hiele perspective, M.T.'s reasoning on this task was at the first phase of Level 1 (incorrect, visual-holistic identification); J. D.'s, at the second phase of Level 1 (correct, visual-holistic identification) but perhaps starting the beginning phase of Level 2; and E. R.'s, at the middle phase of Level 2. Also, the abstractions made by M.T. and J. D. seemed completely situated in the interactive geometry software microworld, whereas that made by E. R. was a more general, formal mathematical conception.

Research shows that appropriately structured sequences of instructional tasks enable students working with the Shape Makers to progress through the first two van Hiele levels and into the third level (Battista 1998, 2001, in press; Yu, Barrett, and Presmeg 2009). The dynamic manipulation facility of the Shape Makers makes certain spatial invariants salient to students, and appropriate guidance by teachers encourages students to invent or use geometric concepts that describe those invariants. For example, appropriately structured activities and class discussions can encourage and support students like M. T. and J. R. to reformulate their informal ideas into formal concepts as E. R. did. Importantly, for instruction to provide the appropriate guidance to students, we must *not* assume (a) that students will automatically "see" the formal geometric properties that are "built into" draggable figures or (b) that students will see a draggable drawing as "representing" some formal geometric concept. Much conceptual work is needed for beginning students to interpret movement invariants in terms of formal geometric properties. In fact, when analyzing draggable drawings, novice students first notice movement *constraints*. Later, and often only with great effort and instructional guidance, they begin to conceptualize movement constraints and regularities in terms of formal geometric properties. As another example, initially many students see the Kite Maker as making shapes with "flapping wings" (Level 1 reasoning); only after they develop an appropriately abstract concept of symmetry can they conceptualize this movement pattern in terms of a geometric concept, reaching Level 2 reasoning (Yu, Barrett, and Presmeg 2009).

Interactive Geometry Software in Middle and High School

Besides having students investigate geometric shapes by manipulating them on the computer screen, two additional kinds of interactive geometry software activities are used with older students. First, students themselves are asked to use the software's tools to construct *draggable* figures, with a construction considered "*valid* if and only if it is *not* possible to '*mess it up'* by dragging." Thus the crucial properties of the construction are maintained when parts of it are dragged. Second, students are asked to *justify* or *prove* conjectures that arise from their interactive geometry software explorations.

To promote an understanding of quadrilaterals in junior high school students, Jones (2000) gave a series of interactive geometry software tasks in which students were asked to construct *draggable* figures. For instance, students were asked to construct a draggable square and explain why the shape remains a square no matter how it is manipulated on the computer screen. As students progressed through the instructional sequence, Jones found that, initially, students emphasized description rather than explanation, relied on perception rather than mathematical reasoning, and lacked precise mathematical language. At an interim period in the sequence, students' explanations became more mathematically precise but were intertwined with the operation of the software. At the end of the instructional unit, students' explanations were stated entirely in terms of formal geometric concepts.

To illustrate, when students were asked to explain why all squares are rectangles, early in instruction one pair of students wrote, "You can make a rectangle into a square by dragging one side shorter ... until the sides become equal" (Jones 2000, p. 76). Later in instruction, students were asked to construct a trapezoid that could be modified to make a parallelogram and explain why all parallelograms are trapezoids. The same pair of students wrote that parallelograms are trapezoids because trapezoids "have one set of parallel lines and parallelograms have two sets of parallel lines" (Jones 2000, p. 76). In the first example the students' explanation is couched in terms of the software's actions, whereas in the second, the students' explanation relies on formal geometric properties. In van Hiele levels, the students' first explanation is reasoning at the second or third stage of Level 2, whereas the second explanation is at the third stage of Level 3.

Moving toward Justification

At the secondary school level, many researchers and teachers see construction activities with interactive geometry software as a bridge to justification and proof. Consistent with this belief, but counter to suggestions that students who have explored a conjecture extensively in computer environments often feel no need for further justification (e.g., Mariotti [2001]), de Villiers (1998) reported that he has been able to encourage deductive thought by asking students to determine *why* interactive geometry software results are true. Similarly, although Marrades and Gutiérrez (2000) found that without prompts for proof, students seemed convinced by empirical evidence provided by interactive geometry software manipulations, with teacher prompting, students were able to use the software in concert with proof. (Editor's note: See Paniati [2009] for another example.)

To illustrate how students can derive a proof from their explorations with interactive geometry software, consider this problem (Battista 2007; Healy and Hoyles 2001):

> Construct a quadrilateral in which the angle bisectors of two consecutive interior angles are perpendicular. Discover, then prove, other properties of this quadrilateral.

Imagine a pair of students who construct quadrilateral *ABCD* with the given angle bisectors, create a dynamic display of the measure of the angle between the bisectors *BEC*, and manipulate the construction to various positions in which angle *BEC* is 90 degrees (see fig. 7.7). On the basis of these manipulations, these students conjecture that when the angle bisectors are perpendicular, sides *AB* and *DC* must be parallel. They empirically verify their conjecture by constructing line *k* through point *A* and parallel to side *DC* and observing that whenever the measure of angle *BEC* is 90 degrees, vertex *B* is on line *k*. To understand and prove this empirical regularity, the students measure various angles in their diagram and discover that whenever the angle bisectors are perpendicular, angles *EHC* and *EBA* have equal measures. The students realize that proving these angles congruent proves that segments *AB* and *CD* are parallel. They then use the given information, relationships among angles in right triangles, and conditions for parallels to write their proof.

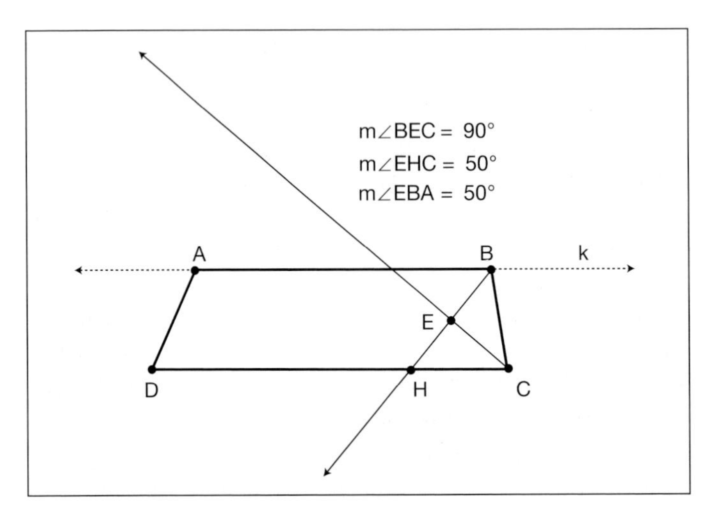

Fig. 7.7. Quadrilateral in which the angle bisectors of two consecutive interior angles are perpendicular

In addition to helping students find regularities and gain insight into why these regularities occur, explorations with interactive geometry software can also help students develop the confidence to proceed with a justification (consistent with de Villiers [1998], when he cited Pólya, "When you have satisfied yourself that the theorem is true, you start proving it"). Unfortunately, instruction infrequently takes advantage of the capability of interactive geometry software to support students' formulation of proofs (Laborde 2001).

Research by Healy and Hoyles (2001), however, suggests an important cautionary note on students' use of constructions with interactive geometry software for exploring and justifying geometric ideas. They found that when students were given the consecutive-angle-bisectors problem just discussed, many students successfully constructed the angle bisectors and manipulated them until they were perpendicular yet failed to see the parallel relationship between the sides of the resulting quadrilateral. Furthermore, all students who were successful on this problem constructed a general quadrilateral with consecutive bisectors and manipulated it to find when the bisectors were perpendicular. Students who attempted to construct a draggable quadrilateral with the given bisector-perpendicularity property—as the researchers had intended—encountered two problems. The first difficulty was constructing the given angle-bisector relation for only a special case, such as a parallelogram. The second difficulty was that constructing a general quadrilateral that satisfied the given conditions proved too difficult. Because they could not use Cabri to construct the quadrilateral, students became frustrated and failed to solve the problem.

Proof and Justification in Geometry

In 1992, Clements and Battista concluded that, in geometry, students are extremely unsuccessful with formal proof and that they are deficient in their ability to establish mathematical truth.[4] More recently, Hershkowitz and her colleagues (2002) drew a similar conclusion: "The teaching of mathematical proof appears to be a failure in almost all countries" (p. 675). In a large study in the United Kingdom, Healy and Hoyles (1998) reported that (a) high-achieving year 10 students showed poor performance in constructing proofs, with empirical verification the most popular form of justification; (b) students' proof performance was considerably better in algebra than in geometry; and (c) students were better at recognizing a valid mathematical argument than constructing one themselves. Hoyles and Jones (1998) reported that (a) students do not distinguish between empirical and deductive arguments and prefer empirical arguments; (b) for many students, deductive proof only *adds evidence* for validity; and (c) most students regard proof as an irrelevant add-on and do not understand its purpose. Similarly, Hershkowitz and her colleagues (2002) stated that students rarely see the point of proving. Battista and Clements (1995) argued that although proof is crucially important to mathematicians because it enables them to establish the validity of their mathematical thought, students do not conceive of proof as establishing validity but instead see it as conforming to a set of formal rules that are unconnected with their personal mathematical activity.

Consistent with earlier research, Koedinger (1998) reported that high school students rarely think that evidence beyond examples is needed to support geometric conjectures. He also found that many students, when considering conjectures, first said that the one example they drew was enough evidence, then, after prompting, that many examples were needed. Most students had to be explicitly asked to write a proof "like they had seen in class" before they attempted to produce one. Also, Koedinger reported that only 10 percent of students successfully formulated proof problems and began to work on them. Interestingly, students found that stating the givens was especially difficult.

---

4. See Harel and Sowder (1998) for a discussion of the proof schemes that students use in all branches of mathematics.

Chazan's (1993) investigation of high school students' views of proof elaborates and clarifies some long-standing issues in the research. He found that five out of seventeen interviewed high school geometry students believed that measuring examples "proved" assertions, seven did not believe that such empirical evidence proved assertions, and five were unclear or uncertain about the issue. Some students who believed that empirical evidence is proof focused on the number of examples examined. Others focused on the type of examples examined. For instance, in considering a proposition about triangles, one student believed that if he checked examples that were right, obtuse, and acute, then the proposition would be proved for all triangles.

In contrast with students who believed that empirical evidence was proof, some students thought that proof was merely evidence (Chazan 1993). For instance, some students believed that counterexamples to a statement were still possible, even if a deductive proof of the statement was given in a textbook. However, Chazan's finding could be due to students' not understanding the logic of proof, or it could be a reasonable reaction to the possibility of incorrect proofs (especially proofs written by students). This finding brings up an important issue. Students generally assume that proofs in mathematics textbooks are correct because they appear in the book, which is taken as the unquestioned authority, rather than because they have critically examined the proofs. This view is very different from how mathematicians work—usually, proposed new theorems and proofs are greeted with skepticism, and often with attempts to find counterexamples. So for some students, skepticism about deductive proofs should not be viewed as a weakness but rather as justified and healthy.

Another reason that students might be skeptical about the generality of a deductive proof occurs when the proof does not really lend insight into why a proposition is true. For instance, one can read a proof and believe in its validity because each step in its argument is valid but still not, in the words of Pólya, "see it at a glance." In this instance, the proof does not provide enough insight to make genuine sense of the proposition, so one might still be skeptical. In such instances, a mathematically healthy and reasonable response is to explore the proof further by examining examples, perhaps reviewing the proof for some of those examples, even developing alternative proofs (think of how many proofs have been given for the Pythagorean theorem). Thus, students' skepticism about proof is not always negative. Too often we cite such skepticism as a lack of understanding of proof, ignoring contexts in which such skepticism in warranted.

In summary, researchers and teachers alike are still grappling with this basic question: Why do students have so much difficulty with geometric proof? Is it because they see little need for proof or because proof is so abstract that it is very difficult to learn, or is it a combination of the two—that because students see proof as difficult and not useful, they do not expend sufficient intellectual effort to master it? As previously discussed, however, the use of interactive geometry software has shown promise in encouraging and supporting the development of students' geometric proof. So we have another instructional tool that can help us help students develop proficiency with mathematical proof.

Empowering students to make personal sense of, and establish the validity of, geometric ideas is a primary goal of geometry instruction. However, although proof is the gold standard for establishing mathematical truth, geometry curricula too often myopically focus on formal proof in ways that disconnect it from students' sense making. Indeed, as the van Hiele levels demonstrate, for students, formal proof comes only at the end a long sequence of prior less-sophisticated ways of reasoning. Although our past failures to teach proof may cause us to view it as an instructional objective that only the best students need to master, we cannot dismiss the process of establishing mathematical truth as an objective for all our students. It is the essence of mathematics. Without it, students cannot do mathematics, they can only noncritically examine what others have done.

Conclusion

Teaching in ways that encourage, guide, and support students' personal sense making in geometry requires a basic understanding of research on the development of students' geometric thinking (Bransford, Brown, and Cocking 1999; Fennema and Franke 1992; Fennema et al. 1996). This article describes several research-based frameworks, along with several important research findings, that can be used to understand and promote students' geometric sense making. Understanding these frameworks and research is crucial for selecting and creating instructional tasks, asking appropriate questions of students, guiding classroom discussions, adapting instruction to students' needs, understanding students' reasoning, assessing students' learning progress, and diagnosing and remediating students' learning difficulties.

REFERENCES

Battista, Michael T. "On Greeno's Environmental/Model View of Conceptual Domains: A Spatial/Geometric Perspective." *Journal for Research in Mathematics Education* 25, no. 1 (January 1994): 86–94.

———. *Shape Makers: Developing Geometric Reasoning with The Geometer's Sketchpad*. Berkeley, Calif.: Key Curriculum Press, 1998.

———. "Fifth Graders' Enumeration of Cubes in 3D Arrays: Conceptual Progress in an Inquiry-Based Classroom." *Journal for Research in Mathematics Education* 30, no. 4 (July 1999): 417–48.

———. “Shape Makers: A Computer Environment That Engenders Students’ Construction of Geometric Ideas and Reasoning.” *Computers in the Schools* 17, no. 1 (2001): 105–20.

———. “The Development of Geometric and Spatial Thinking.” In *Second Handbook of Research on Mathematics Teaching and Learning,* edited by Frank K. Lester, pp. 843–908. Greenwich, Conn.: Information Age Publishing, 2007.

———. “Development of the *Shape Makers* Geometry Microworld: Design Principles and Research.” In *Research on Technology and the Learning and Teaching of Mathematics: Syntheses, Cases, and Perspectives: Vol. 2, Cases and Perspectives,* edited by M. Kathleen Heid and Glendon W. Blume, pp. 131–56. Greenwich, Conn.: Information Age Publishing, 2008.

Battista, Michael T., and Douglas H. Clements. “Geometry and Proof.” *Mathematics Teacher* 88, no. 1 (January 1995): 48–54.

———. “Students’ Understanding of Three-Dimensional Rectangular Arrays of Cubes. *Journal for Research in Mathematics Education* 27, no. 3 (1996): 258–92.

Baulac, Yves, Franck Bellemain, and Jean-Marie Laborde. Cabri 2. Software. Dallas, Tex.: Texas Instruments, 1994.

Beaton, Albert E., Ina V. S. Mullis, Michael O. Martin, Eugenio J. Gonzalez, Diana L. Kelly, and Teresa A. Smith. *Mathematics Achievement in the Middle School Years: IEA’s Third International Mathematics and Science Study (TIMSS)*. Chestnut Hill, Mass.: Boston College, 1996.

Borrow, Caroline. “An Investigation of the Development of 6th-Grade Students’ Geometric Reasoning and Conceptualizations of Geometric Polygons in a Computer Microworld.” Unpublished doctoral dissertation, Kent State University, 2000.

Bransford, John D., Ann L. Brown, and Rodney R. Cocking. *How People Learn: Brain, Mind, Experience, and School*. Washington, D.C.: National Research Council, 1999.

Burger, William F., and J. Michael Shaughnessy.” Characterizing the van Hiele levels of Development in Geometry.” *Journal for Research in Mathematics Education* 17, no. 1 (January 1986): 31–48.

Chazan, Daniel. “High School Geometry Students’ Justification for Their Views of Empirical Evidence and Mathematical Proof.” *Educational Studies in Mathematics* 24, no. 4 (December 1993): 359–87.

Clements, Douglas H. “Teaching and Learning Geometry.” In *A Research Companion to “Principles and Standards in School Mathematics*,” edited by Jeremy Kilpatrick, W. Gary Martin, and Deborah Schifter, pp. 151–78. Reston, Va.: National Council of Teachers of Mathematics, 2003.

Clements, Douglas H., and Michael T. Battista. “Geometry and Spatial Reasoning.” In *Handbook of Research on Mathematics Teaching and Learning,* edited by Douglas A. Grouws, pp. 420–64. Reston, Va., and New York: National Council of Teachers of Mathematics and Macmillan Publishing Co., 1992.

de Villiers, Michael. “An Alternative Approach to Proof in Dynamic Geometry.” In *Designing Learning Environments for Developing Understanding of Geometry and Space,* edited by Richard Lehrer and Daniel Chazan, pp. 369–94. Mahwah, N.J.: Lawrence Erlbaum Associates, 1998.

Fennema, Elizabeth, and Megan L. Franke. "Teachers' Knowledge and Its Impact." In *Handbook of Research on Mathematics Teaching and Learning,* edited by Douglas A. Grouws, pp. 147–64. Reston, Va., and New York: National Council of Teachers of Mathematics and Macmillan Publishing Co., 1992.

Fennema, Elizabeth, Thomas P. Carpenter, Megan L. Franke, Linda Levi, Victoria R. Jacobs, and Susan B. Empson. "A Longitudinal Study of Learning to Use Children's Thinking in Mathematics Instruction." *Journal for Research in Mathematics Education* 27, no. 4 (July 1996): 403–34.

Fuys, David, Dorothy Geddes, and Rosamond Tischler. *The van Hiele Model of Thinking in Geometry among Adolescents. Journal for Research in Mathematics Education* Monograph No. 3. Reston, Va.: National Council of Teachers of Mathematics, 1988.

Greeno, James G. "Number Sense as Situated Knowing in a Conceptual Domain." *Journal for Research in Mathematics Education* 22, no. 3 (May 1991): 170–18.

Harel, Guershon, and Larry Sowder. "Students' Proof Schemes: Results from Exploratory Studies." *CBMS Issues in Mathematics Education* 7 (1998): 234–83.

Healy, Lulu, and Celia Hoyles. *Technical Report on the Nationwide Survey: Justifying and Proving in School Mathematics.* London: Institute of Education, University of London, 1998.

———. "Software Tools for Geometrical Problem Solving: Potentials and Pitfalls." *International Journal of Computers for Mathematical Learning* 6 (2001): 235–56.

Hershkowitz, Rena, Tommy Dreyfus, Dani Ben-Zvi, Alex Friedlander, Nurit Hadas, Tzippora Resnick, Michal Tabach, and Baruch Schwarz. "Mathematics Curriculum Development for Computerized Environments: A Designer-Researcher-Teacher-Learner Activity." In *Handbook of International Research in Mathematics Education,* edited by Lyn D. English, pp. 657–94. Mahwah, N.J.: Lawrence Erlbaum Associates, 2002.

Hoyles, Celia, and Keith Jones. "Proof in Dynamic Geometry Contexts." In *Perspectives on the Teaching of Geometry for the 21st Century*, edited by Carmelo Mammana and Vinivio Villani, pp. 121–28. Dordrecht, Netherlands: Kluwer Academic Publishers, 1998.

Jackiw, Nicholas. Geometer's Sketchpad, version 3.0. Software. Berkeley, Calif.: Key Curriculum Press, 1995.

Johnson-Laird, Philip N. *Mental Models: Towards a Cognitive Science of Language, Inference, and Consciousness*. Cambridge, Mass.: Harvard University Press, 1983.

———. "Imagery, Visualization, and Thinking." In *Perception and Cognition at Century's End,* edited by Julian Hochberg, pp. 441–67. San Diego, Calif.: Academic Press, 1998.

Jones, Keith. "Providing a Foundation for Deductive Reasoning: Students' Interpretations When Using Dynamic Geometry Software and Their Evolving Mathematical Explanations." *Educational Studies in Mathematics* 44, no. 1–2 (2000): 55–85.

Koedinger, Kenneth R. "Conjecturing and Argumentation in High-School Geometry Students." In *Designing Learning Environments for Developing Understanding of Geometry and Space,* edited by Richard Lehrer and Daniel Chazan, pp. 319–47. Mahwah, N.J.: Lawrence Erlbaum Associates, 1998.

Laborde, Colette. Solving Problems in Computer-Based Geometry Environments: The Influence of the Features of the Software. *Zentralblatt fur didaktik der mathematik (International Reviews on Mathematical Education)* 4 (1992): 128–35.

———. "Integration of Technology in the Design of Geometry Tasks with Cabri-Geometry." *International Journal of Computers for Mathematical Learning* 6, no. 3 (2001): 283–317.

Mariotti, Maria A. "Justifying and Proving in the Cabri Environment." *International Journal of Computers for Mathematical Learning* 6 (2001): 257–81.

Marrades, Ramon, and Angel Gutiérrez. "Proofs Produced by Secondary School Students Learning Geometry in a Dynamic Computer Environment." *Educational Studies in Mathematics* 44, no. 1–2 (2000): 87–125.

Mitchelmore, Michael C., and Paul White. "Development of Angle Concepts by Progressive Abstraction and Generalisation." *Educational Studies in Mathematics* 41, no. 3 (March 2000): 209–38.

Mullis, Ina V. S., Michael O. Martin, Albert E. Beaton, Eugenio J. Gonzalez, Diana L. Kelly, and Teresa A. Smith. *Mathematics Achievement in the Primary School Years: IEA's Third International Mathematics and Science Study (TIMSS)*. Chestnut Hill, Mass.: Boston College, 1997.

———. *Mathematics and Science Achievement in the Final Year of Secondary School: IEA's Third International Mathematics and Science Study (TIMSS)*. Chestnut Hill, Mass.: Boston College, 1998.

Paniati, James. "Teaching Geometry for Conceptual Understanding: One Teacher's Perspective." In *Understanding Geometry for a Changing World,* Seventy-first Yearbook of the National Council of Teachers of Mathematics (NCTM), edited by Timothy V. Craine, pp. 175–88. Reston, Va.: NCTM 2009.

Pinker, Steven. *How the Mind Works*. New York: W. W. Norton & Co., 1997.

Presmeg, Norma C. "Generalization Using Imagery in Mathematics." In *Mathematical Reasoning: Analogies, Metaphors, and Images,* edited by Lyn D. English, pp. 299–312. Mahwah, N.J.: Lawrence Erlbaum Associates, 1997.

Senk, Sharon L. "How Well Do Students Write Geometry Proofs?" *Mathematics Teacher* 78 (September 1985): 448–56.

———. "Van Hiele Levels and Achievement in Writing Geometry Proofs." *Journal for Research in Mathematics Education* 20, no. 3 (May 1989): 309–21.

Yerushalmy, Michal, and Daniel Chazan. "Overcoming Visual Obstacles with the Aid of the Supposer." In *The Geometric Supposer: What Is It a Case Of?* edited by Judah L. Schwartz, Michal Yerushalmy, and Beth Wilson, pp. 25–56. Hillsdale, N.J.: Lawrence Erlbaum Associates, 1993.

Yu, Paul, Jeffrey Barrett, and Norma Presmeg. "Prototypes and Categorical Reasoning: A Perspective to Explain How Children Learn about Interactive Geometry Objects." In *Understanding Geometry for a Changing World,* Seventy-first Yearbook of the National Council of Teachers of Mathematics (NCTM), edited by Timothy V. Craine, pp. 109–25. Reston, Va.: NCTM, 2009.

Prototypes and Categorical Reasoning:

A Perspective to Explain How Children Learn about Interactive Geometry Objects

Paul Yu
Jeffrey Barrett
Norma Presmeg

In this article, we share classroom episodes in which middle school students learn geometry while using a computer-based, interactive geometry curriculum. In these episodes, we use the concepts of prototypes and categorical reasoning to describe changes in students' understanding that come as they work in an interactive geometry environment. Our motivation is to understand better how children learn in this context, and to help teachers prepare effective curricular sequences and assessments that reflect an understanding of the complexities of students' reasoning as they use computer-based tools for exploring big ideas in geometry.

Interactive Geometry Software: Shifts in Representations and Learning

In the recent past, rapid developments in computer technology and mathematical software have given students new means by which to visualize mathematical objects and relationships (Yu 2003; Yu and Barrett 2002; Battista 2001). With interactive geometry software, a new class of geometric objects can be created that allows students to explore in new and exciting ways.

For example, one such program, Shape Makers (Battista 2003), uses such interactive geometric objects as the Parallelogram Maker that can be manipulated on screen and morphed into various forms while preserving the defining qualities of a parallelogram. Students can drag its vertices and sides to make a square. Students can, through physical activity and visual perception, conclude that a square is a type of parallelogram because the Parallelogram Maker can make a square. This relationship between a parallelogram and square is traditionally mediated through reasoning about the properties of a parallelogram: since a parallelogram is defined as having two sets of parallel sides, and a square has two sets of parallel sides, then a square must be a type of parallelogram. However, with Shape Makers, a student may understand this relationship between a parallelogram and square without knowing the properties of the shapes. A parallelogram *is always* the shape that the Parallelogram Maker makes, even though the Parallelogram Maker may make a shape that a student does not initially recognize as a parallelogram (see also Battista [2009]).

The use of interactive geometry software allows for the visualization, manipulation, and discussion of geometric objects in a manner that was once impossible with static paper-and-pencil representations. This software does *not* replace the need for students to come to know the properties of shapes. Rather, the software facilitates interactive entry points for constructing prototypes that encapsulate these properties. To help better understand this shift in representations and learning, we rely on a cognitive perspective about prototypes for such geometric ideas as parallelograms.

Category Membership and Prototypes

An important curricular and cognitive activity in mathematics is categorizing geometric shapes according to their properties. Some curricular activities in the early elementary grades include sorting a group of shape tiles according to some holistic, visual form or characteristic, or drawing a particular shape given its name or a list of characteristics. In middle and late elementary grades, students may begin to make sense of the defining characteristics of a category of shapes, such as triangles, quadrilaterals, or polyhedra. In middle and high school, students may explore the hierarchical and class-inclusion relationships for a particular category of shapes on the basis of a definition or list of properties.

From a cognitive perspective, however, the mental processes of categorization are difficult to identify. One perspective, classical categorization theory, might suggest that since all members of a category share the same defining properties, then all members would have equal status as category members (Lakoff 1987). So, for example, since a rectangle can be defined as a quadrilateral with four right angles, then any quadrilateral with four right angles (fig. 8.1) would have equal status as rectangle.

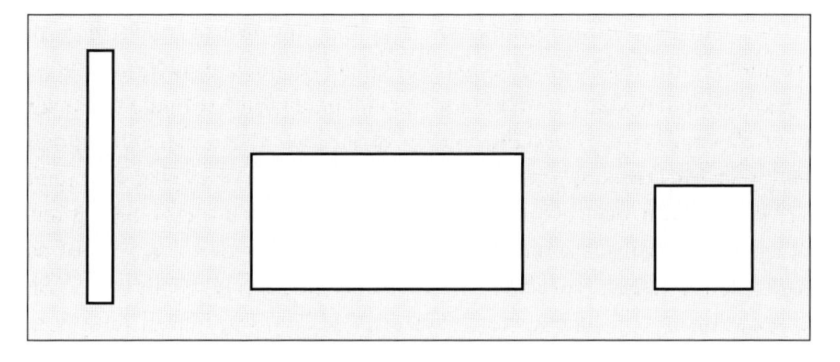

Fig. 8.1. Category of "rectangles"

Consider, however, the following thought experiment. If you were to ask a group of students to draw a rectangle, even those who understand the formal definition of a rectangle, which of the three rectangles do you think would most commonly appear in their drawings? If you picked the middle rectangle, then your guess would support a second cognitive theory of categorization known as prototype effects. Based on the work of cognitive psychologist Eleanor Rosch, the term *prototype effect* refers to a categorization process in which a certain member of a category is judged to be more representative of that category than other members (Lakoff 1987). For example, a sparrow would be considered more representative of the category "bird" than an ostrich. In figure 8.1, the shape in the middle would be considered more "rectangular" than the other two shapes, even though all the shapes have the defining properties of rectangles. To illustrate this effect, Presmeg (1992) gives an example of a student, Crispin, who redrew a right triangle so that one of the legs, rather than the hypotenuse, would be positioned horizontally and the other leg positioned vertically with the right angle in the bottom corner. In this instance, he rearranged the right triangle to conform to his prototypical notion of a right triangle.

Also helpful in our discussion is an understanding of the nature and characteristics of prototypes. First, the prototype effect does not specify that one particular object best exemplifies a category, but rather reflects a person's tendency to identify some members of a category as more representative than others. Extending the bird example, in addition to sparrows, a person might consider robins, larks, and blue jays to be more birdlike than ostriches, penguins, and puffins. Second, the prototype effect for a particular category is not universal for all people, but rather is idiosyncratic depending on each person's experiences (Dörfler 2000). Third, prototypes exist not only for objects but also for actions (e.g. walking, looking, or speaking) and descriptors (e.g. tall, short, or fast). Fourth, prototypes occur in context. For example, a prototype for *tall* may not be activated when comparing a tall man with a tall building. However, within their respective contexts, the basketball player Michael Jordan is tall, as is the Empire State Building. This idiosyncratic and contextual nature of prototypes may help explain situations in which students know a formal definition for a shape but fail to recognize visual examples of that shape because the examples are either out of context or do not match any of the students' prototypes for the shapes.

Dörfler (2000) identifies two types of prototypes that we find relevant to students' investigations in interactive geometry environments. The first are those prototypes that are based on a shape's form or properties. For example, a student's visual recognition of a rectangular shape exemplifies this kind of prototype, as does his or her recognition of a rectangle's properties of parallel sides and congruent angles.

The second type of prototype is based on operations enacted on the shape. For example, a shape such as a trapezoid can be translated, rotated, or reflected to form congruent trapezoids in different orientations. These isometric transformations can be illustrated by teachers with paper cutout shapes that are moved about on the blackboard or overhead projector. Interactive geometry software has the additional capability of nonisometrically morphing and reshaping certain objects to build relationships with other objects. For example, a shape made by Trapezoid Maker[1] can be reshaped to make a figure that is a square. In this article we refer to these prototypes as motion-based (Glass 2001).

We next illustrate how prototypes in students' thinking can be used to understand classroom learning sequences and how children's learning is inherently related to the development of their prototypes (categories of knowledge) for geometric shapes. We find evidence of these prototypes in several aspects of children's work in an interactive geometry–based environment. We illustrate how appropriate instructional tasks can help students form more powerful concepts through more advanced and more integrated prototypes.

1. Shape Maker software embodies an inclusive definition of trapezoid as a quadrilateral with at least one pair of parallel sides.

Classroom Examples of Students' Prototype Development

The classroom episodes that follow are taken from a four-week unit on quadrilaterals using Shape Makers (Battista 2003) and The Geometer's Sketchpad (Jackiw 1991) from two different seventh-grade classrooms in a large middle school in the Midwest. The school used an alternating-day block schedule so that each class lesson lasted eighty minutes. The purpose of the lessons was to facilitate the students' analysis of the characteristics and properties of quadrilaterals and to prompt the development of mathematical arguments about geometric relationships based on visualization, spatial reasoning, and geometric modeling (NCTM 2000) through interactive geometry software. These instructional tasks introduced students to the quadrilateral objects without giving definitions; the students were challenged to describe and form definitions by checking and observing the behavior of the specific quadrilateral objects. The students worked in groups of four. Two computers with software were available for each group. Each lesson began with a recap of the previous day's activities and an introduction to a new activity. The students worked in groups on the computer-based activities for thirty to forty minutes. After the small-group work, the teacher facilitated a whole-class discussion to engage students in comparative analysis of their ideas and to help them form logical justifications for their conclusions.

The following examples describing two students' interactions as they worked with the Kite Maker to investigate questions on an activity sheet titled "Can You Make It?" illustrate the development of their prototypes for kites. On the second and third days of the unit's lessons, the students were asked to match the various Shape Makers shapes to shapes A through G in a clown-face picture by dragging and manipulating one of the seven Shape Makers to fit exactly on top of the fixed shapes A through G (fig. 8.2). For this task, each Shape Maker can be used only one time, so that once a Shape Maker is used to make a lettered shape, it cannot be used to make another shape. Episode 1 illustrates the uncertainty of students' understanding of a kite and the tentative manner in which they worked with the Kite Maker.

Episode 1: Initial Prototypes for the Kite Maker (Day 2)

As Micah and Paula began the activity sheet "Can You Make It?" (fig. 8.2), they started with the Rectangle Maker.

Micah first took the Rectangle Maker and easily manipulated it, placing it on the long rectangle (A) at the top. Micah then tried to put the Kite Maker on the mouth (G), saying, "This one will be kite." He continued to manipulate the Kite Maker unsuccessfully but still replied, "Yep. Definitely Kite." He continued to try until he came to the conclusion, "I don't think we will be able to use Kite

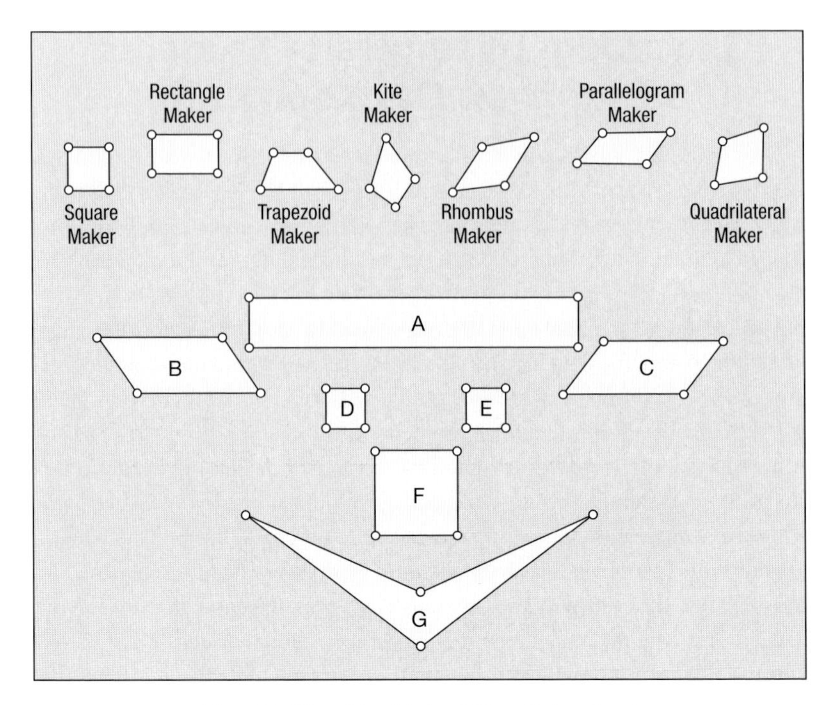

Fig. 8.2. "Can You Make It?" activity sheet

Maker." Micah's inability to make the mouth (G) out of the Kite Maker seemed to be a result of his lack of control of, and familiarity with, the Kite Maker.

"Why not?" asked Paula. It seemed as if other students had been able to fit the Kite Maker on the mouth.

Giving up on the Kite Maker, Micah suggested, "I think we need to use rhombus or something…."

Still curious about the Kite Maker, Paula asked, "What would that [Kite Maker] be, then? Like, what else could that make?"

Micah pointed to the parallelogram (C on the worksheet), "Can it [Kite Maker] make one of those?"

The nature of their inquiry regarding the Kite Maker indicated that at this point, both students had very limited prototypes for the kite shape. Both students expressed some uncertainty as to what shapes the Kite Maker could make.

Later on in the activity, Micah tried to get the Kite Maker to fit on the parallelogram shape, thereby testing his earlier idea. As he morphed the Kite Maker shape, it momentarily took the form of a rhombus (fig. 8.3), which, having two sets of parallel sides, looked very similar to the parallelogram (C) that Micah was trying to make.

However, since the parallelogram on the screen had adjacent sides that were not congruent, he was unable to make the Kite Maker shape fit. As they ex-

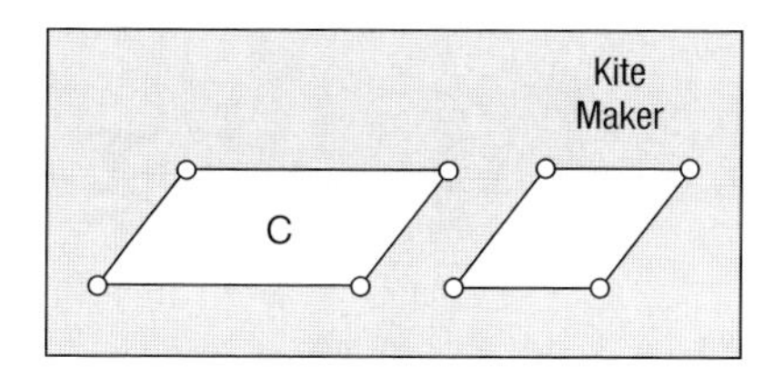

Fig. 8.3. Kite Maker as a rhombus

perimented with the Kite Maker, Paula asked, again expressing her uncertainty, "What is a kite shape, anyway?"

"One side is usually longer than the other, like that," Micah answered, turning the Kite Maker shape into a shape that looked like a prototypical kite, a convex quadrilateral having one pair of congruent sides longer than the other pair (fig. 8.4).

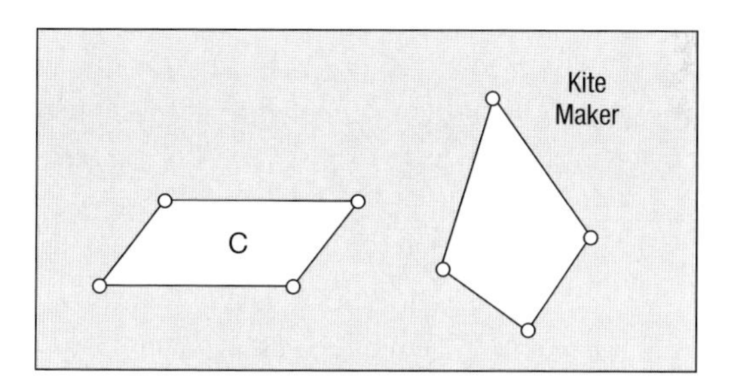

Fig. 8.4. A prototypical kite

This comment was the first meaningful descriptive statement about the kite made by either of the two students. Micah's statement reflected a holistic, yet property-based prototype for the kite. The prototype is holistic in that Micah turned the Kite Maker shape into a form most often associated with the term "kite" but is property-based in that his reasoning attends to the parts or characteristics of the shape. While dragging and morphing the Kite Maker shape, Micah turned it into a nonconvex kite, or chevron, shape (fig. 8.5). This outcome prompted Paula to ask the following question.

"I don't see why we can't make that [the mouth] with the Kite Maker," said Paula.

"Look at that, the tip [of the kite] is slanted," said Micah as he morphed the Kite Maker shape. He again tried unsuccessfully to make the Kite Maker shape fit on the mouth (shape G on the sheet).

"Try and find another one [to make the mouth]," suggested Paula.

"Yeah …, rhombus," Micah said as he put the Kite Maker to the side and took the Rhombus Maker. They then took the Rhombus Maker shape and unsuccessfully tried to make it fit on the mouth.

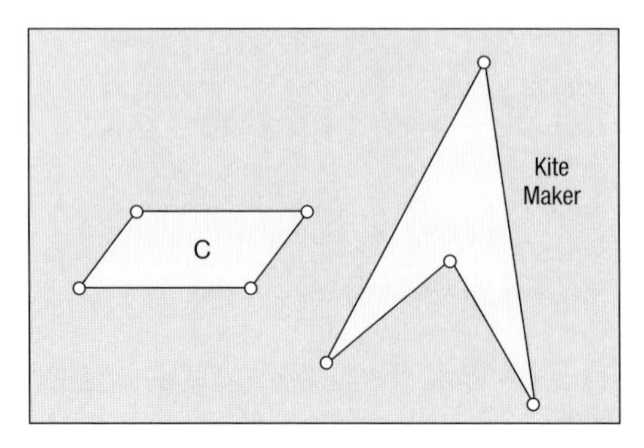

Fig. 8.5. A nonconvex kite

The previous episode serves to illustrate the degree of uncertainty and unfamiliarity that Micah and Paula had with the Kite Maker, the kite shape, or both. Their actions with the Kite Maker tended to be experimental. Their conversation and questioning were speculative. Their concept for kite seemed limited to the prototype, as in figure 8.4. In the next episode, we see the development of a motion-based prototype for a kite based on the kite's symmetry.

Episode 2: Symmetry of the Kite Maker (Day 2)

During the whole-class discussion of "Can You Make It?" one group had indicated that they had made the chevron-shaped mouth (G) with the Kite Maker. This comment prompted Paula to say, "We tried the kite, I don't know if we were doing something wrong, … but it didn't fit exactly … for G [the chevron mouth]. Kite didn't fit, it was a little bit off."

The teacher conducting the class discussion asked the rest of the class, "OK, did anyone get the kite to work exactly for that one [shape G]?" In response to the question, a number of students raised their hands, indicating that it was possible to make shape G with the Kite Maker.

Paula objected, "But it shouldn't work exactly, because it's [shape G] uneven, so it shouldn't work."

Seeking clarification, the teacher asked, "So your thing [shape G] is uneven?"

Paula continued, "In the picture it's a little bit uneven, so … and when you move one of the things [vertices on the Kite Maker], they both move. So they can't be exactly on because they are not even."

"OK, so this idea of even and uneven, let's write this down." Documenting Paula's thoughts, the teacher wrote down her statement on the overhead projector for the rest of the class to consider in their future discussions. The teacher then directed the students to write down any conjectures or queries that they may have had regarding any of the shapes, Shape Makers, or issues raised in the class discussion. After giving the students some time to write down their ideas, the teacher then asked students in the class to share their conjectures and queries. Paula raised her hand to share. "OK, Paula?"

"I just said that the kite has to be equal on both sides."

"OK, the ... kite... has ... to ... be ... equal ... on ... both ... sides," repeated the teacher as he wrote down Paula's statement.

"Yeah, like when you draw a line of symmetry through the middle, both sides should be the same," Paula clarified.

Paula's initial statements about how "when you move one of the things [vertices on the Kite Maker], they both move... so they can't be exactly on because they are not even" reflect a motion-based prototype for the visual symmetry of the Kite Maker shape. Although she did not use the term *symmetry*, her description and use of the terms *even* and *uneven* indicated concepts of symmetry. In an attempt to lead students to articulate their observations better, the teacher prompted the students to write down their conjectures and queries, giving them time to reflect personally on their observations and the classroom discussion. As a result of this conjecture-making activity, Paula was given the opportunity to extend her thinking further and describe the kite as being "equal on both sides" and having a "line of symmetry through the middle," indicating property-based prototypes for the kite.

At the beginning of the unit, Paula and Micah had little familiarity with the shapes. Much of their work was trial and error with little understanding of various shapes' properties or Shape Maker actions. The only shapes they seemed familiar with were the rectangle and square. For example, on this day, while working with the Kite Maker, Paula asked a variety of questions throughout the day's activities, for example, "What would that be, then? What else could it make? What is a kite shape, anyway? Does Kite Maker make a square?" Collectively, her constant inquiry indicated that she had very limited prototypes for the kite.

Micah also showed limited prototype development of the kite and Kite Maker. When Micah was asked by another student, "Does the Kite Maker make a square?" his answer, "No ..., I don't think so ... it might," illustrated the uncertainty that he had regarding the actions and properties of the Kite Maker. Although Micah did attempt to define a kite by saying, "One side is usually longer than the other," his statement seemed to be more a reflection on the flying variety of kite prototype rather than the geometric definition of kite.

In spite of Paula's limited knowledge of a kite or Kite Maker, the day's activities resulted in the initial construction of a prototype for the Kite Maker shape based on symmetry. Furthermore, the symmetrical nature of the Kite Maker shape, as described by Paula, was elicited by the movement of the Shape Maker in which two opposite vertices on the Kite Maker shape always move in tandem, like flapping bird wings, to preserve its symmetry. In the next episode we see Paula and Micah apply their motion-based prototype for a kite.

Episode 3: That One Is Kite Maker (Day 3)

On day 3, the students were given the activity "Hiding Shape Makers." In this activity, all seven on-screen Shape Makers started out temporarily as congruent squares (fig. 8.6). The goal was to determine which Shape Maker made up each square by manipulating each Shape Maker (fig. 8.7) and identifying them by their associated properties, actions, or both.

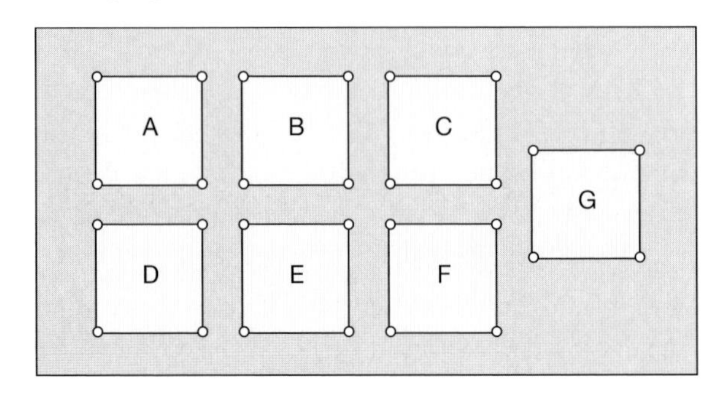

Fig. 8.6. Shape Makers hiding as squares

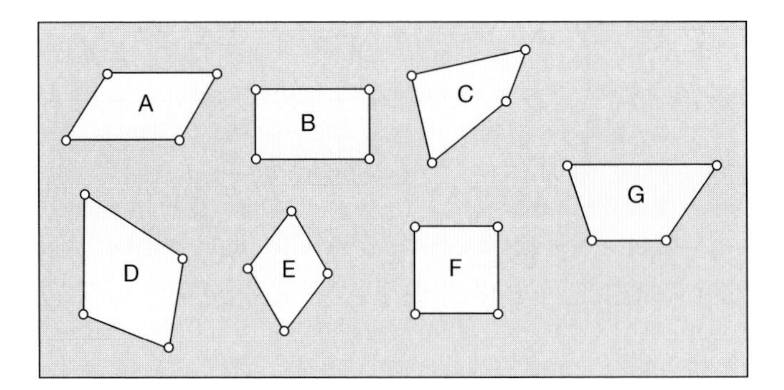

Fig. 8.7. Identifying the Shape Makers

While the researcher was still giving directions, Micah and Paula began the activity by randomly manipulating shapes. Knowing that they should have been listening to the instructions, their initial on-screen actions were tentative. Micah dragged one vertex on shape D, causing two opposite vertices to move together. This coordinated movement was the same movement described by Paula the previous day when she said, "When you move one of the things [vertices on the Kite Maker], they both move."

As Micah continued to morph the shape, Paula said, "That one is Kite Maker."

"Definitely," replied Micah. He took a closer look at the shape as he morphed it and determined, "D is kite."

"How did you figure it out?" asked Paula.

"It moved like a Kite Maker," said Micah.

Note that both Paula's and Micah's identifications of the Kite Maker shape were based on the coordinated movement of two nonadjacent vertices. Paula's recognition and Micah's use of the term *moved like* reflected a motion-based prototype for the Kite Maker shape. Thus, the interactive geometry environment helped Paula and Micah start constructing meaningful prototypes for the kite shape.

Gradual Processes Involved in Prototype Construction

The following episodes revisit the clown-face task and the "Hiding Shape Makers" task with a different pair of students, for the purpose of showing the complexity of the cognitive processes involved in the construction of robust prototypes of shapes. The discussion focuses on the progressive conceptualization of the rhombus for two student participants, Chloe and Lucas. Like most of the students in their classroom, Chloe and Lucas had little familiarity with the rhombus, as indicated by their prestudy assessments and video data taken on the first few days of the study. In general, of all the shapes, the rhombus seemed least familiar to most students. One reason could be that unlike the square, rectangle, and parallelogram, the rhombus tends to have less exposure in the grades K–6 geometry curriculum. Another possible reason is that the name *rhombus* is not associated with any particular characteristic or figure. For example, the name *parallelogram* reflects the two sets of parallel sides that define it, and the kite can be visually associated with the shape of a flying toy kite. In spite of their unfamiliarity, however, Chloe and Lucas were able to establish property-based prototypes for the Rhombus Maker shape by relating it to the more familiar square. The following episode was taken from the second day of the unit during the clown-face activity called "Can You Make It?"

Episode 4: "Can You Make It?" (Day 2)

With only square D left to make and the Rhombus Maker remaining (fig. 8.8), Lucas and Chloe attempted to turn the Rhombus Maker shape into a square. When they succeeded, thus finishing the worksheet, they exclaimed, "We got it! We got it all done!"

Lucas began to highlight the Shape Maker shapes and turn them into different colors. Chloe tried to get him back on task, "We have to fill this [student worksheet 2] in now. Fill this in!"

As they filled in the sheets Lucas asked, "Why is it [square D] rhombus?"

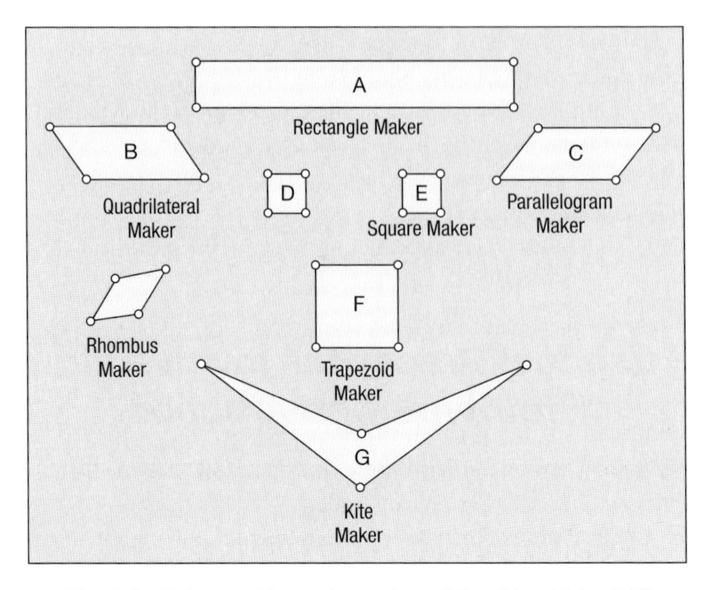

Fig. 8.8. Chloe and Lucas's work on "Can You Make It?"

"Because it [Rhombus Maker shape] always has four equal sides, but not always 90 degrees …, ah, … 60 degrees," replied Chloe.

"No, 90 degrees," corrected Lucas.

"No, 60 …," countered Chloe.

"Because there are four …," began Lucas.

Correcting herself, Chloe said, "I'm thinking about triangles."

"Not always 90," repeated Lucas.

As part of the "Can You Make It?" activity, students were asked to tell why they used each Shape Maker to make a particular shape in the clown face. In her response, Chloe stated, "Because it [Rhombus Maker shape] always has four equal sides, but not 90 degrees," reflecting a property-based prototype for the rhombus. The construction of this prototype for the Rhombus Maker shape appeared to have been supported through the actions the students used to morph the Rhombus Maker shape into a square. Since the Rhombus Maker shape could take on the form of a square, Chloe seemed to notice that like a square, a rhombus had to have four equal sides. However, unlike the square, it could have angles other than 90 degrees, as determined by shifting the Rhombus Maker shape so that it no longer had right angles. The relationship "the Rhombus Maker makes a square" helped the students construct this property-based prototype for the rhombus shape. The students had not internalized the fact that a rhombus has four congruent sides as a defining property. Rather, this property emerged in the context of the Rhombus Maker shape taking the form of a square.

When the context created by the association between the Rhombus Maker shape and a square was removed, both Chloe and Lucas were unable to recognize or correctly identify the Shape Maker. This inability was illustrated by an activity that was done on the next day, "Hiding Shape Makers." In the following episode, Chloe and Lucas were not able to identify the Rhombus Maker shape correctly in spite of their discussion on the previous day, when they had determined that "it [Rhombus Maker shape] always has four equal sides, but not always 90 degrees." The episode suggests that in the absence of the direct relationship between the Rhombus Maker shape and a square, the property-based prototype that a rhombus has four equal sides may have been lost on the students.

Episode 5: Hiding Shape Makers (Day 3)

Lucas grabbed shape E, the Rhombus Maker shape, and began to morph it (fig. 8.9).

"That's Kite Maker," guessed Chloe incorrectly.

"No, I think this is the parallelogram," said Lucas, also incorrectly.

"The Kite Maker, it's always in a diamond," reasoned Chloe.

"Yeah, so is a parallelogram," added Lucas.

"No it's not," objected Chloe.

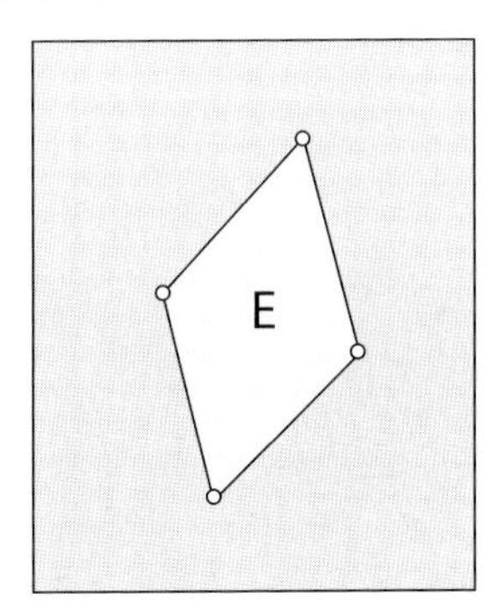

Fig. 8.9. Unknown shape

Lucas continued to check each vertex carefully to determine the actions and nature of the unknown shape E. Disagreeing with Chloe's conjecture that shape E was made by Kite Maker, Lucas pointed out, "With Kite Maker, one side can be short and one side can be long,… that's parallelogram. E is parallelogram." His statement about the lengths of the sides of the Kite Maker shape reflected a figurative prototype for a kite.

Because this day's context or activity did not explicitly draw the students' attention to the lengths of the sides of shape E, neither of the students seemed to realize that all the sides on shape E were congruent. This outcome was in contrast with that of the previous day, when the Rhombus Maker shape was in the form of a square, a context that helped mediate this figurative prototype of the rhombus through its relationship with the square.

Although both students were incorrect in their identification of the Rhombus Maker shape, both were correct in pointing out correct aspects of the Rhombus Maker shape. Chloe's focus was on the "diamond shape" that was often used to describe both the Rhombus Maker shape and the Kite Maker shape and that led her to misidentify shape E as made by the Kite Maker. Lucas's focus was on the two sets of parallel sides that remained invariant with the Rhombus Maker shape that led him to misidentify the shape as made by the Parallelogram Maker. Without realizing it, the students separated out two related properties for a rhombus shape, namely, that a diamond shape with two sets of parallel sides is a rhombus. Furthermore, calling the shapes made by the Rhombus Maker a kite or a parallelogram are in fact correct categorizations when using inclusive definitions, because any shape made by a Rhombus Maker is both a kite and a parallelogram. Rather than continue to investigate the unknown shape E, the Rhombus Maker shape, Chloe moved ahead in the activity by trying to identify some of the other Shape Makers shapes.

Later, toward the end of activity, both Chloe and Lucas then misidentified the unknown shape A as made by the Rhombus Maker (when its correct identification was the Parallelogram Maker shape). Looking over their nearly completed sheets, Chloe concluded by the process of elimination, "[Shape] A has to be rhombus."

"Let's check," suggested Lucas. Turning it into a parallelogram (fig. 8.10), with two long sides and two short sides, Lucas incorrectly concluded, "Yep, it's the rhombus."

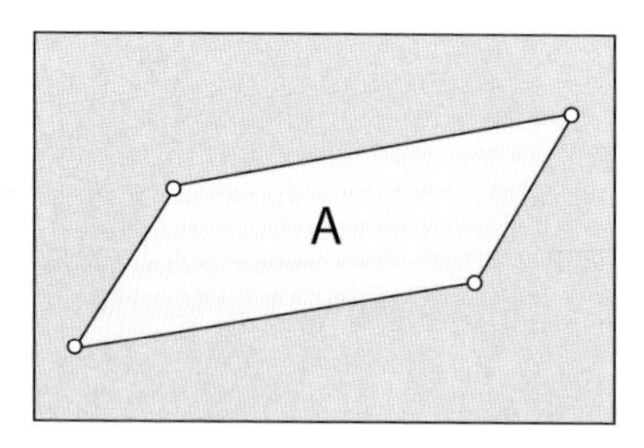

Fig. 8.10. Students' misidentification of the "Rhombus Maker"

When Lucas began to drag one vertex of shape *A,* it initially moved like a Rhombus Maker shape, preserving the congruence of all four sides. When he dragged a second vertex, he turned the unknown Shape Maker shape into a parallelogram with two short sides and two noticeably longer sides, contradicting Chloe's statement from day 2, "it [Rhombus Maker shape] always has four equal sides, but not always 90 degrees." In spite of this fact, neither student changed his or her identification of shape A, keeping the incorrect identification of the Rhombus Maker shape. By not having the square as a referent shape with which to compare the two shapes as on the previous day, the property-based prototype, that a rhombus had to have four equal sides, was lost in this day's activity, resulting in the students' incorrect identification of Shape Maker E and Shape Maker A.

The episode indicates the gradual nature of the students' construction of robust prototypes, with reversals along the way. For a teacher, the episode brings out the need to facilitate these constructions with repeated and varied activities and different contexts.

Shifting Roles of Classroom Discourse and Teachers' Interactions

Many traditional geometry curricula promote instruction with a deductive focus on shape properties as embodied by an axiomatic system with definitions, postulates, and theorems. In contrast, Shape Makers instruction takes the form of an inductive analysis of computer models that preserve defining properties and geometric relationships regardless of how they are manipulated. However, students' interaction with these objects should be done with careful guidance from the instructor (as illustrated in the vignettes of whole-class dialogue presented previously) so that students develop adequate meaning for their prototypes. Some practical suggestions that can be taken from these episodes are as follows:

1. Effective instruction should allow the flexibility and freedom for students to interact with the Shape Makers and use their own initiative to examine the objects and relations among them. The classroom episodes illustrate the investigative nature of this interactive geometry curriculum as students make sense of, visualize, and mentally deconstruct the various Shape Maker shapes. Furthermore, the motion-based nature of manipulating the on-screen shapes helped facilitate a more robust, property-based understanding of the geometric shapes, the kite and rhombus in particular. In this instructional unit, students were given enough time to explore the shapes, make conjectures, test their conjectures, and document their findings. Their experiences with the Shapes Makers expanded their prototypes and mental models for the various quadrilaterals.
2. Effective instruction should allow the flexibility and freedom for students to interact with one another. The classroom episodes also illustrate the robust conversations that resulted from the collaborative work of the pairs of students on the activities. Although prototypes are idiosyncratic, the interactive geometric objects and activities provided

a common context for students to share their ideas. On one hand, the conversation was very collaborative because students worked together to make sense of the tasks assigned to them. On the other hand, the students also worked to communicate with, and defend their ideas to, their peers while using the on-screen shapes to justify or make sense of their conjectures. These peer-to-peer discussions were important because they provided the opportunity for students' participation in more global classroom discussions facilitated by the teacher.

3. Effective instruction should include meaningful interactions with the teacher. Although most of the discussion documented in this article is between students in pairs, the importance of a teacher's guidance must not be understated. In an open, investigative learning environment, misunderstandings do arise, as in the last episode, with Chloe and Lucas's misidentification of the Rhombus Maker shape in "Hiding Shape Makers." The teacher's interaction should be deliberate and precise, yet it should maintain an openness to foster students' generation of ideas. As illustrated by the classroom teacher's interactions with Paula as they discussed her motion-based prototype for the symmetry of a kite, a teacher should help students articulate their ideas instead of giving them ideas. Furthermore, the teacher needs to expose students to proper terminology, make connections between concepts, and raise questions or encourage investigations that will help dispel misconceptions that may arise.
4. Effective instructional design should be deliberate. An important curricular feature of Shape Makers (Battista 2003) is the intentional relationship between the nature of the activities and models for learning in mathematics education.[2] Like writing a lesson with a blank piece of paper and pencil, using interactive geometry software allows for a broad range of options limited only by the imagination of the teacher. Effective instructional design and implementation should be based not only on the imagination of teachers but also on their experiences, their content goals, and their understanding of how students learn. The cognitive perspective of the prototype effect helps educators understand better the nature of object categorization and learning in the area of geometry.

2. Battista (2001, 2003) also provides a different, but complementary theoretical analysis based on the work of Dina van Hiele and Pierre van Hiele.

REFERENCES

Battista, Michael. "Shape Makers: A Computer Environment That Engenders Students' Construction of Geometric Ideas and Reasoning." *Computers in Schools* 17 (2001): 105–20.

———. *Shape Makers*. Emeryville, Calif.: Key Curriculum Press, 2003.

———. "Highlights of Research on Learning School Geometry." In *Understanding Geometry for a Changing World,* Seventy-first Yearbook of the National Council of Teachers of Mathematics (NCTM), edited by Timothy V. Craine, pp. 91–108. Reston, Va.: NCTM, 2009.

Dörfler, Willi. "Means for Meaning. " In *Symbolizing and Communicating in the Mathematics Classroom: Perspectives on Discourse, Tools, and Instructional Design,* edited by Paul Cobb, Erna Yackel, and Kay McClain, pp. 99–131. Mahwah, N.J.: Lawrence Erlbaum Associates, 2000.

Glass, Bradley J. "Students' Reification of Geometric Transformations in the Presence of Multiple Dynamically Linked Representations." Unpublished doctoral dissertation, University of Iowa, 2001.

Jackiw, Nicholas. The Geometer's Sketchpad. Software. Emeryville, Calif.: Key Curriculum Press, 1991.

Lakoff, George. *Women, Fire, and Dangerous Things: What Categories Reveal about the Mind.* Chicago: University of Chicago Press, 1987.

National Council of Teachers of Mathematics (NCTM). *Principles and Standards for School Mathematics.* Reston, Va.: NCTM, 2000.

Presmeg, Norma. "Prototypes, Metaphors, Metonymies and Imaginative Rationality in High School Mathematics. " *Educational Studies in Mathematics* 23 (1992): 595–610.

Yu, Paul. "Prototype Development and Discourse among Middle School Students in a Dynamic Geometric Environment." PhD. diss., Illinois State University, 2003.

Yu, Paul, and Jeffrey Barrett. "Shapes, Actions, and Relationships: A Semiotic Investigation of Student Discourse in a Dynamic Geometric Environment." Paper presented at the Twenty-fourth Annual Meeting of the North American Chapter of the International Group for the Psychology of Mathematics Education, Athens, Ga., October 26–29, 2002.

Conceptions of Angle:

Implications for Middle School Mathematics and Beyond

Christine Browning
Gina Garza-Kling

From their very first geometric experiences, such as playing, drawing, sorting, and naming shapes, children are constructing initial conceptions of angle. These conceptions are integral to the development of higher-level geometric understanding. Both *Principles and Standards for School Mathematics* (NCTM 2000) and *Curriculum Focal Points for Prekindergarten through Grade 8 Mathematics* (NCTM 2006) state that elementary school students should be able to describe, analyze, compare, and classify two-dimensional shapes by their angles and sides and connect these properties with definitions of shapes. By middle school, students should be able to use angles to describe and analyze figures and situations in two- and three-dimensional space. These expectations suggest that the angle concept should be fairly well developed by the end of middle school. Can the assumption be made, then, that preservice elementary and middle school teachers already have sufficient understandings of angle?

Several years ago, following a brief, initial class discussion on angle, we asked our preservice teachers a question we thought would be fairly straightforward: how does measuring an angle of a shape differ from measuring a side?

Collected responses showed that a significant number of them could not articulate well what it meant to measure an angle and chiefly defined an angle by its unit of measure. Concerned by their limited conceptual understanding, we developed activities designed to promote angle understanding for our preservice teachers that we later piloted with a group of middle school students. In this article, we describe these angle activities, examine the results from a study with middle school students, and uncover how both middle school students and preservice teachers think about angle.

Angle Conceptions

When asked to define angle, quite often our preservice teachers reply with such words as "a corner" or "something you measure in degrees." A few mention something involving a vertex and rays or lines. Yet on careful consideration of what an angle actually is, we realized that such descriptions indicate some limitations in our students' understandings. Traditionally, when the topic of angle is first encountered, it is defined in a very "static" manner, such as "where two sides of a polygon meet, they form an angle." Rather than initially be given a formal definition of angle, Keiser (2000) suggests children should instead see various representations of angle, for example, a turn or a wedge, and that they be provided time and opportunities to share and challenge one another's emerging ideas. Numerous other studies with elementary and secondary school students also suggest that more thoughtful attention be given to the development of the concept of angle, particularly with respect to exposing students to a variety of experiences with angle and angle measure (Clements and Battista 1989, 1990; Clements et al. 1996; Keiser 2004; Mitchelmore and White 2000; Scally 1990; Wilson and Adams 1992). Several have examined children's development of the angle concept in a learning environment that included the computer program Logo (Clements and Battista 1989, 1990; Kieran 1986; Noss 1987; Scally 1990). Their findings show that appropriately designed Logo activities can facilitate students' exploration and their development of concepts of angle and angle measure.

Thus research supports the supposition that angle is a complex idea, best understood from a variety of perspectives. These perspectives generally fall into three classes, namely a pair of rays meeting at a common vertex, the region formed by the intersection of two half-planes (more informally, the space or spread between two rays or an infinite wedge), and finally, a more dynamic idea of angle as a representation of a turn (Mitchelmore and White 2000; Keiser 2004). These diverse conceptions of angle appear in a variety of contexts. For example, when focusing on the angle formed by a slice of pizza (a context readily suggested by both middle school students and preservice teachers), thinking of an angle as a wedge is more useful than thinking of an angle as a turn. In contrast, when looking for the angle to describe the rotation that a snowboarder may make when completing a complex trick, such static notions as space or rays and a vertex may limit us from seeing that an angle's measure can exceed 360 degrees.

Research then leads us to two conclusions: students benefit from a broad conception of angle, and for students to develop this broad angle concept, they need opportunities to wrestle with multiple notions of angle. In an attempt to provide preservice teachers with such opportunities, we developed lessons that focus on angle and angle measure and incorporate the use of hands-on activities, graphing calculator applications, and computer software. From these lessons, three particular activities have been implemented for several years in a college geometry course designed for preservice elementary and middle school teachers. These activities have also been pilot tested in a sixth-grade classroom (Browning, Garza-Kling, and Sundling 2007). The first activity, the "wedge activity," makes use primarily of a region or space conception of angle; the other two activities involve handheld and desktop computer technology and illustrate the dynamic turn conception of angle.

Angle Activities

We start our angle explorations by simply asking our preservice teachers to describe what an angle is. Some suggest the idea of an angle as the space or spread between two rays; others express hesitation to accept that concept, since "space" implies to them a bounded region to be measured in square or even cubic units. The informal term *space* is indistinguishable for some preservice teachers from *area*. To attend to this idea of the "space or spread between two rays" representing something different from a "space in a bounded region," we introduce a "wedge activity" designed to have students consider what type of unit can be used to measure an angle and how that unit differs from those used to measure area or volume.

Wedge Activity

In the wedge activity, students are given a piece of hamburger patty paper and are asked to "invent" a device that would allow them to measure several angles (adapted from Wilson 1990). To prepare them for the wedge investigation, students may be prompted to think about how they measure length, area, and volume. We might choose a common item, such as a pencil, and ask, "Would you measure the area of the room with this pencil as a unit of measure?" Students invariably say "no" because the idea of a covering, the *area*, is not as prominent a feature of the pencil as it is of, say, a sheet of paper. However, when asked to find the length of a desk, the pencil could be used because it has the characteristic of "longness" or length, the attribute we are trying to measure. Therefore, as students think about measuring angles, they need to create a unit that embodies the attribute they are trying to measure.

To help students keep their focus on what they are actually measuring, we ask them to pretend that neither the protractor nor the idea of degrees has been invented. Doing so is a struggle for both preservice teachers and middle school students alike, who are already familiar with, and attached to, using degree measures, and we find ourselves repeatedly reminding them not to use the "d word" during this activity. We encourage our students to develop their own unit of angle measure and a measuring tool from their square piece of patty paper. Their ideas vary from folding their papers into fan shapes, to making a series of folds all of which meet in the center when opened (resembling a 360-degree protractor), to labeling their folds to match a clock. Many students fold their papers into roughly triangular "wedges" of various sizes, so chosen for how well a wedge mirrors the attribute of an angle that we are attempting to measure (the "space" between two rays sharing a common vertex), and then count the number of wedges needed to fill each angle they are measuring. The various wedge sizes represent nonstandard units and prompt discussion of how the same angle can be associated with different measures, allowing students to realize that the size of the unit is relevant to determining the measure of an angle.

Through this activity students can also begin to realize that the apparent length of an angle's sides is, in contrast with the idea described above, an *irrelevant* attribute when using wedges. For example, when given two angles of identical measure, one of whose sides are represented by shorter segments than the other, students will find that the same number of wedges "fill" both as they concentrate on filling the space near the vertex as opposed to the entire area "outlined" by the angles' sides. This experience encourages them to conclude that the "length" of the sides has no relation to the amount of spread or space measured by the wedge. The observation of the irrelevance of "side length" in angle measure can also promote a discussion regarding the difference between the words *side* and *ray,* in particular that the two rays meeting at a common endpoint to form an angle are in fact infinitely long, and that when an angle is drawn, only a finite piece of each ray is shown.

Further, the creation of their varied units prompts a discussion of the rationale behind having a common or standard unit to measure angles. The idea of degree emerges in this discussion as a traditionally accepted unit, viewed now by the students as a very tiny, standard wedge. Overall, the wedge activity attends to several objectives: providing an experience with a nonstandard unit of measurement (a "wedge"), attending to relevant attributes being measured when finding an angle, and developing the idea of a degree.

Angle Estimation

The next major activity is angle estimation. We use the SmileMath application on the Texas Instruments TI-73 Explorer calculator (see fig. 9.1), which allows students either to guess the measure of a given angle or to attempt to "draw" an angle of a given measure. In both instances the angle begins with a measure of 0 degrees and is formed as the terminal ray turns away from the initial ray much in the way that clock hands work. Thus students experience the idea of angle as a representation for a turn. Students can choose to work with angles that range from either 0 to 90 degrees or 0 to 360 degrees, thus introducing them to angles with measures greater than 180 degrees. In one version of the activity, students guess the measure of the angle once the terminal ray stops moving. In the other version, students are given a particular angle measure in degrees and have to freeze the terminal ray (by pressing a key) when they believe it has made an angle having the given measure. In both instances, five angles are drawn for each game and a "report card" is given at the end, providing students feedback about the accuracy of their estimations.

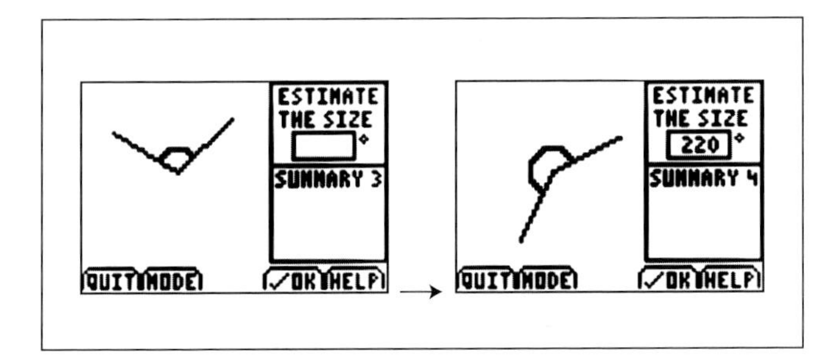

Fig. 9.1. Sample screens from the SmileMath application.

SmileMath invokes several interesting ideas about angle, including the idea of angle as a representation of a turn and the importance of knowing and using such benchmark angle measures as 45, 90, 180, or 270 degrees when estimating. Also, because the position of the initial ray varies each time a new angle is drawn, students begin to realize that the angle's orientation need not look "typical," with the initial ray along the horizontal. Students then realize many angles that "look different" actually have the same measure. Having a variety of orientations is not trivial; even our preservice teachers struggle with identifying 90-degree angles without a horizontal side.

Logo Light

Of the various conceptions of angle, the concept of an angle as a representation of a turn seems to be the least familiar to our students. SmileMath provides one venue to experiment with this concept. To further develop the idea of angle as turn, we introduce the computer program Logo with a scaled-down version designed for the TI-73 Explorer called "Logo Light." In Logo, a small turtle icon is present on the screen, and the user draws a shape by typing in commands, such as "rt 72" or "fd 20" to make a right turn of 72 degrees or a forward move of 20 "turtle steps," respectively. Once our students are familiar with the turtle, we challenge them to create specific figures. Producing these figures successfully requires knowledge of fundamental properties of shapes. Further, these activities lead students to discover the relationship between the exterior and interior angles of a polygon.

For example, when students are asked to create an equilateral triangle, many apply their knowledge of these triangles and initially attempt the construction using turns of 60 degrees. However, they soon discover the turtle "walks" around the outside of the shape as it is formed, and so asking it to turn 60 degrees will form an interior angle of 120 degrees. Thus the three sides do not form a closed figure, as shown in figure 9.2.

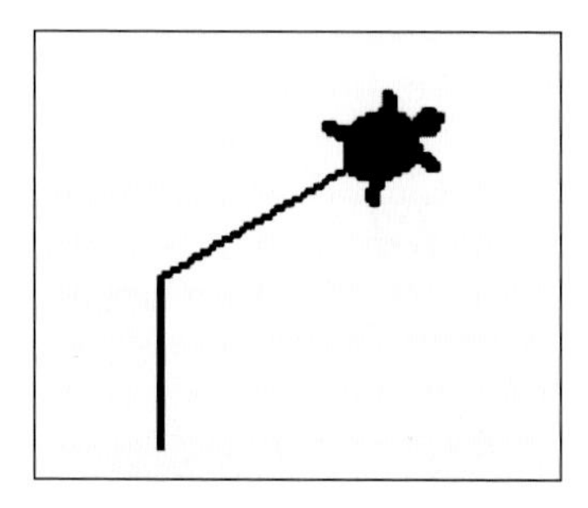

Fig. 9.2. Turtle executing "fd 60 rt 60 fd 60"

To resolve this problem we have a student volunteer to "be the turtle" and physically walk out the path of the triangle. This action promotes discussion regarding the amount of turn needed at each vertex to create the 60-degree interior angles, the notion of straight angles (180-degree measures), and the relationship between interior and exterior angles of polygons. Thus Logo provides an environment in which students can further their understanding of two-dimensional shapes and their properties while specifically focusing on angles as being created by "turtle turns." Some studies have shown that young students may struggle with seeing the angle in the turtle's turn (Kieran 1986; Mitchelmore 1998). Our experiences with both preservice teachers and middle school students, however, suggest that using SmileMath prior to the students' investigations with Logo can facilitate their visualization of the angle in the turtle environment.

Angle Understanding

In theory, the activities described in the foregoing have potential for both broadening and solidifying students' conceptions of angle, but validating this claim requires a way to measure growth in the students' angle concepts. The theoretical framework that we chose to examine students' conceptions is based on the work by the van Hieles, which characterized students' geometric understanding into five levels, the first three of which—the visual, descriptive, and relational levels—are most applicable to school mathematics (van Hiele 1986). See Crowley (1987) or Battista (2009) for an introduction to the van Hiele levels.

Previous work by Scally (1990) related the first three van Hiele levels to the angle concept by developing a set of level indicators that focused specifically on angle. For example, if a student is asked to explain why one angle is larger than another, a response such as "because it looks bigger" would suggest thinking at the visual level, for it gives no indication of what the student sees that makes the angle "look bigger." In contrast, arguing "it is larger because the lines are more spread apart" would demonstrate understanding of the appropriate property for comparing angle sizes and thus indicate the descriptive level of thinking. Given that in the van Hiele model, understanding is strongly connected with learning experiences, preservice teachers (grades K–8) and middle school students may be likely to operate at similar van Hiele levels and have similar understandings of the angle concept. Accordingly, Scally's indicators may be effective for analyzing students' work at either level. Thus we decided to work with both preservice teachers and sixth graders, hypothesizing that both groups of students could potentially benefit from the previously described angle activities. In particular, we now describe a study we conducted with a class of sixth-grade students and attempt to ascertain growth in their angle concepts as measured by Scally's indicators.

Angle Study

Our study was conducted in a rural Midwestern middle school classroom with sixth-grade students who were described by their teacher as "very typical" in their mathematical abilities (Browning, Garza-Kling, and Sundling 2007). Twenty-one of the twenty-two students in the class participated in providing data. The students had already done some work with angle that year in their study of the "Shapes and Designs" unit of the Connected Mathematics Project curriculum (Lappan et al. 1996), which involved using goniometers, or "angle rulers"; estimating angle measures; and comparing angles of different measures (fig. 9.3). They had also investigated other geometric ideas, such as properties of polygons, area, perimeter, and circumference. We had developed a seven-day sequence of angle lessons that were chiefly taught by the classroom teacher with the two co-authors each leading one class session and assisting with the others. The lessons integrated the three aforementioned activities with periodic whole-class discussions, often in regard to the development and revision of the class's working definition of angle.

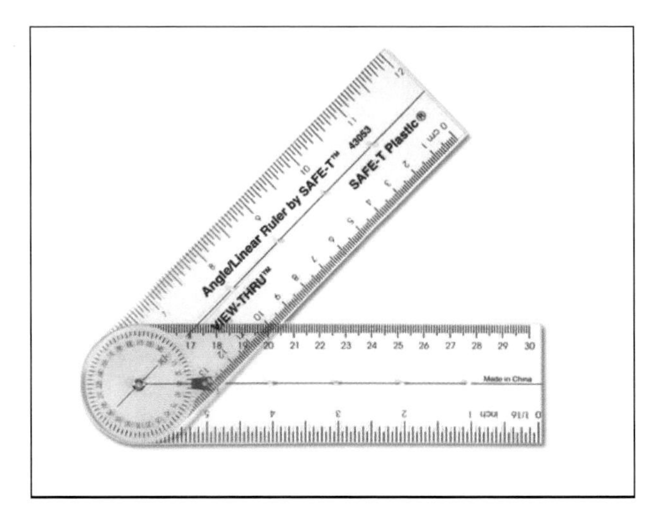

Fig. 9.3. An angle ruler

Before engaging in the angle lessons, the students took a seven-item pretest consisting of angle questions (sample items appear in fig. 9.4). The items on the tests were taken primarily from the work of Clements and Battista (1989, 1990) and from earlier work with our preservice teachers. During the unit, all the work of students participating in the study was collected, including activity sheets, Logo tasks, and "exit" cards (three-by-five-inch note cards of students' summarizing thoughts that were completed before students could "exit" to the next class). The coauthors also made observational notes during each of the lessons. A posttest identical to the pretest was administered two months after the completion of the lessons. (This delay in administration was due to the tight curriculum schedule the classroom teacher needed to follow, in which sufficient time for the posttest was available only near the end of the school year.) No further angle activities were conducted with the students during the interim. We discuss the overall results of the pretests and posttests, with a particular focus on the four items shown in figure 9.4.

For the pretest and posttest, the van Hiele level indicators described by Scally (1990) were used to determine both initial angle understanding and overall growth. Pretest results showed that the understanding of the majority of the students (15 of 21) was at a visual level of understanding for angle. When answer-

- What is an angle?
- Draw an angle. Why are you convinced what you have drawn is an angle?
- Draw a larger angle. What is it about this angle that convinces you that it is larger than the one you drew in item #2?
- How large is the largest possible angle? Why is that the largest angle?

Fig. 9.4. Sample items from the pretest and posttest

ing the question "What is an angle?" the most common response was "a corner of a shape." However, many of the students could use the language of *obtuse, acute,* and *right* and could classify angles appropriately, which might suggest some attention to properties or emergent descriptive level thinking. Yet at times this knowledge appeared to be from memory alone. For example, when asked to draw an angle and indicate how they knew it was an angle, many students drew a right angle. The students' supporting rationale for why they knew it was an angle was consistently "because I know it is a right angle." Further, when asked to draw an angle that was larger than their initial one and explain how they knew it was larger, many students drew an obtuse angle ("obease," as one student called it) and claimed that such angles are larger than right angles. Such responses suggest that students are able to recall terminology they have been taught without necessarily understanding what the words mean, thus giving a false impression of their true understanding.

We did not interview the students, in an attempt to minimize the overall classroom "interruption" of the study. Nevertheless, our observational notes of classroom dialogues support our belief that these students were using vocabulary that suggests a higher level of angle thinking than what they really understood. A possible explanation for this phenomenon may lie with the increased demands placed on classroom teachers to "cover" a given curriculum in a fixed time frame. Teachers are pressed to provide students with a summary of big ideas, such as angle classifications, rather than take time to allow the summary to emerge from the students' integration of ideas on the basis of their personal experiences. "[S]uch a shortened learning process never can lead to real understanding of a concept; the pupils will get the use of a verbal network of relations not based on any experience of action" (van Hiele 1986, p. 180).

Thus many students came into this sixth-grade class equipped with some angle-classification vocabulary but able to describe angles only as corners—an apparent mismatch in levels of understanding. We would suggest that when angles are initially introduced in earlier grades, time be taken to investigate what makes an angle, how an angle increases or decreases in size, and how angles can be measured. Angle classifications will make more sense to students later if they first have a better understanding of what they are classifying.

Posttest results from the first three items presented in figure 9.4 showed growth in students' understanding. More students defined an angle either as "two lines that meet at a vertex" or "the measure between two lines that form a vertex" and included their definition as support for their drawn angle. When asked why the second angle drawn was larger, students indicated that "there was more space inside"; "the space in between the rays has more room"; or "I added another 90-degree point (sic): 90 + 90 = 180 degrees," when a student drew a straight angle made up of two right angles with a common side and used what the authors describe as an additive argument. On the pretest, only two of the twenty-one students talked about the space between the rays as a means of describing larger angles, whereas fifteen students used either a "space" or an additive argument on the posttest. Thus posttest responses show that students had integrated more of the space concept into their notion of angle, suggesting that they were moving toward a broader understanding of angle properties.

The fourth item of interest asked the students to describe the largest possible angle and justify their response. On the pretest, nine students responded 180 degrees, eight gave 360 degrees, and the remaining four said 120 degrees, 100 degrees, or 179 degrees as the measure of the largest angle. We believe that this limited notion for largest angle measure is related to an initial static definition of angle, such as "the corner of a shape." In contrast, posttest results showed four students responding 180 degrees, twelve students giving 360 degrees, and three suggesting that no largest angle exists. The two remaining responses suggested that students were referring to a previous posttest item and thus their responses were noninformative. Although we were pleased that many more students thought of angles beyond 180 degrees, we had hoped their experience using Logo would have led more students to decide that no largest angle exists. A class discussion had also occurred on how to mark angles that were greater than 360 degrees, but the posttest showed that only three students had actually integrated that idea into their understanding. Because the SmileMath activity had students investigating angles up to 360 degrees in measure, we argue those experiences moved more students to see angles with measures up to 360 degrees. However, these results suggest that teachers may need to provide more time for discussion and integration of the limitless aspect of angle measure.

Many of the students' responses on the pretest and posttest were informative, both in helping us understand the types of angle definitions that students bring with them to middle school and in helping us determine a student's general level of angle understanding. Overall, sixteen of the students exhibited significant growth in understanding, determined by an increase in van Hiele level; the remaining five showed very little to no growth. Thus the angle lessons would seem to have helped students reach higher levels of angle understanding, and the activities involving technology in particular seem to have promoted the incorporation of varying ideas of angle.

Conclusion

When we consider the ways that we classify and define shapes, describe such transformations as rotations, look for rotational symmetry in objects, or discuss how such objects as lines on the plane or in space relate to one another, we are struck with how fundamental the idea of angle is and how widely it is used throughout geometry and later in trigonometry and calculus. The fact that many of our preservice elementary and middle school teachers appear to struggle to articulate what an angle is or to describe what a degree measures is disconcerting. After observing similar difficulties exhibited by middle school students, we realized that both preservice elementary school teachers and middle school students might begin their studies of angle with roughly the same levels of conceptual understanding. Thus our study with middle school students would seem to have the potential to guide the teaching and mathematics content choices not only of elementary and middle school teachers but of teacher educators as well. As the sixth-grade students participated in the angle unit, the three main activities and frequent class discussions about their emerging angle conceptions appear to have made an impact on their understandings of angle. The posttest results showed that the majority of the class reached higher van Hiele levels of understanding with respect to angle.

Overall, the responses of the sixth graders on the posttest included various components of angle. For example, when asked simply to provide an angle definition, the static representation consisting of rays and a vertex was given by the majority of students. However, students typically referenced ideas related to the wedge aspect when describing the size of angles, frequently responding that "there is more space between the lines." Lastly, the item asking for the largest angle evoked ideas relating to turn, evidenced by statements of "you can turn 360 degrees, 720 degrees, and so on" or "because it goes around in one whole complete rotation."

We find these results to be quite encouraging and also consistent with what we have informally observed happening with our preservice teachers. Since the introduction of these angle activities several years ago, we have seen our preservice teachers move from limited initial conceptions of angle to those that incorporate more aspects, such as "the space turned between two rays that meet at a common vertex." We believe that our findings validate the general idea that teachers at all levels can learn much by carefully examining our students' understandings of fundamental geometric concepts. If we incorrectly assume too much about students' fundamental knowledge, then building on a shaky foundation will likely lead to future problems, frustrations, and misconceptions. Strong geometric knowledge must be rooted in rich conceptual understanding of fundamental ideas—ideas such as angle. Only by first developing these deep roots can we expect our students to reach higher levels of geometric understanding.

REFERENCES

Battista, Michael T. "Highlights of Research on Learning School Geometry." In *Understanding Geometry for a Changing World,* Seventy-first Yearbook of the National Council of Teachers of Mathematics (NCTM), edited by Timothy V. Craine, pp. 91–108. Reston, Va.: NCTM 2009.

Browning, Christine A., Gina Garza-Kling, and Elizabeth Hill Sundling. "What's Your Angle on Angles?" *Teaching Children Mathematics* 14 (December 2007/January 2008): 283–87.

Clements, Douglas H., and Michael T. Battista. "Learning of Geometric Concepts in a Logo Environment." *Journal for Research in Mathematics Education* 20 (November 1989): 450–67.

———. "The Effects of Logo on Children's Conceptualizations of Angle and Polygons." *Journal for Research in Mathematics Education* 21 (November 1990): 356–71.

Clements, Douglas, Michael Battista, Julie Sarama, and Sudha Swaminathan. "Development of Turn and Turn Measurement Concepts in a Computer-Based Instructional Unit." *Educational Studies in Mathematics* 30 (June 1996): 313–37.

Crowley, Mary L. "The van Hiele Model of the Development of Geometric Thought." In *Learning and Teaching Geometry, K–12,* 1987 Yearbook of the National Council of Teachers of Mathematics (NCTM), edited by Mary M. Lindquist, pp. 1–16. Reston, Va.: NCTM, 1987.

Keiser, Jane M. "The Role of Definition." *Mathematics Teaching in the Middle School* 5 (April 2000): 506–11.

———. "Struggles with Developing the Concept of Angle: Comparing Sixth-Grade Students' Discourse to the History of the Angle Concept." *Mathematical Thinking and Learning* 6 (2004): 285–306.

Kieran, Carolyn. "LOGO and the Notion of Angle among Fourth- and Sixth-Grade Children." In *Proceedings of the 10th PME International Conference,* edited by University of London Institute of Education, pp. 99–104. London: University of London, 1986.

Lappan, Glenda, James Fey, William Fitzgerald, Susan Friel, and Elizabeth Phillips. *Shapes and Designs: Two-Dimensional Geometry.* Connected Mathematics Project. Palo Alto, Calif.: Dale Seymour Publications, 1996.

Mitchelmore, Michael. "Young Students' Concepts of Turning and Angle." *Cognition and Instruction* 16 (1998): 265–84.

Mitchelmore, Michael, and Paul White. "Development of Angle Concepts by Progressive Abstraction and Generalisation." *Educational Studies in Mathematics* 41 (2000): 209–38.

National Council of Teachers of Mathematics (NCTM). *Principles and Standards for School Mathematics*. Reston, Va.: NCTM, 2000.

———. *Curriculum Focal Points for Prekindergarten through Grade 8 Mathematics*. Reston, Va.: NCTM, 2006.

Noss, Richard. "Children's Learning of Geometrical Concepts through Logo." *Journal for Research in Mathematics Education* 18 (November 1987): 343–62.

Scally, Susan. "The Impact of Experience in a Logo Learning Environment on Adolescents' Understanding of Angle: A van Hiele Based Clinical Assessment." Ph.D. diss., Emory University, 1990.

van Hiele, Pierre. *Structure and Insight*. Orlando, Fla.: Academic Press, 1986.

Wilson, Patricia S. "Understanding Angles: Wedges to Degrees." *Mathematics Teacher* 83 (April 1990): 294–300.

Wilson, Patricia S., and Verna M. Adams. "A Dynamic Way to Teach Angle and Angle Measure." *Arithmetic Teacher* 39 (January 1992): 6–13.

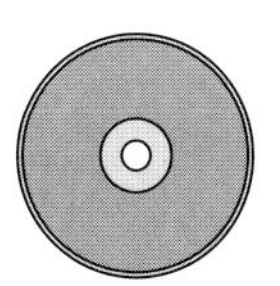

A link to Smile Math and Logo Lite is found on the CD-ROM disk accompanying this Yearbook.

Developing the Spatial Operational Capacity of Young Children Using Wooden Cubes and Dynamic Simulation Software

Jacqueline Sack
Retha van Niekerk

THE development of spatial thinking forms "a major element of practical preparation for many occupations (engineering, architecture, construction, topography, drafting, machine operation, etc.)" (Yakimanskaya 1991, p. 7). In their review of research on spatial reasoning, Clements and Battista provide evidence that visual processing involves manipulation and transformation of visual representations (1992). The National Council of Teachers of Mathematics' *Principles and Standards for School Mathematics* (NCTM 2000, p. 43) supports this view.

> Beginning in their early years of schooling, students should develop visualization skills through hands-on experiences with a variety of geometric objects and through the use of technology that allows them to turn, shrink, and deform two- and three-dimensional objects. Later they should become comfortable analyzing and drawing perspective views, counting component parts, and describing attributes that cannot be seen but can be inferred. Students need to learn to physically and mentally change the position, orientation, and size of objects in systematic ways as they develop their understandings about congruence, similarity, and transformations.

In this article, we present a framework and representative sample of activities that align with the vision of the NCTM Geometry Standard. Our work is part of an ongoing collaboration among educators in South Africa, the United States, and Holland. Its focus is to develop the spatial knowledge of children of eight to twelve years of age through an extensive program of hands-on activities and a simulated computer environment. The theoretical framework, the Spatial Operational Capacity (SOC) model (Wessels and van Niekerk 2000), combines the approach described by Yakimanskaya (1991) and the research of van Niekerk (1995a, 1995b, 1996, 1997). The framework forms the basis of the geometry curriculum for grades 1–9 in Gauteng Province, South Africa (Wessels 2001; GICD 1999). Parts of this model were also adapted by the Rice University School Mathematics Project in collaboration with Houston Independent School District as part of a geometry initiative to develop grades K–12 teachers' content and pedagogical knowledge. In the next section, we present an overview of the theoretical framework, the SOC model.

Spatial Operational Capacity Model

Our model stems from the belief that children should be exposed to activities that require them to act on a variety of physical and mental objects and transformations to develop the skills necessary for solving spatial problems (Yakimanskaya 1991). The instructional design based on the SOC model uses—

- *full-scale* models (or scaled-down models) of large objects that can be handled by the child;
- *conventional-graphic* models, two-dimensional graphic (2-D) representations that bear resemblance to the real, three-dimensional (3-D) objects; and
- *semiotic* models, which are abstract, symbolic representations that usually do not bear any resemblance to the actual objects. Examples include view and floor-plan diagrams.

Children should develop competence using physical and mental processes with all three visual representation modes in addition to *verbal* descriptions, regardless of the representation given in any particular problem (see fig. 10.1).

Our carefully designed instructional activities move among these different representations, initially using wooden cubes and Soma pieces (Weisstein 1999), then progressing to 2-D, conventional graphic and semiotic models (Freudenthal 1991), verbal representations, and a dynamic computer simulation, Geocadabra (Lecluse 2005).

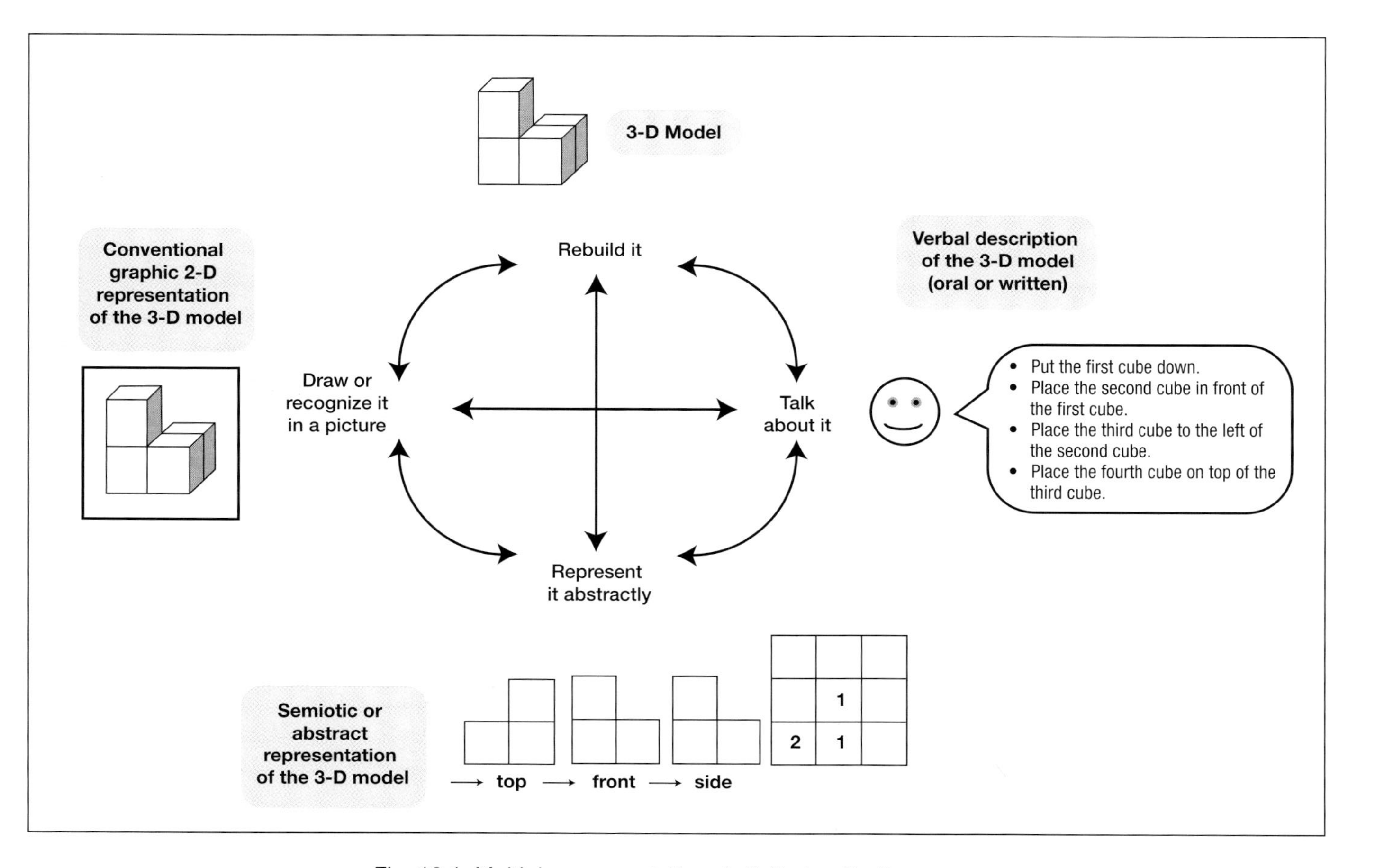

Fig. 10.1. Multiple representations in 3-D visualization

Intervention Program

An early version of our intervention program was carried out in rural South Africa (van Niekerk 1997) with twenty-one children of six to seven years of age. Although English was the language of instruction, eleven of the children did not speak English as a first language. The intervention program was introduced after the children had completed six weeks of compulsory instruction that included activities to develop fine and gross motor skills, visual skills, auditory skills, spatial skills, and form perception. The children received instruction presented by one of the authors for one hour a week for the remainder of the academic year. All instruction periods were video-recorded and transcribed for subsequent analysis. In this article, we include sample activities and important results from this intervention.

The revised intervention program, based on van Niekerk's (1997) work, addresses Presmeg's (2006, p. 226) research question, "How can teachers help learners to make connections between visual and symbolic inscriptions of the same mathematical notions?" The activities we share can be flexibly integrated into an elementary school mathematics curriculum. Ideally, we suggest that the program be introduced to children eight years of age and paced for completion within two to four years. However, regardless of the age of the children when the program is started, the introductory activities should be used. We also suggest that the program activities be interspersed over the course of the whole school year rather than compacted into short-term blocks of time.

The intervention program uses same-size, same-color unit cubes and Soma pieces (see fig. 10.2) and dynamic computer simulation software, Geocadabra (Lecluse 2005).

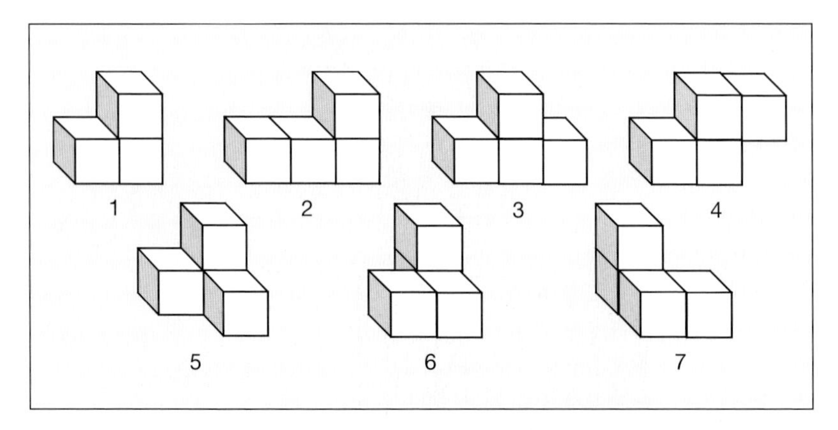

Fig. 10.2. The Soma set can be made by gluing unit cubes together.

Identification and orientation activities introduce children to real 3-D models, conventional 2-D pictures, and verbal representations (see fig. 10.1). Semiotic representations are introduced at a later stage. These introductory activities provide opportunities for children to develop and use appropriate vocabulary while focusing on position and orientation of 3-D figures and their component parts. Such terms as *on top, above, over, below, at the bottom, beside, next to, left of, to the left, right of, to the right, in front, behind, at the back, top, bottom,* and *beside* are used throughout to develop spatial vocabulary. Examples of identification and orientation activities follow.

- The teacher builds a structure using three to twelve loose cubes. Children reproduce what they see (3-D stimulus to 3-D product) from their locations in the room using loose cubes. The children then describe what they have built and why their constructions are different, depending on what they can see from their locations in the room (3-D stimulus to verbal product). The children also build a 3-D figure to correspond to a 2-D stimulus.
- The teacher shows one of the Soma pieces in a specific orientation. Each child selects the appropriate piece from his or her own set of Soma pieces and places it in the same orientation as the teacher's piece (3-D stimulus to 3-D product). Each child then draws the figure (3-D stimulus to 2-D product). Alternatively, the teacher provides each child with a set of cards illustrating the individual Soma pieces in different orientations. Each child then selects the card that matches the teacher's Soma piece (3-D stimulus to 2-D product).
- Children use loose cubes to build individual Soma pieces from verbal directions read by the teacher or from verbal directions on cards (verbal stimulus to 3-D product). Children write or state their own directions for building or describing Soma pieces or simple combinations of Soma pieces (2-D or 3-D stimulus to verbal product). As a variation, in pairs or small groups, one child hides a Soma piece and describes it while the others try to build it using loose cubes, or other students ask descriptive questions to try to identify the hidden piece (verbal stimulus/product to 3-D product/stimulus).

Spatial development activities gradually increase in complexity through the four representations that children encounter in the program (see fig. 10.1). Activities using the dynamic computer simulation Geocadabra should be interspersed throughout the program to develop children's abilities to relate to multiple representations through technology. The computer interface is addressed later in the article. We next describe sample spatial development activities with students' accompanying responses from the revised intervention program.

Sample Verbal-Descriptive Activity

In small groups, one child selects one of the Soma pieces and hides it from the other members of the group. He or she describes the hidden piece verbally while the other members ask clarifying questions to try to build a copy of the hidden piece using loose cubes. For an example, see figure 10.3.

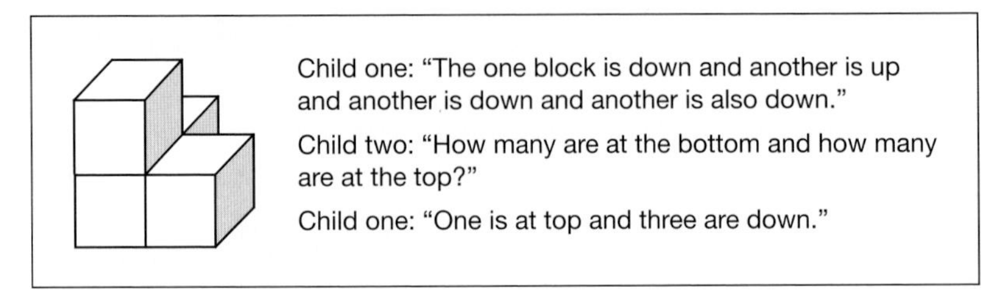

Fig. 10.3. A child's verbal explanation of the position of the top block

In general, children at all levels struggled with such terms as *left, right, in front,* and *to the back.* The children concluded that "left" and "right" are not determined by one's position in space but by the direction from which one faces the object. Some children used their right hands to guide themselves through this experience.

Initially, children described the Soma pieces using holistic comparisons with known shapes. However, for Soma pieces 5, 6, and 7, they resorted to analytic descriptions based on the cubes' positions within the object (see fig. 10.4). Over time, they moved toward analytic descriptions for most of their verbal representations. The class realized the importance of using common terminology to communicate effectively. Over time, the ease with which second-language learners progressed increased noticeably through extended practice requiring verbal descriptions (van Niekerk 1997, pp. 163–75).

Sample 2-D Products

Several activities have been developed to move between 3-D and 2-D representations. For example, children draw individual pieces or combinations of Soma pieces in various orientations. Figure 10.5 displays children's drawings for various three-dimensional cube figures observed in van Niekerk's earliest study with rural South African six- to seven-year-old children (1997). The drawings were assigned to categories, indicated by columns in figure 10.5, on the basis of similar attributes found in the children's responses. The order in which the categories appear should not be regarded as a developmental hierarchy. Very similar results were observed in informal trials with ten-year-old children in South Africa and with six-year-old children in the United States who were asked to draw figures i, ii, and iii.

Soma Piece	Holistic Description	Analytic Description
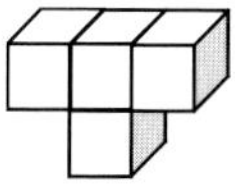	"It looks like a T."	"Three blocks next to one another and one underneath in the middle"
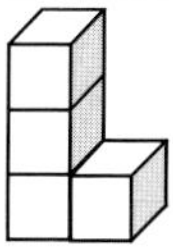	"It looks like an L."	"Three blocks stacked on one another with one block next to the stack of three"
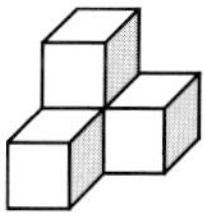	"It looks like a chair."	"One block in front with one block behind that one, and one block next to that one, and one block in the corner on the top"

Fig. 10.4. Holistic and analytic descriptions for the Soma pieces

The line drawings for figures i through v in category A were made by three children at the beginning of the research project. These children reverted to other categories of representation for figures vi through viii. Most of the children drew pictures in categories B and C. Some children made mistakes in the number of objects represented, as in category B. These children attempted to make front-view drawings of figures i through iv, but they all used other forms of representation for figures v through viii. Drawings in category D were made by the children who struggled most with this activity. The children who represented hidden objects as in category F were from a range of ability levels. The drawings in category G, made by children with strong spatial ability, used different heights to represent depth in the individual components of the figure. Only one child attempted to make drawings represented by category H.

Sample 3-D Representations from 3-D or 2-D Stimuli

Evaluating children's understanding of given tasks was easiest through the 3-D construction medium. None of the children had difficulty working with 3-D constructions when the stimulus was a 3-D object. Initially, the children constructed simple Soma-like figures with a maximum of four cubes. They later progressed toward complicated combinations of Soma pieces in different orientations. Initially, those children who had difficulty when the stimulus was a 2-D

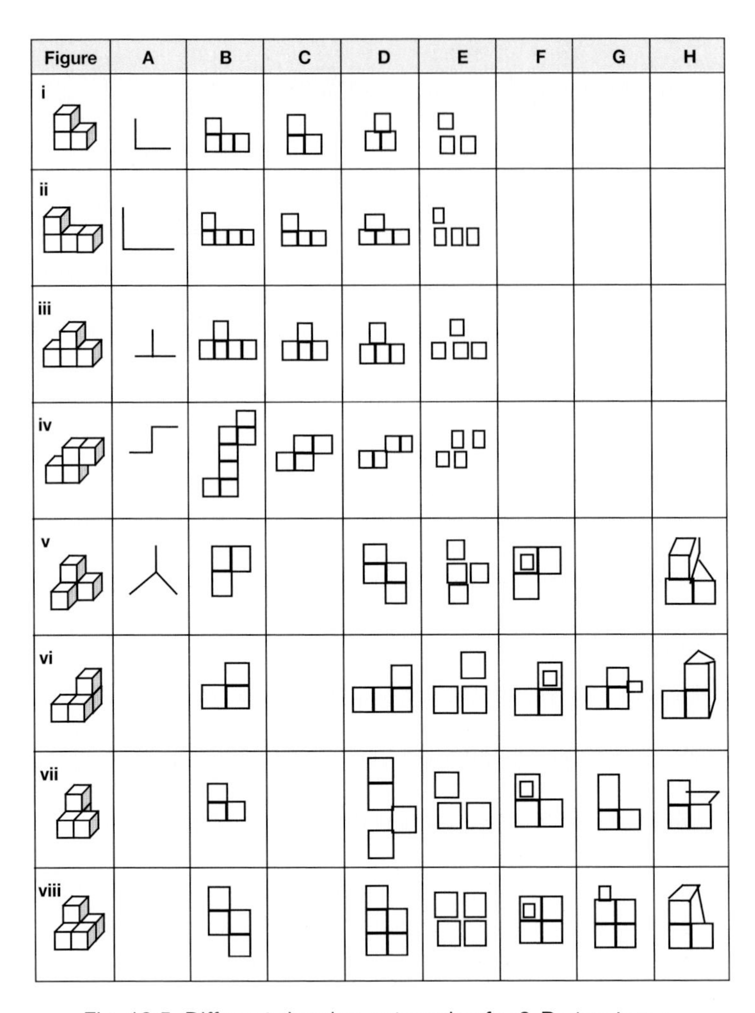

Fig. 10.5. Different drawing categories for 3-D structures, from an analysis of students' work

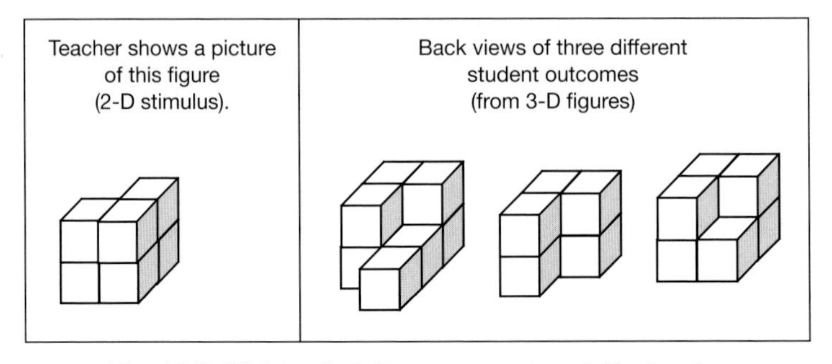

Fig. 10.6. Children's 3-D responses to a 2-D stimulus

picture understood after experiencing several similar activities that some cubes might not be visible in a 2-D picture (for example, see fig. 10.6).

The following example of a 2-D assembly-construction stimulus (fig. 10.7a) for a 3-D product shows how some children initially thought about sameness in objects that were mirror images of each other (fig. 10.7b).

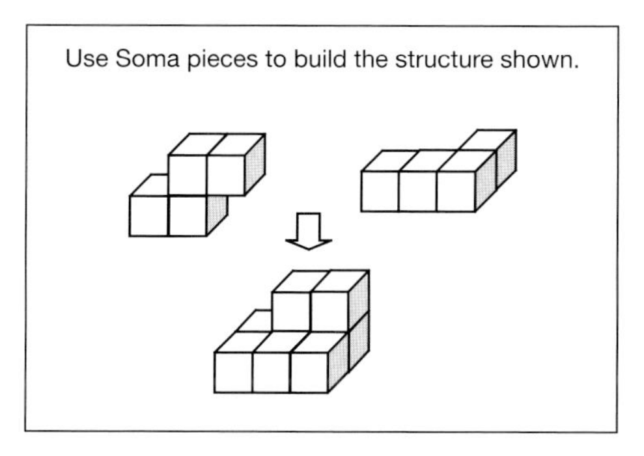

Fig. 10.7a. 2-D task card for a 3-D assembly task

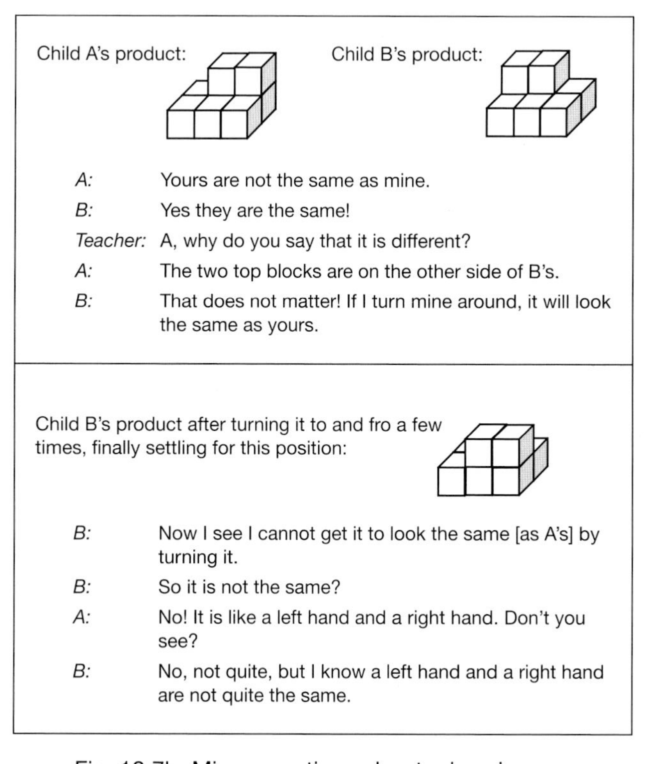

Fig. 10.7b. Misconceptions about mirror images

From this discussion, Child B is clearly convinced that the two products are different, but she is not sure why. Although children had previously constructed Soma pieces 6 and 7, which are mirror images of each other, reverting back to the verbal construction task cards (fig. 10.7c) helped Child B restructure her understanding.

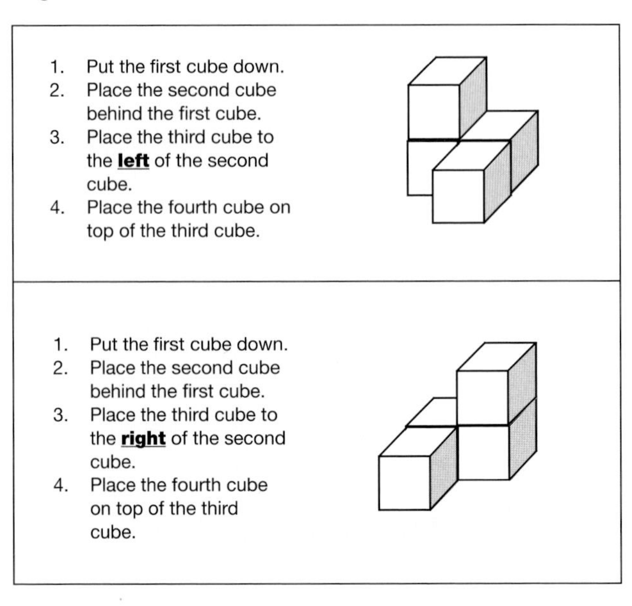

Fig. 10.7c. Verbal task cards for constructing Soma pieces 6 and 7 using loose cubes

Over time, children developed competence with 2-D stimuli that increased in complexity. Children's mental processes were examined through questioning. Figure 10.8 illustrates an example of two children's thinking as they constructed the figure shown on the task card. Each child recognized simpler, but similar substructures within the whole but mentally separated the parts along different axial directions.

The Geocadabra software interface

Lecluse (2005) customized the Geocadabra software interface to align with the SOC intervention program providing a digital model with dynamic capabilities. The Geocadabra Construction Box screen includes a four-by-four grid and a drop-down menu (see fig. 10.9). Using the keypad (a semiotic model), the child creates a conventional-graphic model consisting of cubes, which appears on the grid together with its front, side, and top views. All views or individual views can be shown or hidden. In addition, the child can view the structure dynamically from different perspectives by manipulating the view controls (see fig. 10.10).

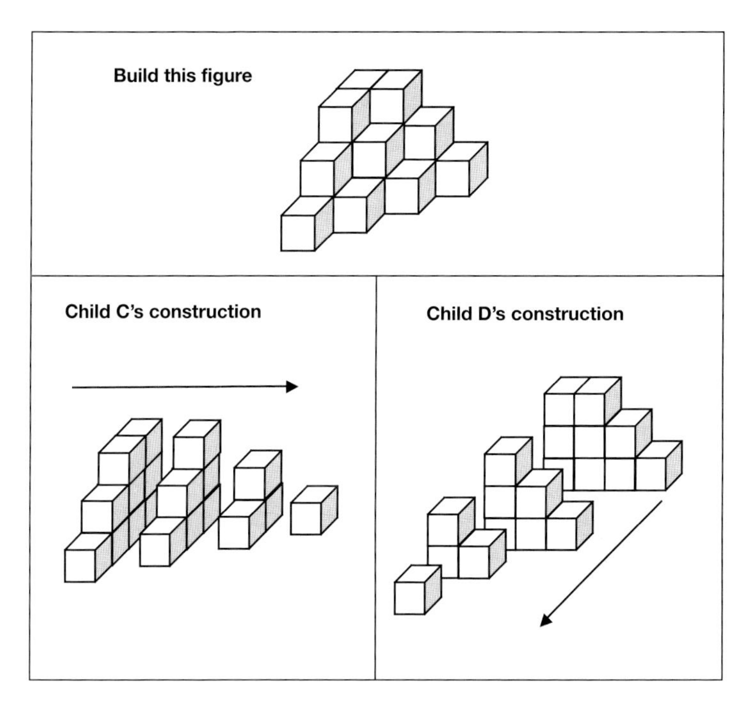

Fig. 10.8. Two children's processes in the construction of a 3-D figure from a 2-D stimulus

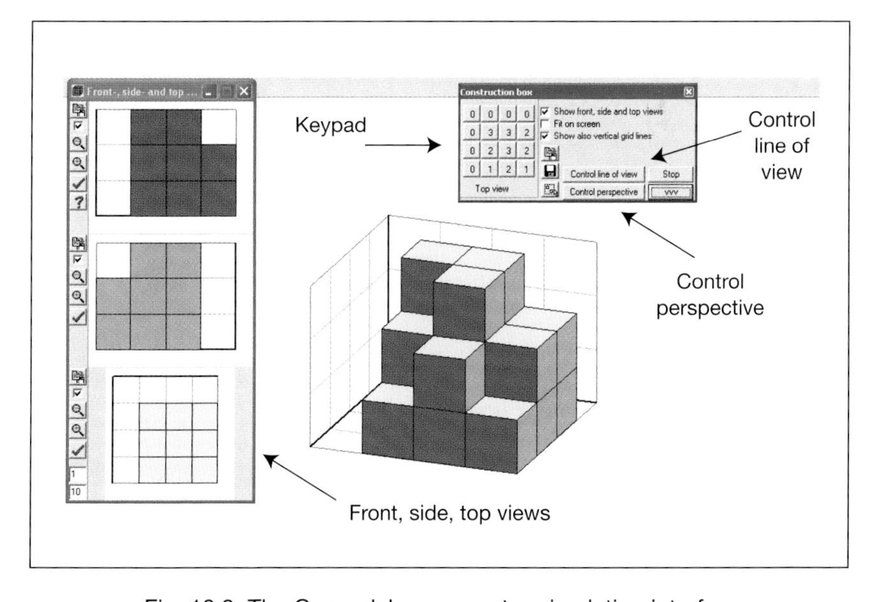

Fig. 10.9. The Geocadabra computer-simulation interface

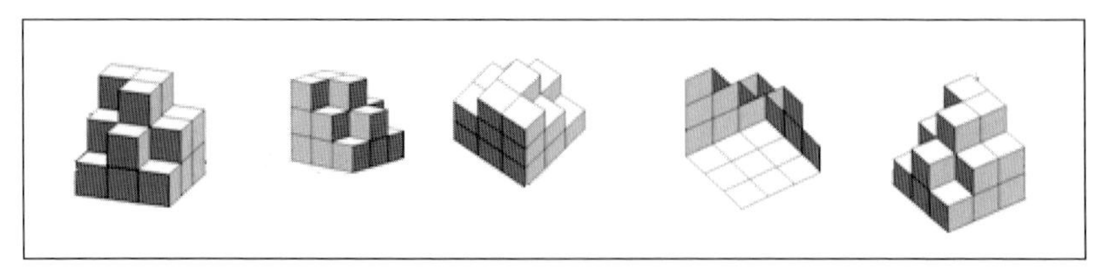

Fig. 10.10. A sequence of views of the same structure

Using Geocadabra, the child is able to create computer-generated copies of full-scale models, reproduce full-scale-model copies of computer-generated figures, and verify paper-and-pencil drawing tasks by selecting or deselecting the different activation buttons (see fig. 10.11).

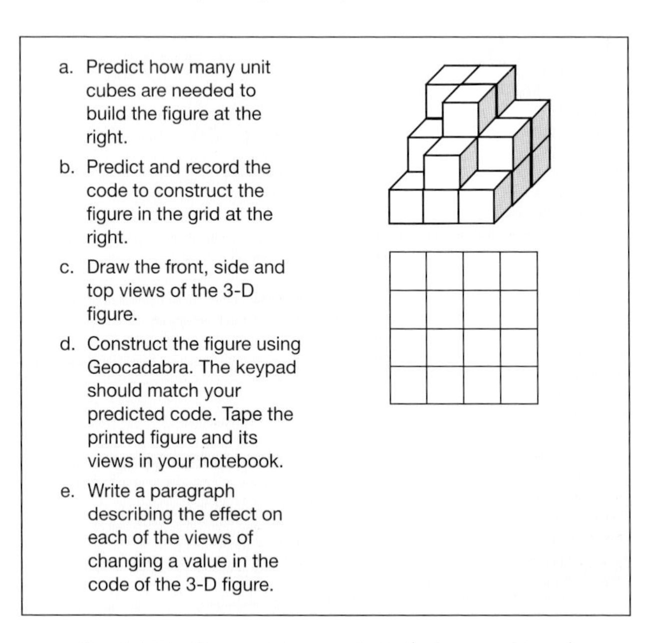

a. Predict how many unit cubes are needed to build the figure at the right.
b. Predict and record the code to construct the figure in the grid at the right.
c. Draw the front, side and top views of the 3-D figure.
d. Construct the figure using Geocadabra. The keypad should match your predicted code. Tape the printed figure and its views in your notebook.
e. Write a paragraph describing the effect on each of the views of changing a value in the code of the 3-D figure.

Fig. 10.11. Geocadabra activity (advanced level)

Conclusion

The SOC model serves as the theoretical framework for an intervention program designed to develop the spatial operational capacities of children of eight to twelve years of age. Wooden cubes and Soma pieces, and the dynamic geometric software interface, appeal to children of various abilities and ages. Initially, children develop their spatial operational capacities with rigid transformations through the use of concrete 3-D objects and then progress to conventional-graphic and semiotic representations. This program deliberately moves among the three visual representations and the verbal representation for 3-D visualization. In this context, children's spatial operational ability levels can be tracked on the basis of their successes with activities requiring manipulation of objects using these different representations. The computer interface integrates 3-D and 2-D representations through the Construction Box and through face views of complex multicube objects. The interface also has the capacity to transform prisms, pyramids, and figures created by sectioning prisms and pyramids into their 2-D net equivalents. Faces of 3-D objects serve to connect one's 3-D world with abstract 2-D geometric figures. The authors and software collaborator intend to use the framework presented in this article to design programs to develop the spatial operational capacities of children beyond the age of twelve.

REFERENCES

Clements, Douglas H., and Michael T. Battista. "Geometry and Spatial Reasoning." In *Handbook of Research on Mathematics Teaching and Learning,* edited by Douglas A. Grouws, pp. 420–64. New York: Macmillan, 1992.

Freudenthal, Hans. *Revisiting Mathematics Education: China Lectures*. Dordrecht, Netherlands: Kluwer Academic Publishers, 1991.

Gauteng Institute for Curriculum Development (GICD). *Mathematics Literacy, Mathematics, and Mathematical Sciences Draft Progress Map: Foundation, Intermediate and Senior Phases Levels 1–6 for Grades 1–9*. Johannesburg, South Africa: GICD, 1999.

Lecluse, Ton. Geocadabra. Computer software. 2005. www.geocadabra.nl/setupeng.exe.

National Council of Teachers of Mathematics (NCTM). *Principles and Standards for School Mathematics*. Reston, Va.: NCTM, 2000.

Presmeg, Norma. "Research on Visualization in Learning and Teaching Mathematics." In *Handbook of Research on the Psychology of Mathematics Education: Past, Present, and Future*, edited by Angel Gutierrez and Paolo Boero, pp. 205–36. Rotterdam, Netherlands: Sense Publishers, 2006.

van Niekerk, H. M. (Retha). "From Spatial Orientation to Spatial Insight: A Geometry Curriculum for the Primary School." *Pythagoras* 36 (1995a): 7–12.

———. "4-Kubers in Africa." Paper presented at the Panama Najaarsconferentie Modellen, Meten en Meetkunde: Paradigmas's van Adaptief Onderwijs, Netherlands, 1995b.

———. "4-Kubers in Africa." *Pythagoras* 40 (1996): 28–33.

———. "A Subject Didactical Analysis of the Development of the Spatial Knowledge of Young Children through a Problem-Centered Approach to Mathematics Teaching and Learning." Ph.D. diss., Potchefstroom University for CHE, South Africa, 1997.

Weisstein, Eric W. "Soma Cube." 1999. From MathWorld, a Wolfram Web Resource. mathworld.wolfram.com/SomaCube.html

Wessels, Dirk C. J. "Building the Spatial Operational Capacity (SOC) of the Primary School Child through Rich Learning Experiences: A Geometry Curriculum for Gauteng Province in South Africa." *Tydskrif vir Christelike Wetenskap* 31, nos. 1–2 (2001): 91–113.

Wessels, Dirk C. J., and H. M. (Retha) van Niekerk. "Semiotic Thinking and the Development of Secondary School Spatial Knowledge." *Tydskrif vir Christelike Wetenskap* 36, nos. 3–4 (2000): 1–16.

Yakimanskaya, I. S. *The Development of Spatial Thinking in Schoolchildren*. Edited and translated by Patricia S. Wilson and Edward J. Davis. Vol. 5 of Soviet Studies in Mathematics Education. Reston, Va.: National Council of Teachers of Mathematics, 1991.

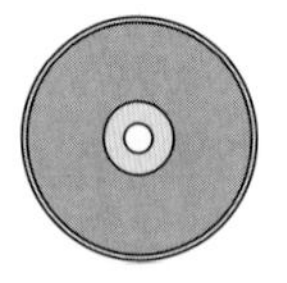

Video clips that support this article are found on the CD-ROM disk accompanying this Yearbook.

Fostering Geometric Thinking in the Middle Grades:

Professional Development for Teachers in Grades 5–10

Mark Driscoll
Michael Egan
Rachel Wing DiMatteo
Johannah Nikula

Broadly speaking I want to suggest that geometry is that part of mathematics in which visual thought is dominant whereas algebra is that part in which sequential thought is dominant. This dichotomy is perhaps better conveyed by the words "insight" versus "rigour" and both play an essential role in real mathematical problems. The educational implications of this are clear. We should aim to cultivate and develop both modes of thought. It is a mistake to overemphasise one at the expense of the other and I suspect that geometry has been suffering in recent years.

—*Michael Atiyah*

During a recent meeting with a group of mathematics coaches, we were exploring geometric thinking using a set of relatively quick exercises. After each exercise, a few volunteers explained how they thought about it. From their reflections, we were trying to capture and describe the variety in thinking. One such exercise asked, "In two minutes or less, can you draw a quadrilateral that has two right angles, but has no pair of parallel sides?" After a couple of minutes, three respondents put words to their thinking about the question:

Person A: I just started drawing, and pretty quickly realized that the two right angles couldn't be adjacent to each other, or else you'd have parallel sides. So they are opposite each other, and the other two angles, opposite each other, also have to add to 180 degrees. From putting two right angles opposite each other, it wasn't long before I had the other two angles and all four sides.

Person B: Right angles always suggest right triangles to me, so I drew one. Then I reflected it through the hypotenuse. I knew that, as long as the original triangle wasn't isosceles, the resulting quadrilateral fits the description. (See fig. 11.1.)

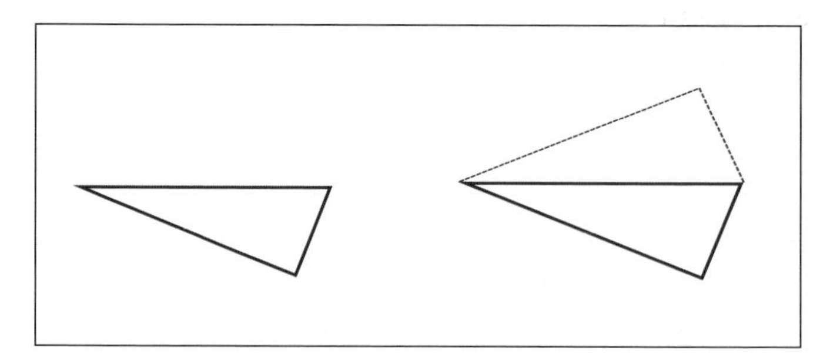

Fig. 11.1. Person B's attempt to draw a quadrilateral with two right angles but no sides parallel

Person C: Believe it or not, when I hear "right angle," I think of circles—or semicircles, actually. So, I drew a circle and diameter, picked a point *A* on one side of the diameter, then a point *B* on the other side. I completed the right angles by connecting *A* and *B* to the ends of the diameter, and there was the quadrilateral I wanted. Then I realized that I could get an infinite number of them by moving *A* along its semicircle, or by moving both *A* and *B*. I guess by doing that, you'd get all the ones with a fixed diagonal connecting the nonright angles. (See fig. 11.2.)

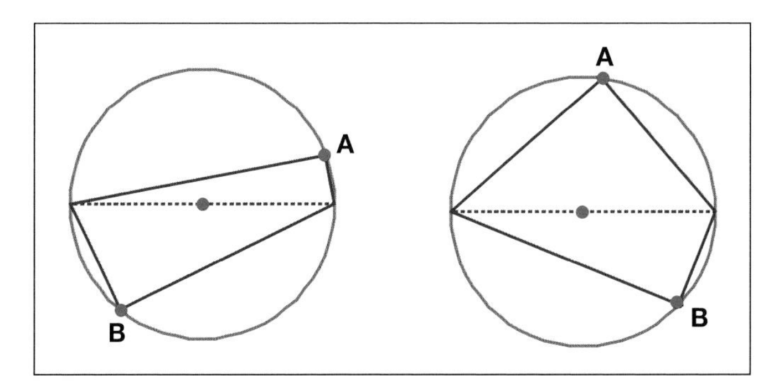

Fig. 11.2. Person C's drawing of quadrilaterals with two right angles but no sides parallel

A Focus on Geometric Thinking

In revealing the rich variety of thinking that a relatively simple, although open, geometry challenge can produce, the incident described above is far from being atypical. Something about asking people to visualize, construct, and reason in one, two, or three dimensions seems to invite different ways of thinking. Of course, some ways are more productive than others. We have a project, Fostering Geometric Thinking in the Middle Grades[1](FGT), which is dedicated to describing productive geometric thinking in ways that help teachers understand the development of geometric thinking and foster their students' geometric thinking. We place a high value on analyzing evidence of geometric thinking—both adults' and students'—to identify lines of thinking that make geometric problem solving successful, as well as common conceptual obstacles to productive geometric thinking. To explore lines of thinking, let us take the three examples above.

Person A, after some experimentation with drawing and some visualizing, takes quick stock of what he has learned from experimenting and then focuses mainly on the relationship that the angles in the desired figure must have with one another. This attention to, and reasoning about, relationships in and among geometric figures is fundamental to productive geometric thinking.

Person B also considers properties and relationships but adds some motion—reflecting the triangle through the hypotenuse. Because she knows something about what stays invariant under reflection (angles, for one thing), she is

1. Fostering Geometric Thinking in the Middle Grades is a collaborative effort between Education Development Center and Horizon Research, Incorporated, spanning the years 2004–2008. Its products include a multimedia book (Heinemann 2007) and a set of professional development materials (Heinemann 2008). The project is funded by the National Science Foundation (NSF) under grant number NSF EHR-0353409. The NSF does not assume responsibility for findings or interpretations produced through this project.

able to assure herself that by starting with a nonisosceles triangle, she has not created a figure with four right angles. This consideration of the invariance of angles under reflection—however tacit it may have been for this teacher—marks another primary feature of productive geometric thinking.

Person C also made use of dynamics—points moving around the circle—to go beyond the problem's bounds and look for some general information—first, "How can I get more of them?" and then "How can I describe all of them?" This transfer of attention from one object to a whole class containing it is indicative of generalization, a core thinking process for mathematics and for geometry, in particular.

Later, we will return to the thinking of Persons A, B, and C to illustrate a framework for four geometric habits of mind used in the Fostering Geometric Thinking project.

Fostering Geometric Thinking in the Middle Grades (FGT): Background

In 2007, FGT completed a two-year field test of a professional development curriculum for teachers in grades 5–10, comprising twenty sessions of two hours each. For mathematics content in the materials, we chose the three areas in geometry in which the literature shows the most challenging conceptual hurdles for middle graders. Accordingly, roughly a third of the materials address each of (1) properties of geometric figures; (2) geometric transformations; and (3) measurement of length, area, and volume.

Two frameworks guided the development of the FGT materials. One, the Structured Exploration Process, underlies the organization of work; the other, Geometric Habits of Mind (GHOMs), guides the approach to geometric thinking.

The Structured Exploration Process

The Structured Exploration Process (Kelemanik et al. 1997) is an essential component of the FGT materials. The process allows teachers to see mathematical thinking from different points of view. It is a cyclical process that repeats each time teachers engage with a new mathematics problem. The cycle involves five stages:

Stage 1: *Doing mathematics.* Teachers work together with colleagues to explore and solve mathematics problems they will later use with their students.

Stage 2: *Reflecting on the mathematics.* Using an explicit conceptual framework (for FGT, the Geometric Habits of Mind [GHOMs]), teachers discuss the mathematical ideas and their thinking about the problem.

Stage 3: *Collecting students' work.* Teachers use the problems in their own classes and collect students' work.

Stage 4: *Analyzing students' work.* Teachers bring students' selected work back to the study group to analyze and discuss with colleagues.

Stage 5: *Reflecting on students' thinking.* Once again using an explicit conceptual framework (for FGT, the GHOMs), teachers discuss students' mathematical thinking, as revealed in the students' work, and ways to elicit more productive thinking in future classes.

FGT departs from previous efforts to employ the Structured Exploration Process, in which "students' work" has primarily meant written records of work. Because of the dynamic nature of much geometric problem solving, our materials are providing, and requiring teachers to provide, videotaped examples and transcript examples of students' work, along with written samples. The video examples are of two types: (1) brief snippets useful in discussing a single issue, such as an apparent misconception by students, a struggle with mathematical language, or an example of a geometric habit of mind; (2) longer pieces, which reveal how students' thinking can shift over the course of working on a problem.

Mathematical Habits of Mind

Mathematical habits of mind are productive ways of thinking that support the learning and application of formal mathematics. A major premise of the FGT materials is that learning mathematics is as much about developing these habits of mind as it is about understanding established mathematical results. Goldenberg, Cuoco, and Mark (1998, p. 39) highlighted this notion of habits of mind, equating it with mathematical power:

> Mathematical power is best described by a set of *habits of mind.* People with mathematical power perform thought experiments; tinker with real and imagined machines; invent things; look for invariants (patterns); make reasonable conjectures; describe things both casually and formally (and play other language games); think about methods, strategies, algorithms, and processes; visualize things (even when the "things" are not inherently visual); seek to explain *why* things are as they see them; and argue passionately about intellectual phenomena.

Building on earlier work around mathematical habits of mind (Driscoll 1999; Driscoll et al. 2001), FGT offers a framework highlighting productive mental habits geared specifically toward geometric thinking. Selecting GHOMs for the FGT framework has been an extended as well as iterative process, with revisions driven by several forces: conversations with project advisors (both mathematicians and mathematics educators) and with pilot-test and field-test teachers; examinations of the research literature on geometric thinking; and analyses of artifacts of students' work on the problems that have been used in our pilot and field tests. Throughout, we have been guided by four criteria:

- *Each GHOM should represent mathematically important thinking.* We aspire to have our framework reflect important lines of geometric thinking, particularly as they contribute to geometric problem solving.
- *Each GHOM should connect with helpful findings in the research literature on the learning of geometry and the development of geometric thinking.* We want to point teachers toward insights gained by researchers into the development of such thinking as well as insights into common hurdles faced by learners in that development.
- *Evidence of each GHOM should appear often in our pilot-test and field-test work.* We want to ensure that the lines of geometric thinking we choose to emphasize will show up with some frequency in the work of students in grades 5–10, even if the appearance may be unpolished and underdeveloped.
- *The GHOMs should lend themselves to instructional use.* Our core interest is in helping teachers foster geometric thinking in and among their students. Each GHOM must point the way toward helpful instructional strategies—for example, productive questions to ask students and clues toward problem design and adaptation.

To constitute our framework, we have selected four GHOMs, which we elaborate in the next section. They are *reasoning with relationships, generalizing geometric ideas, investigating invariants,* and *balancing exploration and reflection.* These habits of mind resonate with existing notions of effective geometric thinking found in the research literature and, hence, address the second criterion listed above. Barrett and his colleagues (2006) have explored how learners develop understanding of linear measurement, and their account helped highlight for us the role that attending to relationships among geometric objects (such as the congruence of opposite sides of rectangles) plays in geometric problem solving. Mitchelmore (2002, p. 161) points out that "[g]eneralisations ... are at the core of school mathematics—numerical generalisations in algebra, spatial generalisations in geometry and measurement, and logical generalisations everywhere." With Mitchelmore, we believe that school children are afforded too few opportunities to locate general results through investigation; rather, the norm tends to be that students are expected to learn and apply results obtained by others. The importance of considering both change and invariance in geometric contexts has surfaced in research related to the use of interactive geometry tools in school mathematics (see, for example, Hoyles and Jones [1998]).

Finally, our fourth habit of mind, *balancing exploration and reflection,* resembles effective geometric problem-solving processes identified by Herbst (2006) and Harel and Sowder (2005).

Fostering Geometric Thinking (FGT) Materials

The habits-of-mind framework and students' work are the central influences on the three main FGT efforts: Geometric Habits of Mind (GHOMs), materials development, and attention to teachers' learning. This section describes how these efforts have evolved.

The Geometric Habits of Mind (GHOMs) framework

With our four criteria as guides, we have settled on the following four GHOMs:

- *Reasoning with relationships:* Actively looking for relationships (e.g., congruence, similarity, parallelism, and so on) in and among geometric figures in one, two, and three dimensions, and thinking about how the relationships can help your understanding or problem solving. Internal questions would include the following:
  - "How are these figures alike?"
  - "In how many ways are they alike?"
  - "How are these figures different?"
  - "What would I have to do to this object to make it like that object?"
  - "Have I found all the ones that fit this description?"
  - "What if I think about this relationship in a different dimension?"

  Rationale: To solve many geometry problems, the ability to recognize how geometric objects in one, two, and three dimensions can be related to one another on the basis of geometric properties is advantageous. Often, the relationships can be exploited to arrive at exact solutions to the problem; without reasoning based on relationships, solvers are usually reduced to finding approximate solutions, at best.

  Indicators: Basic indicators of this GHOM include the identification of figures presented in a problem and correct enumeration of the properties of the figures. More advanced indicators include relating multiple figures in a problem through proportional reasoning and reasoning through symmetry.

  Example: In our opening example, Person A seemed to be reasoning with an awareness of relationships determined by parallel lines, and Person B seemed to be reasoning with symmetry relationships.

- *Generalizing geometric ideas:* Wanting to understand and describe the "always" and the "every" related to geometric phenomena. Internal questions would include the following:
  - "Does this happen in every case?"
  - "Why would this happen in every case?"
  - "Have I found *all the ones* that fit this description?"
  - "Can I think of examples when this is not true, and, if so, should I then revise my generalization?"
  - "Would this apply in other dimensions?"

  Rationale: Generalizing—shifting attention from a given set of objects to a larger set containing the given one—has been a driving force in the history of mathematics and, indirectly, in the history of science, as well. Even in the narrower frame of school mathematics, young solvers of mathematics problems need to learn that, quite often, finding just one solution, or even a finite number of solutions, to a posed problem is not adequate.

  Indicators: The habit of generalizing can start to develop early on when a solver uses one problem solution to generate another (e.g., through reflection in the plane), or when the solver intuits that he or she has not found all the solutions but may not know how to identify them. More advanced geometric generalizers can generate all solutions and make a convincing argument why no more are possible. Another more advanced indication of this GHOM is the habit of wondering what happens if a problem's context is changed—for example, to a higher dimension.

  Example: Person C, in our opening example, used his knowledge of circles not only to create one quadrilateral but also to extend his awareness to an infinite class of examples.

- *Investigating invariants:* An *invariant* is something about a situation that stays the same, even as parts of the situation vary. This habit of mind shows up, for example, in analyzing which attributes of a figure remain the same and which change when the figure is transformed in some way (e.g., through translations, reflections, rotations, dilations, dissections, combinations, or controlled distortions). Internal questions include the following:

- "How did that get from here to there?"
- "What changes? Why?"
- "What stays the same? Why?"

  Rationale: In advanced mathematics, the idea of invariants under transformations is fundamental in distinguishing one kind of geometry from another. In many kinds of engineering, understanding what stays invariant under change is crucial. Even young mathematics learners need to appreciate the role of mathematics in analyzing change, to understand that geometry is useful in analyzing change in space, and to realize that looking for what *does not* change under geometric transformation is often essential to geometric problem solving.

  Indicators: At a basic level, an indication of this GHOM appears when a solver decides to try a transformation of figures in a problem without being prompted to do so (as Person B did in our opening example) and considers what has changed and what has not changed. At more advanced levels, solvers naturally consider extreme instances for what is being asked by a problem—for example, asking, "If I let this vertex go out to infinity, will the area stay the same, and what happens to the perimeter?"

  Example: In our opening example, we suggested that Person B was implicitly capitalizing on the knowledge of what changes and what stays the same under reflections. A less subtle example of this GHOM may be the following: Knowing that a square's diagonals are perpendicular to each other, someone might do a thought experiment and imagine the square collapsing into flatter and flatter rhombi (as, for example, in fig. 11.3), wondering what changes and what stays the same. Area changes as the shape varies, but perimeter does not. And neither does the angle of intersection between the diagonals.

• *Balancing exploration and reflection:* Trying various ways to approach a problem and regularly stepping back to take stock. This balance of "what if ..." with "what did I learn from trying that?" is representative of this habit of mind. Internal questions include the following:

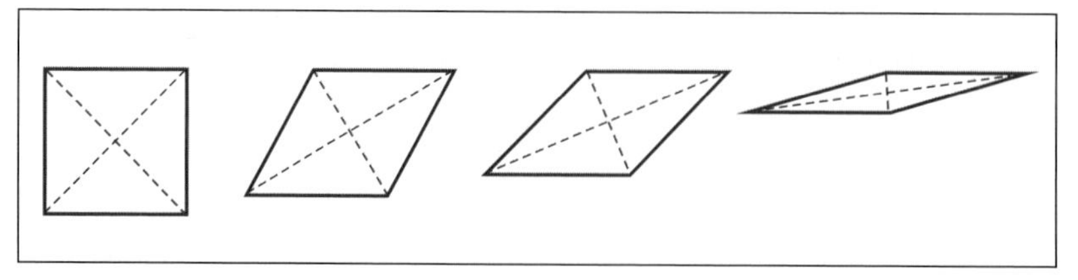

Fig. 11.3. Imagining a square collapsing into flatter and flatter rhombi to consider what changes and what stays the same

- "What happens if I (draw a picture, add to or take apart this picture, work backward from the ending place, and so on)?"
- "What did that action tell me?"
- "How can my earlier attempts to solve the problem influence my approach now?"
- "What if I already had the solution? What would it look like?"

Rationale: One characteristic of successful problem solvers is their metacognitive capacity to balance exploration with taking stock of the productivity of their explorations, then deducing where to take their exploration next. Habituating this appreciation of exploration in mathematics—often quite playful exploration—balanced with metacognitive monitoring, is the essence of this GHOM.

Indicators: Indications that this GHOM is developing include solvers' drawing, playing, or exploring, with occasional (although maybe not consistent) stock-taking. Later indications can include approaching a problem by imagining what a final solution would look like, then reasoning backward; or making what Herbst (2006) calls "reasoned conjectures" about solutions, with strategies for testing the conjectures.

Example. In our opening example, Person A's drawing, visualizing, and taking stock are representative of this GHOM.

Materials development

A major challenge in putting together our materials is describing and exemplifying the GHOMs so they are both understandable and useful to teachers. For one thing, we make no claim that the four-point framework represents an exhaustive list of productive modes of geometric thought. Nor do we view the four habits of mind as mutually exclusive. Quite the opposite: a problem solver is likely to draw on several of these conceptual tools while approaching a problem, and multiple GHOMs might possibly be used to describe the same process. For example, a student who "discovers" pi by investigating the measurements of several circles has reasoned with the relationship between circumference and diameter, has found an invariant ratio between the two, and might even go so far as to generalize this relationship to all circles.

One strategy that has proved useful in meeting this challenge is to appeal to our third criterion for GHOM selection (*evidence of each GHOM should appear often in our pilot-test and field-test work*), and to cull helpful indicators of the different habits of mind from examples we have seen in teachers' and students' work on FGT problems. We hope that, in the end, these indicators will serve teachers not only to understand our meanings but also to sharpen their own thinking as they analyze their students' work on geometric problems, even inviting teachers to add to the list of indicators.

We can illustrate this strategy by considering the current list of indicators for the GHOM *generalizing geometric ideas*. Simply put, mathematical generalization is "passing from the consideration of a given set of objects to that of a larger set, containing the given one" (Pólya 1954, p. 12). In geometry, we are interested in the generalizing of procedures ("Will your dissection method work for all parallelograms?") as well as geometric figures ("Is that true of all trapezoids?"). Quite often, the whole set we are interested in is infinite, either discrete (e.g., points on a lattice), continuous and bounded (e.g., points on a circle or line segment; points in the region bounded by a circle), or continuous and unbounded (e.g., points on a ray or straight line; points in an infinite region of the Cartesian plane).

Many of our FGT problems prompt thinking in generalization by asking solvers to make a convincing argument that they have found all the solutions that fit the problem. In the course of tabulating responses, we have noted the following indicators of generalizing geometric ideas in efforts to make convincing arguments:

- looks only at special cases—for example, right triangles, equilateral triangles;
- looks beyond special cases to a finite class of examples that fit;
- tries generating new cases by changing features in cases already identified, including applying reflections, rotations, and other transformations;
- intuits that unlimited examples are possible but does not know how to generate them;
- realizes that the given conditions work for an infinite class but considers only a discrete class (e.g., using only points on a graph that have integer coordinates);

- sees an infinite, continuously varying class of cases that work but limits the class (e.g., by bounding it in space) or jumps to the wrong conclusion about the class (e.g., by representing the class with the wrong geometric shape);
- sees the entire class and can explain why no more cases are in that class;
- explores the same problem in a broader context—for example, in three dimensions instead of two.

To see several of these indicators in a real context, consider a few solutions to a problem we have used with both teachers and students in our field test.

> ***Perimeter Problem***
> Two vertices of a triangle are located at (4, 0) and (8, 0). The perimeter of the triangle is 12 units. What are all possible positions for the third vertex? How do you know you have them all?

Because we have the work of students across grades 5–10, along with teachers' work in those grades, we have many examples of different ways to think about the problem, and the range of responses covers all the bullets in our indicator list. In particular, at the level of greatest generalization, many solvers realize that the possible positions of the third vertex lie on and constitute an ellipse with foci (4, 0) and (8, 0), and they do so with a variety of techniques (e.g., using a piece of string pinned at the two points) and reasoning (e.g., representing the set with the appropriate equations in x and y). Examples of indicators at other levels of generalizing are plentiful:

1. Examples of the first bullet, consideration of special cases only, generally resemble the drawing in figure 11.4. In this drawing, an indication

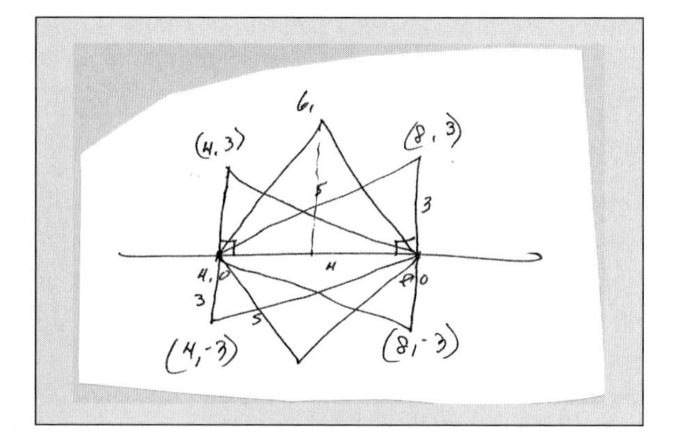

Fig. 11.4. Considering special cases in answer to the Perimeter Problem

of the third bullet also appears to be evident, in that we can infer that the solver expanded the original set of three triangles to six by reflecting through the x-axis.

2. The third and, arguably, the fourth bulleted indicators are represented in the work of a solver who started with the realization that the lengths of the other two sides must add to 8, then used that fact to try to generate new examples. In a nice example of making thinking transparent, he or she wrote, "Will all values between 1 and 8 work, as long as they add up to eight? Or is there a range in which it still will be a triangle? In other words, would 0.25 and 7.75 still make a triangle with 4? No, the short side couldn't connect with the third vertex. How can I possibly find them all? How do I know what the angles are without measuring?" (By the way, the seeming absence here of knowledge of the triangle inequality underscores for us the importance of the relationship among sides that the inequality represents, and reinforces our desire to attend to the GHOM *reasoning with relationships.*)

3. Representative of the third and sixth bullets is the sample in which the solver described his or her strategy as "Find the first point like I did [authors' note: the first point was found by an approximation] and then reflect it over the x-axis. Now I have 2 points. Find another point and then reflect it over the x-axis. Continue this process until I notice that my points are forming a circle." This solver arrived at a faulty conclusion; however, in seeing that the final set of points is symmetric with respect to the x-axis, he or she was also exemplifying one of the indicators we are paying attention to for the GHOM *balancing exploration and reflection*—"describes what the final state would look like (e.g., to find a way to reason backward)."

Research has suggested that teachers who are versed in a structured framework describing modes of students' thinking are better prepared to attend to students' promising ideas and respond productively to those ideas (see, for example, Franke and Kazimi [2001]). On the basis of this research and anecdotal evidence from our field testing, we believe that the GHOM indicators help teachers see lines of potential in which their questions, challenges, and other instructional moves can foster students' geometric thinking. For example, the listed indicators for *generalizing geometric ideas* show gaps of thinking defined by limited attention to whole-number solution sets, or to finite solution sets, or to discrete instead of continuous solution sets. With these gaps in mind when they attend to their students' geometric thinking, teachers can use classroom questioning to assess the extent to which these gaps exist in students' thinking and to advance students' thinking over the gaps. Furthermore, guiding students to think in greater generality about geometric objects can influence teachers' selection and adaptation of geometry problems they chose for their students.

Similarly, we believe that greater acuity when it comes to the other GHOMs can help teachers foster their students' geometric thinking. For example, taking a cue from examples similar to 2 and 3 above, teachers can ask such questions as "To make a triangle, what relationship do three segments need to have with one another?" and "Suppose you had the set of points you are looking for. Is there anything you can say about the shape of that set?"

Attention to teachers' learning

Besides developing and testing the materials, FGT also has a component investigating the impact of the materials on teachers' knowledge of geometry for teaching. For that effort, we have relied on the recent growth in interest in "mathematics knowledge for teaching."

Ball, Lubienski, and Mewborn's (2001) synthesis of the research on requisite knowledge for teaching mathematics revealed two main approaches in the field. One approach focuses on teachers' knowledge of mathematical content, typically measured through academic coursework, test scores, and so on. The other emphasizes pedagogical content knowledge (Shulman 1987), or teachers' knowledge of effective means of presenting content to make it accessible to students. Ball and her colleagues (Ball, Hill, and Bass 2005) have found value in both approaches, and have proposed a third model building on the strengths of the earlier work. Specifically, they called on researchers and teacher educators to consider teachers' use of content and pedagogical content knowledge in the context of teaching. Powell and Hanna (2006, p. 370) suggest how researchers might uncover such knowledge in the field:

> Researchers can infer teachers' mathematical knowledge for teaching by analyzing their practice in action, including interactions with students, questions they ask, issues they make salient to students, student artifacts they use, as well as post-session analyses they perform of their actions, plans, and students' work.

The FGT professional development curriculum actively promotes all three forms of knowledge identified by Ball, Lubienski, and Mewborn (2001). FGT invites teachers to expand their knowledge of geometry through problem solving. The FGT problem set includes challenging forays into geometric properties, transformations, and geometric measurement. Teachers publicly display their work, explain their reasoning, and justify their findings. Finally, teachers are prompted to reflect on their own geometric thinking and that of their colleagues, using the GHOM framework as a metacognitive lens.

We believe that such reflection by teachers into their own geometric thinking readies them to be more receptive to students' thinking as they solve problems.

Focused attention on students' thinking is central to effective pedagogical content knowledge for mathematics. Powell and Hanna (2006, p. 370) offer "that strictly speaking teachers cannot convey or communicate knowledge to students. Instead, teachers can invite students to engage with a mathematical task and discursively connect with students to understand their emergent mathematical ideas and reasoning, as they build their knowledge." The FGT materials promote such an approach to teaching, and the GHOM framework provides teachers a lens through which to interpret students' emerging mathematical ideas.

The FGT approach to professional development reflects Powell and Hanna's (2006) vision of mathematical knowledge in the context of teaching, attending particularly to "questions [teachers] ask, issues they make salient to students, student artifacts they use, as well as post-session analyses they perform ... of students' work" (p. 370). In highlighting the productive mental processes of experienced geometers, we believe that teachers versed in the GHOMs are better equipped to highlight important features of a geometry problem without "telling" students how to solve it, and such highlighting is accomplished through strategic questioning. FGT prompts teachers to consider numerous and varied artifacts of students' work and provides a model for analyzing students' work that moves beyond evaluation and toward insight into students' thinking. These students' work artifacts are collected from the teachers' own classrooms, grounding teachers' analyses in the context of their own work.

An educational bonus

The field test of FGT materials included groups of teachers who taught significant numbers of English language learners (ELLs). Our engagements with these teachers and their students pinpointed many of the language challenges in middle-grades geometry, which are particularly troublesome for ELLs. At the same time, the experience demonstrated for us that geometric problem solving, with its invitation to use diagramming, drawing, colloquial language, and even gesturing to complement academic mathematical communication, affords teachers opportunities to attend to three primary points of emphasis drawn from research on ELLs learning mathematics, namely, the importance of the following:

1. *integrating* content and academic language development in classroom instruction. (see, e.g., Garrison, Amaral, and Ponce [2006]);
2. attending to *cognitive demand* in the mathematical work done by all students, but especially by ELLs (see, e.g., Henningsen and Stein [1997]);
3. creating learning environments that use *multimodal mathematical communication*—speaking, writing, diagramming, and so on—to

reinforce the learning of mathematical language (see, e.g., Khisty and Chval [2002]).

We began this article with an argument that geometry and geometric problem solving deserve greater emphasis in middle-grades classrooms so that students can complement their algebraic thinking with powerful geometric thinking. We believe that argument is strong. The apparent language benefits to ELLs in geometric problem solving make the argument for greater emphasis even more compelling.

REFERENCES

Atiyah, Michael. “What Is Geometry?” In *The Changing Shape of Geometry: Celebrating a Century of Geometry and Geometry Teaching,* edited by Chris Prichard, pp. 24–30. Cambridge: Cambridge University Press, 2003.

Ball, Deborah L., Heather C. Hill, and Hyman Bass. “Knowing Mathematics for Teaching: Who Knows Mathematics Well Enough to Teach Third Grade, and How Can We Decide?” *American Educator* 29, no. 3 (Fall 2005): 14–17, 20–22, 43–46.

Ball, Deborah L., Sarah T. Lubienski, and Denise S. Mewborn. “Research on Teaching Mathematics: The Unsolved Problem of Teachers’ Mathematical Knowledge.” In *Handbook of Research on Teaching,* 4th ed., edited by Virginia Richardson, pp. 433–56. Washington, D.C.: American Educational Research Association, 2001.

Barrett, Jeffrey E., Douglas H. Clements, David Klanderman, Sarah-Jean Pennisi, and Mokaeane V. Polaki. “Students’ Coordination of Geometric Reasoning and Measuring Strategies on a Fixed Perimeter Task: Developing Mathematical Understanding of Linear Measurement.” *Journal for Research in Mathematics Education* 37, no. 3 (May 2006): 187–221.

Driscoll, Mark. *Fostering Algebraic Thinking: A Guide for Teachers Grades 6–10.* Portsmouth, N.H.: Heinemann, 1999.

Driscoll, Mark, Lynn Goldsmith, James Hammerman, Judith Zawojewski, Andrea Humez, and Johannah Nikula. *The Fostering Algebraic Thinking Toolkit.* Portsmouth, N.H.: Heinemann, 2001.

Franke, Megan L., and Elham Kazemi. “Learning to Teach Mathematics: Focus on Student Thinking.” *Theory into Practice* 40, no. 2 (Spring 2001): 102–9.

Garrison, Leslie, Olga Amaral, and Gregorio Ponce. “UnLATCHing Mathematics Instruction for English Learners.” *NCSM Journal of Mathematics Education Leadership* 9, no. 1 (Spring 2006): 14–24.

Goldenberg, E. Paul, Albert A. Cuoco, and June Mark. “A Role for Geometry in General Education.” In *Designing Learning Environments for Developing Understanding of Geometry and Space*, edited by Richard Lehrer and Daniel Chazan, pp. 3–44. Mahwah, N.J.: Lawrence Erlbaum Associates, 1998.

Harel, Guershon, and Larry Sowder. “Advanced Mathematical Thinking at Any Age: Its Nature and Its Development.” *Mathematical Thinking and Learning* 7, no. 1 (January 2005): 27–50.

Henningsen, Marjorie, and Mary Kay Stein. "Mathematical Tasks and Student Cognition: Classroom-Based Factors That Support and Inhibit High-Level Mathematical Thinking and Reasoning." *Journal for Research in Mathematics Education* 28, no. 5 (November 1997): 524–49.

Herbst, Patricio G. "Teaching Geometry with Problems: Negotiating Instructional Situations and Mathematical Tasks." *Journal for Research in Mathematics Education* 37, no. 4 (July 2006): 313–47.

Hoyles, Celia, and Keith Jones. "Proof in Dynamic Geometry Contexts." In *Perspectives on the Teaching of Geometry for the Twenty-first Century,* edited by Carmelo Mammana and Vinicio Villani, pp. 121–28. Boston: Kluwer Academic Publishers, 1998.

Kelemanik, Grace, Susan Janssen, Barbara Miller, and Kristi Ransick. *Structured Exploration: New Perspectives on Mathematics Professional Development.* Newton, Mass.: Education Development Center, 1997.

Khisty, Lena Licón, and Kathryn Chval. "Pedagogic Discourse and Equity in Mathematics: When Teachers' Talk Matters." *Mathematics Education Research Journal* 14, no. 3 (December 2002): 154–68.

Mitchelmore, Michael C. *The Role of Abstraction and Generalisation in the Development of Mathematical Knowledge.* Sydney, Australia: Macquarie University, 2002. ERIC Document Reproduction No. ED 466 962.

Pólya, George. *Mathematics and Plausible Reasoning.* Vol. 1. Princeton, N.J.: Princeton University Press, 1954.

Powell, Arthur B., and Evelyn Hanna. "Understanding Teachers' Mathematical Knowledge for Teaching: A Theoretical and Methodological Approach." In *Proceedings of the Thirtieth Conference of the International Group for the Psychology of Mathematics Education (PME): Mathematics in the Centre,* Vol. 4, edited by Jarmila Notvotná, Hana Moraová, Magdalena Krátká, and Nad'a Stehliková, pp. 369–76. Prague, Czechoslovakia: PME, 2006.

Shulman, Lee S. "Knowledge and Teaching: Foundations of the New Reform." *Harvard Educational Review* 57, no. 1 (February 1987): 1–22.

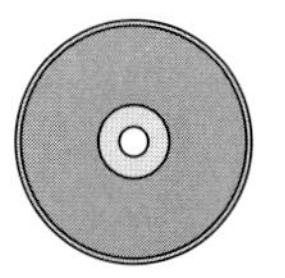

A link to the Fostering Geometric Thinking Web site and a link to a PDF file containing Do Math activities 4 (Dissecting Shapes) and 5 (Analyze Student Work) are found on the CD-ROM disk accompanying this Yearbook.

Part III

Teaching Geometry for Understanding

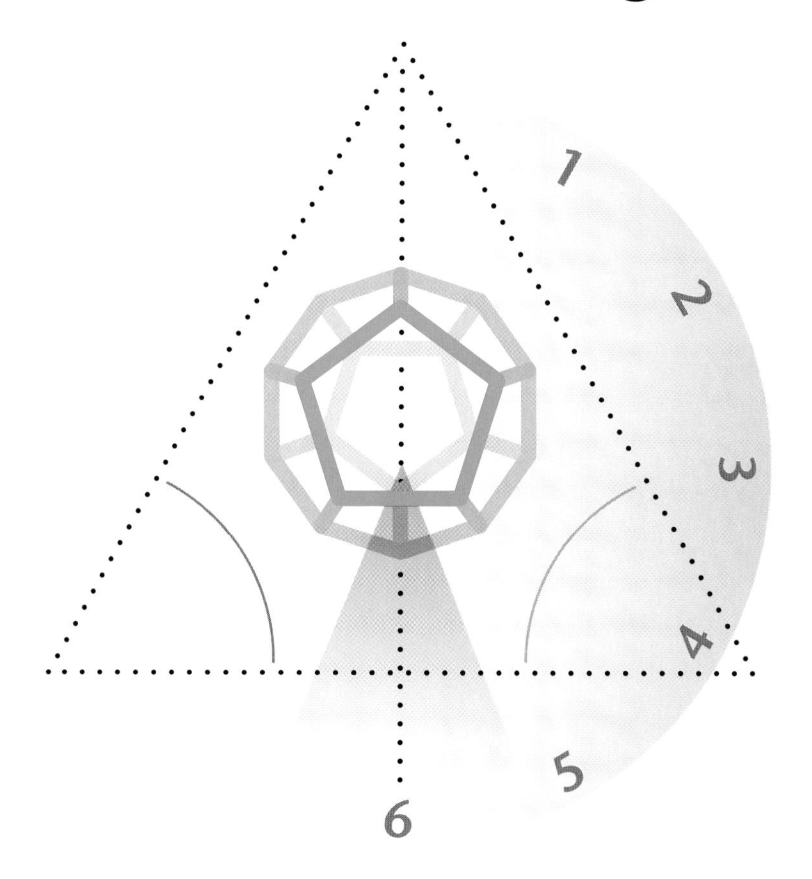

Teaching Geometry for Conceptual Understanding:

One Teacher's Perspective

James Paniati

What instructional approaches help students understand geometric concepts, use sound reasoning, and become insightful problem solvers? How do teachers design classroom environments to foster exploration, discovery, and generalization of ideas? What role does students' verbal and written discourse play in developing mathematical understanding and reasoning? How do teachers build connections within mathematics so students develop a sense of mathematical power? How do teachers assess students' understandings of mathematical concepts so that they can make informed decisions to guide instruction? These are all tough questions. Yet over my twenty-plus years of teaching, including ten years as the head of a high school mathematics department, I have come to believe that best practices do exist in the teaching of mathematics. My department members and I have implemented those practices with much success. The underlying theme of those best practices is teaching for understanding.

Principles and Standards for School Mathematics (NCTM 2000) identifies the following two principles:

1. Teaching Principle: Effective mathematics teaching requires understanding what students know and need to learn and then challenging and supporting them to learn it well (p.16).

2. Learning Principle: Students must learn mathematics with understanding, actively building new knowledge from experience and prior knowledge (p. 20).

Our interpretation of these two principles is that we should teach for conceptual understanding in a challenging learning environment.

In *The Teaching Gap* James Stigler and James Hiebert (1999) explore results of the Third International Mathematics and Science Study (TIMSS), which compared teaching practices and students' performance in different countries. One conclusion that may be drawn is that the standard teaching model of lecture and practice widely used in the United States needs to change in favor of an inquiry or problem-solving approach, similar to the one used in Japan. The TIMSS video study of eighth-grade classrooms (Stigler et al. 1999, p. 71) found that in Japan, students spend about 44 percent of their classroom time on practice tasks, 15 percent of their time on application tasks, and 41 percent of their time on inventing tasks or thinking tasks. In contrast, in the United States students spend about 96 percent of their classroom time on practice tasks, 3.5 percent of their time on application tasks, and less than 1 percent of their time on inventing tasks or thinking tasks. The members of my mathematics department are striving to achieve the balance the Japanese maintain among the three activities (practice, application, invention or thinking) as we implement a discovery-based approach centering on conceptual understanding. In this article I share my evolution as a teacher and the efforts I have made to achieve that balance.

Teaching and Coaching

I became a teacher partly because of my experience coaching kids in youth sports while in college. In coaching kids the first rule of survival is to get the kids active. They learn by doing, even if they know very little at the start. You do not talk to them for forty minutes while they sit in silence listening to you. Rather, you plan a series of activities that teach the skills you want them to learn. At times you may want to stop and lecture, which is perfectly appropriate, but those times do not take up the whole session. The kids are always playing the game in some way, shape, or form. They talk to one another, observe one another, and learn from one another. Since they do not know much when they start, they do things wrong all the time. In fact, if they were not willing to do it wrong, they could not learn to do it right. We accept this reality as a necessary part of learning in youth sports, but often we do not when it comes to teaching mathematics.

Because of this coaching experience, I realized that if students were to develop a deep understanding of mathematics, I would have to teach in the same way as I coached. This approach is not the way I was trained to teach, which was basically to lecture and demonstrate. Therefore I needed some time to develop the methods of discovery-based instruction in which students investigate ideas and concepts and develop the mathematics through those investigations. To make this outcome possible, I knew I had to move my students out of rows and into cooperative groups. This arrangement opened the door for conversations among my students and opportunities for them to teach one another the content. My role as coach was to prepare activities to challenge my students and lead them to the concepts they needed to learn. This approach was a much more exciting way to teach, but also much more daunting. However, when it succeeded, I had a wonderful feeling as I began to see that my students were understanding the concepts underlying the mathematics and not just memorizing procedures.

Early on, a turning point came when I started using *Discovering Geometry* by Michael Serra (1989). The book showed how to teach geometry through a discovery-based approach, and it opened a floodgate of possibilities. I set about learning how to make any type of lesson discovery-based. Sometimes the lesson involved only a small discovery component; at other times the whole lesson was based on discovery. Over time I began transforming all my instruction by converting standard topics in other areas, such as algebra, so that they, too, would have an investigative component. In the end I came to realize that any lesson I wished to teach could be developed as a discovery-based or problem-solving-based investigation to some extent. Furthermore, through collaboration other members of my department have adopted the discovery approach to varying degrees.

Three Formats for Discovery Lessons

The discovery approach places emphasis on students' involvement in constructing mathematics. Like a coach, the teacher selects or creates tasks centering on significant mathematics. The ratio of student-to-teacher involvement in a lesson should approach at least 50:50. Involvement for students means the students are talking about and investigating mathematics in a challenging, problem-solving environment. I have found three ways to implement a discovery or problem-solving lesson:

1. The entire lesson is discovery based. A rich task is needed. The students must be able to enter into the task on their own, and the teacher's role is to monitor and assist them. Students need to work in groups. The ratio of student-to-teacher involvement is about 75:25. See example 1.
2. The teacher starts the class with a discovery problem that consumes about half the class period. Students work in groups. The students present solutions to the problem. The teacher then instructs for the rest of the period using the discovery problem as the focus of the lesson. The ratio of student-to-teacher involvement is about 50:50. See example 2.

3. The teacher starts with an opening problem that takes ten to fifteen minutes. Students work in pairs on this introductory part of the lesson. The teacher then instructs for a few minutes and gives the class another problem to work on. After a few minutes the class discusses the solutions to the problem and then the teacher instructs some more. This process continues to repeat. The ratio of student-to-teacher involvement is about 50:50. See example 3.

I illustrate these three different ways of teaching discovery-based lessons by examples related to geometry or trigonometry. Assessment considerations are included with each example.

Example 1: An Entirely Discovery-Based Lesson

Early in my geometry course, I take students to the computer lab and give them an assignment called the Construction Game. Prior to the advent of interactive geometry software programs such as The Geometer's Sketchpad (GSP) (Jackiw 1990), I had always taught compass-and-straightedge constructions, but I never understood their purpose, other than to acquaint students with the way figures had been drawn a long time ago. Students never got much out of learning constructions, as they had little connection with other work we had done in the geometry class. Now, with geometry software, constructions could be a vehicle for improving students' problem-solving skills, a way to develop their geometric language, and a way for them to make connections with the other topics, including the need for proof. So I created the Construction Game. The rules of this game allow the use of only the straight line and circle tools to construct geometric figures; the more complex constructions from the pull-down menu on GSP are not permitted. We consider a construction "valid" if it can be repeated with these tools and without any guessing. In other words, students must develop a well-defined procedure for the construction.

I then paired the students up for this assignment. Working in pairs is essential so students are able to bounce ideas off each other. With interactive geometry software, students can try many ideas quickly and can move forward and backward by using the undo and redo keys. This level of trial and error is much too time-consuming and inhibiting with paper-and-pencil constructions. Here are the first five constructions I give students:

1. A perpendicular to a line from a point not on the line
2. A midpoint (see fig. 12.1)
3. An equilateral triangle
4. An angle bisector
5. A square

I found that students of all abilities can successfully play this game. Students work at different paces, some taking longer than others to complete a construction. Students make mistakes over and over again, which, as I said previously, is an essential part of learning. Eventually students start to think in a systematic way and begin to reason logically. I check the completed constructions and then require them to write in detail, with proper labels, the steps required to reproduce their construction with a compass and straightedge. Stronger students who quickly solve a construction are given additional constructions or challenged to find another way to do the construction. Even more challenging is to ask them to try to justify why their construction works. Thus they begin to develop deductive arguments, which ultimately lay a foundation for proof.

To help students who are struggling, I may pick a pair of students who were successful with a construction the previous day and have them demonstrate their construction to everyone else using the projection unit. In other words, they become the teachers of the class. The students who do not understand a construction at first can observe the other students' thinking processes and are often able to apply what they learned in the next construction

Students find this game challenging and rewarding. They constantly need to think, solve problems, reflect on what they know, and then apply their knowledge to new situations. Their ability to write and reason in a geometric way grows as they are challenged to justify their work. Connections are made to geometric properties that students will explore throughout the rest of the year. For example, the typical solution to the midpoint-construction problem leads later to a discussion of rhombus properties and a proof involving congruent triangles (see fig. 12.1).

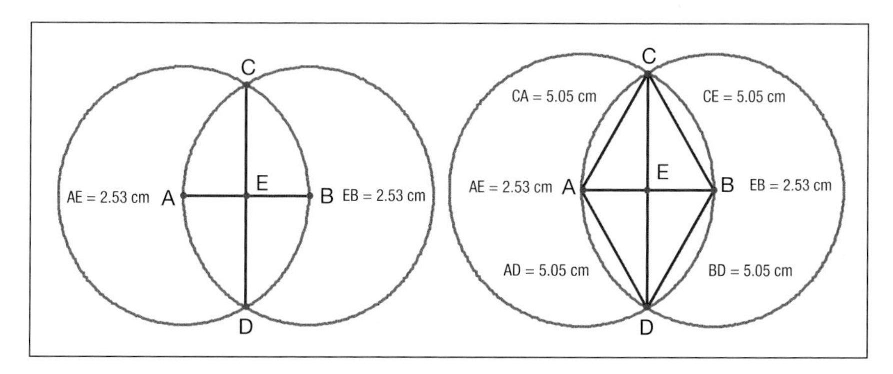

Fig. 12.1. Midpoint construction with rhombus

The informal assessment of this game is multileveled. First, students are constantly self-assessing through the use of the dragging feature of the software, which allows them to see whether the constructed property remains invariant.

Second, through software support, students measure their figures to determine whether they have met the conditions outlined in the definition. As the teacher, I am able to check their work through the undo and redo features of the software. I also assess their written description of the steps needed to reproduce their construction. In this, I am evaluating the correct use of geometric language and symbols and the students' ability to describe mathematics in writing. As a teacher I can observe students' growth in their problem-solving skills by how they make the constructions, how quickly they are able to transfer knowledge from one construction to another, and how well they can begin to reason why their construction works. The "why" is the hardest, yet one of the most important parts. As students build their geometric knowledge throughout the year, they return to these constructions and develop stronger and stronger support for the "why," ultimately leading to a formal proof (see fig. 12.2).

Given: Segment *AB*
Circle *A* passing through point *B*
Circle *B* passing through point *A*
Circle *A* and *B* intersect at points *C* and *D*
Segment *CD* intersects segment *AB* at point *E*

Prove: *E* is the midpoint of segment *AB*

Statements	Reasons
1. Segment *AB* Circle *A* passing through point *B* Circle *B* passing through point *A* Circle *A* and *B* intersect at points *C* and *D* Segment *CD* intersects segment *AB* at point *E*	1. Given
2. Circle *A* ≅ Circle *B*	2. Circles with equal radii are congruent.
3. Construct segments *AC, AD, BC, BD*	3. Only one unique segment exists between two distinct points.
4. $\overline{AC} \cong \overline{AD} \cong \overline{BC} \cong \overline{BD}$	4. Radii of congruent circles are congruent.
5. $\overline{CD} \cong \overline{CD}$	5. Reflexive Property
6. Triangle *CAD* ≅ Triangle *CBD*	6. SSS Postulate
7. Angle *ACD* ≅ Angle *BCD*	7. Corresponding parts of congruent triangles are congruent (CPCTC).
8. $\overline{CE} \cong \overline{CE}$	8. Reflexive Property
9. Triangle *CAE* ≅ Triangle *CBE*	9. SAS Postulate
10. $\overline{AE} \cong \overline{BE}$	10. CPCTC
11. *E* is the midpoint of *AB*	11. Def. of Midpoint

Fig. 12.2. Proof of midpoint construction;students have written proofs similar to this one.

Example 2: A Discovery-Based Opener

Figure 12.3 illustrates a worksheet I give to my precalculus class to help them discover the law of sines. Preceding this problem, the students have solved only right triangles using basic trigonometric functions. I start class by distributing this problem to students in their cooperative groups. Without much prodding, they are able to develop the law of sines on their own.

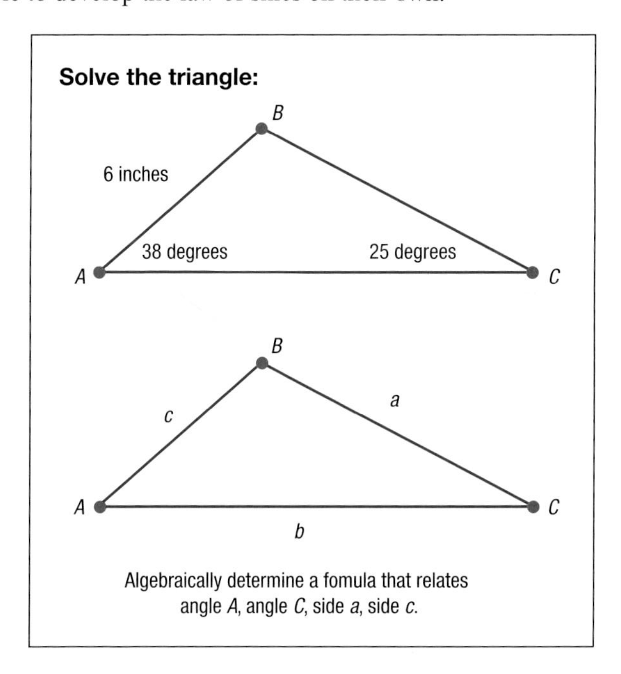

Fig. 12.3. Law of sines worksheet

The first example, done numerically, gets students to think about how to solve a nonright triangle for its missing parts. Most groups quickly realize that they need to draw an altitude from point B to side AC and use right-triangle trigonometry. That strategy leads them into the second example, in which they apply the process learned in the first example in an algebraic way to discover a general formula. They again need to draw an altitude from point B to side AC, but this time label it with a variable such as h. By applying the thinking process used for the first triangle when they had numeric values, they can set up ratios using sines for the two triangles with altitude h. From there they can set expressions for h equal to each other and generate the law of sines (see fig. 12.4). Once they have the formula, they are expected to check it by applying it to the first example to see if they get the previous result.

Next students are asked to explain why the formula will hold for any pair of opposite sides and angles in the triangle. We then turn to a discussion about

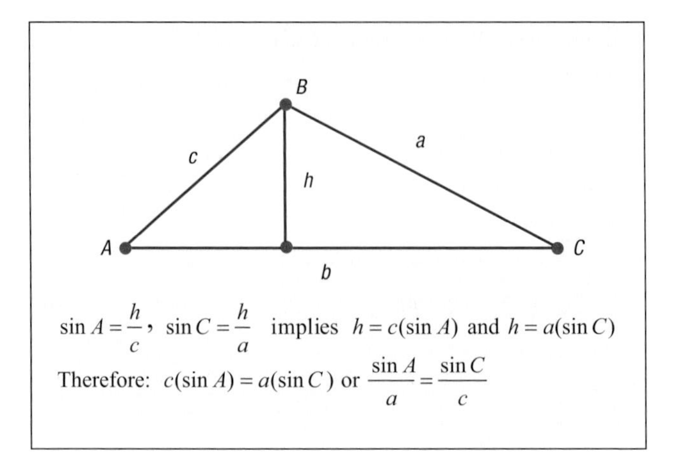

Fig. 12.4. Derivation of the law of sines

what minimum information must be known about a triangle to be able to solve it with the law of sines, making a connection with their earlier work in geometry on congruent-triangle postulates. All these discussion questions are posed to students in their groups, and any student in the group can be called on to summarize the group's answer. Students then practice applying the law of sines in different triangles and word problems. Finally, this practice leads to an investigation of the ambiguous case, which is a result of the earlier discussion on triangle-congruence shortcuts and why side-side-angle does not prove congruence.

Informal assessment of students' knowledge and understanding is built right into the opening exercise. Students are able to solve the initial nonright triangle because the task is set up to make a connection with right-triangle trigonometry. By brainstorming in their group, students find the foundational idea of drawing an altitude from one vertex to create two right triangles. Students can then calculate a numeric solution to the problem and feel confident that their procedure is correct. This numeric example leads naturally to the generalized algebraic example. As the students attack this second problem, they are constantly self-assessing their work on the basis of the process they developed in the previous example, giving them confidence in the validity of the formula for the law of sines they have discovered. Students can then verify their new formula by checking it through the numeric problem they previously solved. In this whole exercise students do not need to rely on the teacher to know that they have developed something significant. I play the essential role of creating the tasks that allow students to begin developing a conceptual understanding of the law of sines. As my students and I work this way in tandem, they become more independent learners. Finally, the class discussion after this opener leads them toward deeper connections between the law of sines and previously learned geometric facts.

Students are now ready to go on and explore the ambiguous case with a clearer understanding of why it exists.

Example 3: A Back-and-Forth Approach

"Back and forth" is illustrated by a lesson in which I introduce radian measure. Radian measure is a concept with which many students struggle; I know this was true for me when I first learned it. The previous lesson is about angles in standard position on a coordinate plane measured only in degrees. I begin the new lesson with the question "What is common to all circles?" Typical answers are (a) the degree measure around their center is 360, (b) all circles are round, (c) the circumference of all circles can be found by the formula $2\pi r$, and (d) the area of all circles is πr^2. I then tell the students that radians are another way to measure an angle other than degrees, just as metric measurements are an alternative to customary units, such as inches, for measuring length. Then I ask the class to explain what happens if you divide both sides of the circumference formula by r. The students write $\frac{c}{r} = 2\pi$. We discuss the idea that since π is always the ratio of the circumference of a circle to its diameter, then 2π is always the ratio of a circle's circumference to its radius. So just as in every circle the degree measure around its center is a consistent 360 degrees, it is also true that in every circle the ratio of circumference to radius is 2π, another feature common to all circles. I tell students, therefore, that 360 in degrees is equivalent to 2π radians, so 180 degrees is equivalent to π radians. Then I give them the following definition for radian measure: "Radian measure is the ratio of the length of an arc of a circle to its radius." At this point I know students really do not have a conceptual understanding of radian measure, but we continue.

Next I have them imagine that we want to construct a degree ruler. Even though no such thing exists, let us imagine it anyway. What values would we put on it? Students quickly realize that we would put 360 degrees on the end and break it into smaller units, such as every 90 degrees (see fig. 12.5). I ask them if the ruler could be extended to have values past 360. Most students are able to understand that, just as we can extend a distance ruler beyond a foot if we want to,

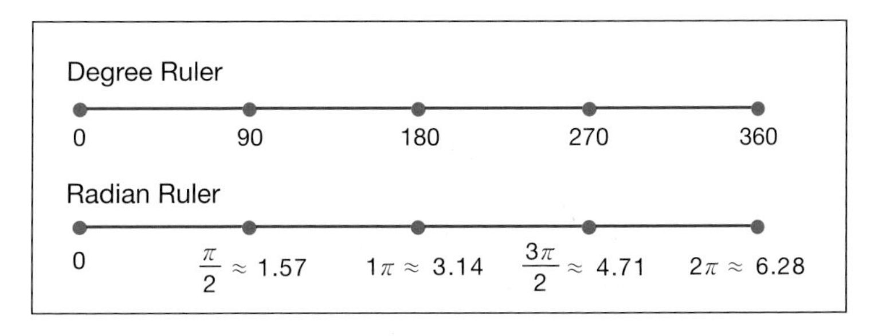

Fig. 12.5. Corresponding degree/radian rulers

we can do the same with the degree ruler. This consideration leads to a discussion of a 450-degree angle and its similarity to 90 when drawn in standard position, thus setting the stage for a later discussion on coterminal angles.

Next I ask the students to construct a radian ruler. They quickly realize that it would end at 2π and could be broken into four divisions of $\pi/2$ (see fig. 12.5). Similarly they realize, through class discussion, that $5\pi/2$ exists and coincides with $\pi/2$ in standard position. These rulers also set the stage for graphing trigonometric functions on the coordinate plane, which occurs later in the course.

At this point my students have developed some sense of radians and their size, but now it is time to challenge their understanding of the definition. I begin by drawing three circles on the board (see top row of fig. 12.6). I ask for a volunteer to draw a 30-degree angle in standard position in each circle using my large protractor. (See middle row of fig. 12.6.) This is a simple task. I ask the students to observe what is the same in all three circles. The students generalize that the 30-degree angle cuts a sector that is the same proportion of the circle no matter what size the circle is. We draw this conclusion visually as well as logically from the fact that all circles are similar. This similarity idea lays the groundwork for understanding why the trigonometric functions are equivalent in any size circle for a given angle and why the unit circle is the most convenient one to use.

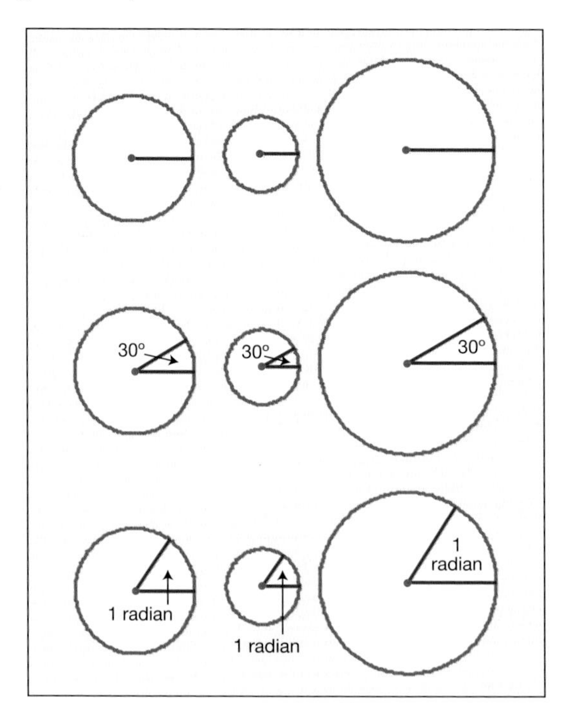

Fig. 12.6. Illustrations used for radian measure investigation

Now comes the challenge question. I ask my students to come to the board and draw an angle of one radian in standard position on each circle. In this instance, however, the only tool they are allowed to use is a piece of string. When I do this exercise, my students inevitably are stumped. This question challenges their conceptual understanding of what a radian represents. I then ask the class to brainstorm in their groups and try to find a method. Often I have to give the students the following hint: "You must not yet understand what radian measure really is, because you are unable to answer this question Perhaps you need to review the definition." After a few minutes, this hint gives some students an idea. A volunteer goes to the board, usually with a partner, and uses the string to measure the radius of the circle and then wraps the string along the circumference to generate an angle of one radian in standard position (see bottom-left circle of fig. 12.6). This demonstration is an "aha" moment for many other students. I then ask another pair of students to come to the board and repeat this process on the other circles and explain why it works. I ask the students to observe all three circles and tell me what looks the same. They realize that the angles cut a sector that is the same proportion of the circle no matter the size of the circle, just as the 30-degree angle did. Students come to understand that radian measure has similar characteristics to degrees in the same way that customary and metric units can be related to length. This task can be followed by students' determining an angle of two radians or π radians using only the string. Finally, this lesson ends with a comparison of the measures of different angles in radians and degrees and the development of a formula to convert from one to the other.

Informal assessment in this type of lesson is built into the design of the activities. The beginning of the lesson connects previous knowledge in geometry with new knowledge about radian measure through a series of scaffoldings. Questions built into the lesson give me a sense of where the students are in their understanding. The activity of developing a ruler for angular measure is a way to bring in some important concepts at a simple level, and it creates a bridge to future concepts. The string activity is an assessment of their conceptual understanding of what a radian is and how it relates to measuring in degrees. Finally, to solidify their understanding, I often ask students to write a paper explaining radian measure to an "imaginary" fellow student who missed our discussion. In those papers I can see how well the students comprehend the concept of radian measure.

Reflections

One of the primary elements underlying all these approaches to designing a discovery-based lesson is that *students are talking and investigating mathematics in a challenging, problem-solving environment.* I draw on my sports experience in assessing the classroom environment. The level of tension that should exist in a mathematics classroom should be equivalent to playing a 2–2 tie game in soccer. The classroom environment should be challenging enough so that students have to push themselves to reason and think at a high level, but not so challenging that they do not have an entry point into making the important connections needed to gain understanding. The classroom environment should never be like a 5–0 game, in which they have an easy win. In such an environment students do worksheet after worksheet of repetitive problems that they have already mastered, and they do not have to think very hard. Neither should the classroom environment be 0–5, in which students are so overwhelmed that they do not have opportunities to enter successfully into the mathematical conversation that leads to conceptual understanding.

In the descriptions of the different ways to apply discovery learning, I have also included the prescription that the teacher should be talking no more that half the time, with the ideal being less than a quarter of the time. This criterion helps us move ourselves away from being the sole provider of knowledge and toward the role of facilitator in the development of mathematical knowledge and understanding. In this approach we do have the opportunity to lecture, but not for the entire period and not on a continuous basis day after day.

At the beginning of this article I asked some tough questions that I hope I have answered, to some degree, in this article.

- *What instructional approaches help students understand mathematical concepts, use sound reasoning, and become insightful problem solvers?* My answer: A discovery-based, problem-solving approach to teaching in which the teacher becomes the facilitator of learning, not the distributor of it.
- *How do teachers design classroom environments to foster the exploration, discovery, and generalization of geometric ideas?* My answers: moving students out of rows and into cooperative groups so they can talk to one another; designing challenging, discovery-based lessons that teach for understanding as students explore the mathematical concepts; and creating an environment in which risk taking and wrong answers are accepted as natural parts of the learning process.
- *What role does students' verbal and written discourse play in developing their mathematical understanding and reasoning?* Students must constantly be put in situations in which they have to explain the underlying ideas or concepts of the mathematics. Therefore the design of the lesson needs to build this discourse through challenging tasks and questions that require students to respond both verbally and in writing.
- *How do teachers build connections within mathematics so that stu-*

dents develop a sense of mathematical power? Lessons need to be connected rather than disjointed. They must lead to conceptual understanding rather than just to rote skill. The content of the mathematics needs to be made relevant to the students, and students need to see real applications of its use. Students get empowered by mathematics when they are challenged and given a doorway through which they can arrive at understanding.

- *How do teachers assess students' understandings of mathematical concepts so that they can make instructional decisions?* Teachers must build informal assessment into the tasks and questions of the lesson. Along the way they need to include markers that students must complete to give the teacher clues that students are developing conceptual understanding. Those markers should be both verbal and written and occur regularly throughout the lesson. This consideration leads to formal assessment, which should be as diverse as the informal assessment. Examples of formal assessments include written responses to problems, tests, performance-based projects, written papers on mathematical topics, portfolios of students' work, and others.

Stigler and Hiebert (1999, p. 179) conclude *The Teaching Gap* with the following statement:

> The star teachers of the twenty-first century will be those who work together to infuse the best ideas into standard practice. They will be teachers who collaborate to build a system that has the goal of improving students' learning in the "average" classroom, who work to gradually improve standard classroom practices. In a true profession, the wisdom of the profession's members finds its way into the most common methods. The best that we know becomes the standard way of doing something. The star teachers of the twenty-first century will be teachers who work every day to improve teaching—not only their own but that of the whole profession.

I truly believe that the authors are right on target with this statement, that we who teach mathematics need to make those "best practices" commonplace in our teaching, and that we need to work together to do so. My next level of work will be continuing to collaborate with department members in sharing our learning and most effective practices.

REFERENCES

Jackiw, Nicholas. The Geometer's Sketchpad. Software. Emeryville, Calif.: Key Curriculum Press, 1990, 2001.

National Council of Teachers of Mathematics (NCTM). *Principles and Standards for School Mathematics*, Reston, Va.: NCTM, 2000.

Serra, Michael. *Discovering Geometry: An Inductive Approach*. Emeryville, Calif.: Key Curriculum Press, 1989.

Stigler, James W., Patrick Gonzales, Takako Kawanaka, Steffen Knoll, and Ana Serrano. *The TIMSS Videotape Classroom Study: Methods and Findings from an Exploratory Research Project on Eighth-Grade Mathematics Instruction in Germany, Japan, and the United States.* Washington, D.C.: National Center for Educational Statistics, 1999.

Stigler, James W., and James Hiebert. *The Teaching Gap: Best Ideas from the World's Teachers for Improving Education in the Classroom*. New York: The Free Press, 1999.

ADDITIONAL READING

National Center for Educational Statistics. *Pursuing Excellence: A Study of U.S. Eighth-Grade Mathematics and Science Teaching, Learning, Curriculum, and Achievement in International Context.* Washington, D.C.: U.S. Department of Education, 1996.

U.S. Department of Education (USDE), Educational Publication Center. *Before It's Too Late: A Report to the Nation from the National Commission on Mathematics and Science Teaching for the Twenty-first Century*. Washington, D.C.: USDE, 2000.

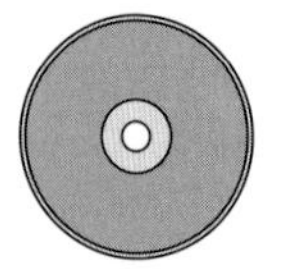

A blackline master of the Law of Sines worksheet that supports this article is found on the CD-ROM disk accompanying this Yearbook.

Defining in Geometry

Michael de Villiers
Rajendran Govender
Nikita Patterson

DEFINITIONS are important in mathematics. They are tools for communication, for reorganizing old knowledge, and for building new knowledge through proof. But definitions present many challenges to both learners and teachers. Mathematical definitions are very concise, contain technical terms, and require an immediate synthesis into a sound concept image. Too often, however, definitions are presented before students' concepts have evolved naturally from existing knowledge. Consequently, students often resort to meaningless memorization. Over time many educators have thought about better ways to teach defining.

Early in the twentieth century the German mathematician Felix Klein strongly opposed the practice of presenting mathematical topics as completed axiomatic-deductive systems. He argued instead for the *biogenetic* principle in teaching, which has also been advocated by Wittmann (1973), Pólya (1981), Freudenthal (1973), and many others. The biogenetic approach takes the stand that the student should retrace, at least in part, the path followed by the original discoverers or inventors, or alternatively, retrace a path by which knowledge could have been discovered or invented. Human (1978, p. 20) calls this the *reconstructive approach*:

> With this term [reconstructive] we want to indicate that content is not directly introduced to pupils (as finished products of mathematical activity), but that

> the content is newly reconstructed during teaching in a typical mathematical manner by the teacher and/or the pupils [freely translated from Afrikaans].

The reconstructive approach allows students to *participate actively* in the development of the content and the related mathematical processes of defining, axiomatizing, conjecturing, and proving. Content is not presented as a finished, prefabricated product, but rather the teacher focuses on the processes by which the content can be developed or reconstructed. Note that a reconstructive approach does not necessarily imply learning by discovery, for it may just involve a reconstructive explanation by the teacher or the textbook.

Mathematicians and mathematics educators alike have often criticized the direct teaching of geometry definitions with no emphasis on the underlying process of defining. For example, as early as 1908 Benchara Blandford (quoted in Griffiths and Howson [1974, pp. 216–17]) wrote,

> To me it appears a radically vicious method, certainly in geometry, if not in other subjects, to supply a child with ready-made definitions, to be subsequently memorized after being more or less carefully explained. To do this is surely to throw away deliberately one of the most valuable agents of intellectual discipline. The evolving of a workable definition by the child's own activity stimulated by appropriate questions, is both interesting and highly educational.

Hans Freudenthal (1973, pp. 417–18) strongly criticized the practice of directly providing geometric definitions on similar grounds. Ohtani (1996, p. 81) argued that the practice of simply telling definitions to students functions to justify the teacher's control over students, to attain a degree of uniformity, to avoid having to deal with students' ideas, and to circumvent problematic interactions with students. Vinner (1991) and others including Battista (2009) have presented arguments and empirical data supporting the thesis that knowing the definition of a concept does not guarantee understanding. For example, students who have been told, and are able to recite, the standard definition of a parallelogram may still not consider rectangles, squares, and rhombi as parallelograms if their concept image (i.e, their private mental picture) of a parallelogram is one in which not all angles or sides are allowed to be equal.

Students often meet mathematics only through the structure presented in formal mathematics textbooks. This structured approach can lead to a common, but false conception that only one correct definition exists for each defined object in mathematics. In fact, several different, correct definitions may exist for a particular concept, so we have a certain amount of freedom in our choice of definitions. Thus definitions can be considered "arbitrary" (Linchevsky, Vinner, and Karsenty 1992, p. 48). For example, a rectangle may be defined as a parallelogram with a right angle, a quadrilateral with three right angles, or a quadrilateral with lines of symmetry through opposite sides. But too often textbooks give the impression that a rectangle can and must be defined only one way.

Furthermore, the formal approach promotes the misconception that mathematics always starts with a definition, and that definitions of mathematical objects are given *a priori.* Students then do not realize that definitions are not discoveries, but human "inventions." Students rarely understand that a main purpose of definitions is to promote accurate mathematical communication. (See also Blair and Canada [2009] regarding developing definitions with students.)

The ideas shared in this article stem from years of consideration of issues related to mathematical definitions and to a variety of efforts to engage secondary school students and prospective teachers in meaningful work in this domain (de Villiers 1986, 1994, 1998, 2003, 2004). If teachers are to engage students in a reconstructive approach in which they create and critique their own definitions, then teachers themselves must first understand the subtle distinctions among various types of definitions. Those distinctions are the major topic of this article.

We first consider the distinction between partitional and hierarchical methods of classification. Next we examine the characteristics of a correct definition based on necessary and sufficient conditions. Then we look at economical definitions and develop criteria for good definitions. Finally, we discuss how these distinctions can help teachers plan and implement strategies to develop their students' mathematical thinking.

Types of Definitions

Partitional and Hierarchical Definitions

A mutual relationship exists between classifying and defining. The classification of any set of concepts implicitly or explicitly involves defining the concepts involved, whereas defining concepts in a certain way automatically involves their classification. For example, defining a rectangle as *a quadrilateral with two axes of symmetry through opposite sides* will then imply that a square is classified as a special rectangle. However, if a rectangle is defined as *a quadrilateral with two axes of symmetry through opposite sides, but no other lines of symmetry,* then squares are clearly excluded from the set of rectangles.

We describe a definition such as the first as *hierarchical* (i.e., inclusive) because it allows the inclusion of more particular concepts as subsets of the more general concept. The latter we call a *partitional* (i.e., exclusive) definition because the concepts involved are considered disjoint from each other (i.e., squares are not considered rectangles).

If students are given the opportunity to create definitions of their own for such geometric concepts as the quadrilaterals, the result can be a lively and fruitful class discussion of why hierarchical definitions are generally preferred in mathematics. De Villiers (1994, p. 15) identifies the following important advantages of hierarchical classification over partitional classification. A hierarchical definition—

- leads to more economical definitions of concepts and formulations of theorems,
- simplifies the deductive system and the derivation of properties of more special concepts,
- often provides a useful conceptual schema during problem solving,
- sometimes suggests alternative definitions and new propositions, and
- provides a useful global perspective.

Nevertheless, in some situations we need partitional definitions to distinguish concepts clearly. For example, we have little choice but to create a partitional classification of convex and concave quadrilaterals, because it is not meaningful to view a concave quadrilateral as a special kind of convex quadrilateral, or vice versa (de Villiers 1994).

Furthermore, from a historical perspective, hierarchical definitions have not always been favored. For example, in Book 1 of *Elements,* Euclid partitioned quadrilaterals into five mutually exclusive categories: *square* (both equilateral and right-angled), *oblong* (right-angled but not equilateral), *rhombus* (equilateral but not right-angled), *rhomboid* (opposite sides and angles equal to one another but neither equilateral nor right-angled), and *trapezium* (any other quadrilateral). Similarly, Euclid did not consider an equilateral triangle to be a special case of an isosceles triangle.

Some of the challenges that definitions create for students are illustrated in the interviews and experiences with children in grades 9 to 12 over several years reported in de Villiers (1994). Here is one example (*I* = interviewer, *S* = student):

I: If we define a parallelogram as any quadrilateral with opposite sides parallel, can we then say that a rectangle is a parallelogram?

S: Yes, … because a rectangle also has opposite sides parallel.… But I don't like this definition of parallelograms.… I know we are taught this definition at school and that squares and rectangles are parallelograms (pulls face), but I don't like it.…

I: How would you define parallelograms instead?

S: As any quadrilateral with opposite sides parallel, but not all angles equal.

I: What about rhombi then? … Would you say a rhombus is a parallelogram?

S: Hmm … according to my definition, yes, … but I don't like that

> either.... I would therefore rather say a parallelogram is a quadrilateral with opposite sides parallel, but not all sides or angles equal.

Clearly this eleventh-grade student (who happened to be a top student at his school) had no problem with drawing correct conclusions from given definitions and making hierarchical class inclusions but preferred not to do so. Moreover, this student clearly exhibited the ability to formulate a definition. But the definitions he preferred were exclusive ones. Battista and Clements (1992, p. 63) have similarly reported two cases of students who were able to follow the logic of a hierarchical classification of quadrilaterals of squares and rectangles but had difficulty accepting it. Quite often the origin of this problem can be traced back to the elementary school, and to direct provision of exclusive definitions or static exclusive images, which then become so fossilized in students' minds that by the time they reach the high school, the students are very resistant to change.

Another challenge for students arises when they struggle to define a more general figure from a more specific one. For example, students may say, "A rhombus is a square with not all angles equal." On the one hand, students are trying to define a rhombus hierarchically as a special square. Yet on the other hand, they are ending up with "A rhombus is a quadrilateral with equal sides, but with not all angles equal," which partitions rhombuses from squares. The authors have frequently observed this type of problem and labeled it as an "inverse hierarchical-partition" definition because students are trying to be hierarchical but instead are partitioning. Textbook authors and teachers often use this approach without realizing the conceptual difficulties it creates. For example, consider the following introduction to a rhombus: "The next shape we are going to be looking at is called a *rhombus*. We can think of this figure as a square that has been pulled out of shape." This approach encourages students to view a rhombus incorrectly as a special kind of square instead of viewing a square correctly as a special kind of rhombus.

Correct Definitions

For our students to participate actively in the construction of definitions, they must know what qualifies as a correct definition. A definition that contains conditions (properties) that are both necessary and sufficient is said to be correct. For a condition in a given description to be necessary, it must apply to all elements of the set we want to define. (The concept implies the property, so the property is necessary for the concept.) However, for a condition to be sufficient, it must ensure that whenever it is met, we obtain all the elements of the set we want to define. (The property implies the concept, so the property is sufficient for the concept.)

It is helpful to recall that logically in the biconditional $p \Leftrightarrow q$, the condition p is viewed as necessary and sufficient for the condition q, meaning that one can conclude that q follows from p, and vice versa. The defining conditions for a set must be both necessary and sufficient. For example, consider the following candidate for a definition: "A rectangle is any quadrilateral with opposite sides parallel." Below are some drawings of a quadrilateral complying with the condition "opposite sides parallel."Certainly the statement does apply to elements of the set we want to define (see fig. 13.1(a).). Therefore we can say that *"opposite sides parallel"* is a necessary condition for a rectangle. However, looking at the three drawings, we notice the existence of elements (figs. 13.1(b) and 13.1(c)) that do not belong to the set we want to define. Thus "opposite sides parallel" is not sufficient to guarantee that a particular quadrilateral is a rectangle. Hence we say that "opposite sides parallel" is a necessary but not a sufficient condition for rectangles.

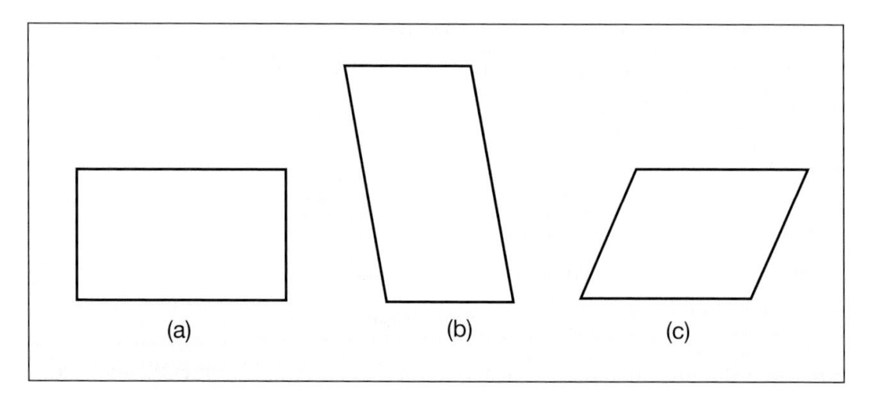

Fig. 13.1. Example and nonexamples of a rectangle

However, having congruent diagonals that are perpendicular bisectors of each other is a sufficient condition for a rectangle (in fact, sufficient for a square). But this condition is not necessary, because it does not apply to many rectangles, including the one in figure 13.1(a).

Next consider the following: "A rectangle is a quadrilateral with opposite sides parallel and with one interior angle equal to 90 degrees." This property applies to every rectangle (making it a necessary condition), and any figure we construct with it will be a rectangle (making it a sufficient condition). Therefore this statement provides a necessary and sufficient condition for rectangles.

Incorrect Definitions

A definition is incorrect if it contains insufficient or unnecessary properties. For example, consider the following:

1. "An isosceles trapezoid is any quadrilateral with perpendicular diagonals."

2. "A kite is a quadrilateral with perpendicular diagonals."

The first statement is incorrect because it contains an unnecessary property, in that isosceles trapezoids do not in general have perpendicular diagonals. The second statement is also an incorrect definition because it does not contain sufficient properties to define a kite and is therefore incomplete. For example, we can construct a diagonal and another one perpendicular to it, and then connect the endpoints, obtaining a quadrilateral as shown in figure 13.2, which is clearly not a kite. In general, to show that a definition is incomplete, it suffices to give a counterexample that meets the purported definition but is not an example of the set of figures one wants to define.

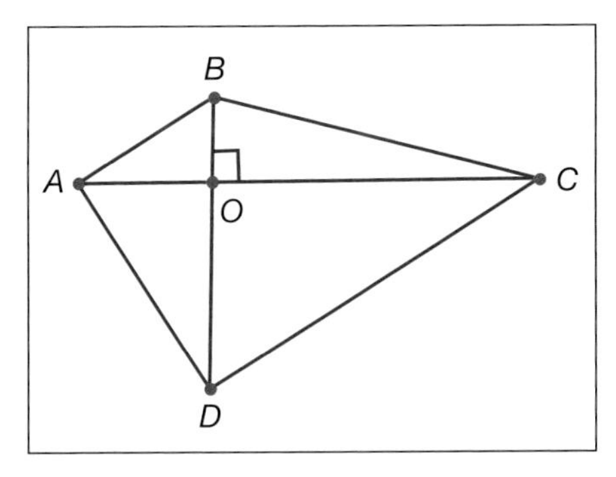

Fig. 13.2. Nonexample of a kite

Note that the statement "A kite is a quadrilateral with perpendicular diagonals" is a correct statement about a property of kites, but it contains too little information to be used as a definition. We therefore say that having "perpendicular diagonals" is a necessary but not a sufficient condition for kites.

The authors have observed that one common difficulty students have in producing correct counterexamples to incomplete definitions is that they often try to refute a definition with a special case. For example, for the incorrect definition "A rectangle is any quadrilateral with congruent diagonals," some students will provide a square as a counterexample. But obviously a square is not a valid counterexample, because a square *is* a rectangle.

Therefore, students should already have developed a sound understanding of a hierarchical (inclusive) classification of quadrilaterals *before* being engaged in formally defining the quadrilaterals themselves (Craine and Rubenstein 1993). This development can be fostered by using interactive geometry software, figures created with flexible wire, or paper-strip models of quadrilaterals. For example, if students use an interactive geometry tool to construct a quadrilateral with opposite sides parallel, then they may notice that they can drag it into the shape of a rectangle, rhombus, or square as shown in figure 13.3. Students can then be asked to describe what this outcome tells them. Teachers should help students realize that this demonstration shows that rectangles, rhombuses, and squares are special cases of a parallelogram.

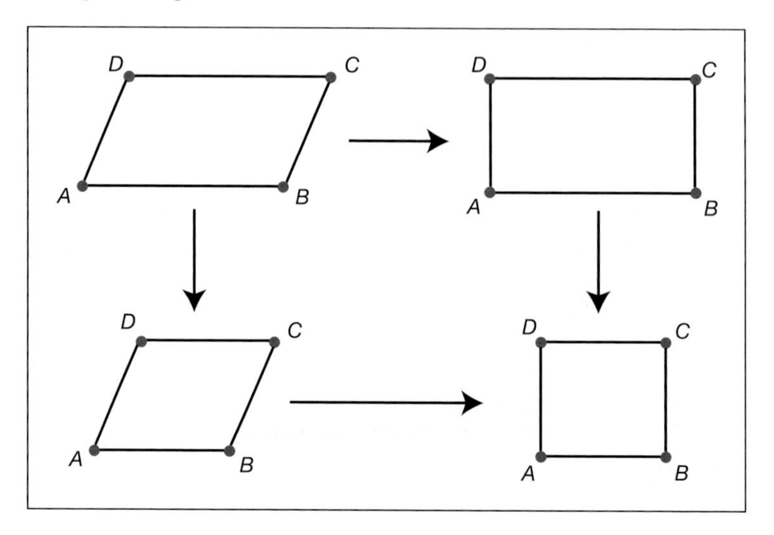

Fig. 13.3. Dynamic transformation of a parallelogram

Economical Definitions

A correct definition can be either economical or uneconomical. An economical definition has a minimal set of necessary and sufficient properties; that is, it has no superfluous information. Conversely, an uneconomical definition contains redundant properties.

For example, consider the following candidates for definitions of a kite:

1. "A kite is a quadrilateral with two pairs of congruent adjacent sides and one pair of opposite angles congruent."
2. "A kite is a quadrilateral with perpendicular diagonals with one being bisected by the other."

The first one is correct, but uneconomical because it contains too much information. In other words, the conditions are necessary and sufficient, but not all of them are required. But which condition can be left out? When students evaluate these conditions, they may realize that if they were to leave out "two pairs of congruent adjacent sides," they would obtain an incorrect definition because it is possible to construct a quadrilateral with one pair of congruent, opposite angles that is not a kite (fig. 13.4).

However, we can construct a kite according to the condition "two pairs of congruent adjacent sides" by placing a compass first at *A* and then at *D* and drawing circular arcs as shown in figure 13.5. In addition, we can easily show that this

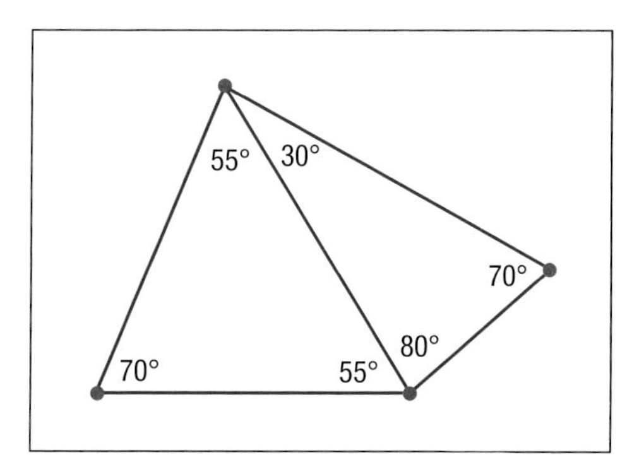

Fig. 13.4. Nonexample of a kite

condition logically implies that "one pair of opposite angles are congruent" because triangles ADB and ADC are clearly congruent (by side-side-side congruence), and therefore $m\angle B = m\angle C$.

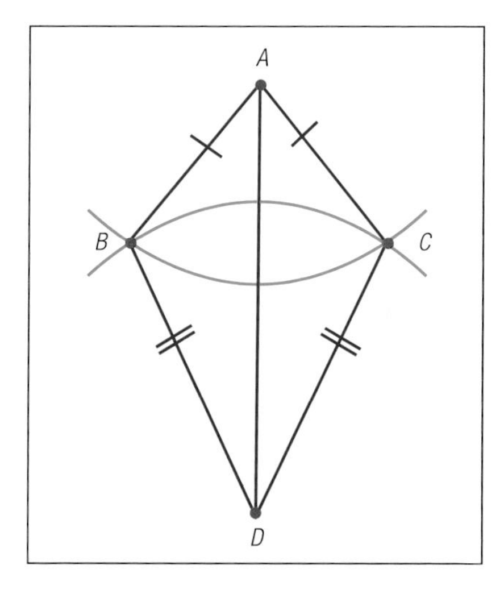

Fig. 13.5. Testing a definition of a kite

The second candidate is a correct, economical definition because no superfluous information is supplied. In other words, we cannot leave out any of the properties, and we can construct a kite from the given conditions. We can demonstrate the latter, as shown in figure 13.6, by first constructing two perpendicular lines, then placing the compass at O to cut off equal distances on the one line, and picking two arbitrary points on the other to determine the vertices. In addition, students can show that the other properties of a kite can be logically deduced from the given conditions. For example, in figure 13.6, triangles *AOB* and *AOC* are clearly congruent (by side-angle-side congruence) and therefore $AB = AC$. In the same way they can show that $DB = DC$.

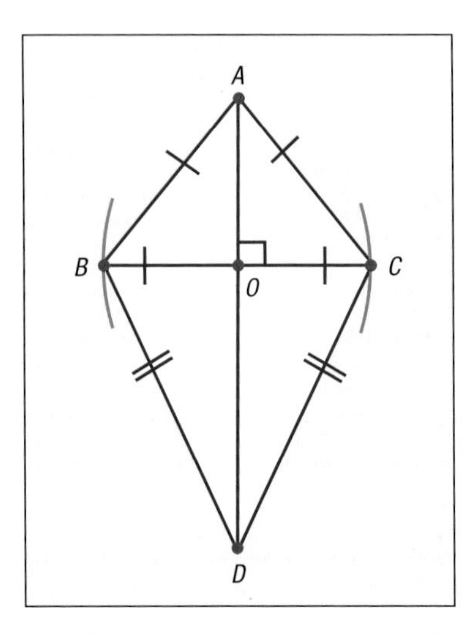

Fig. 13.6. Testing another definition of a kite

This minimality principle—that is, that definitions should be economical—is a crucial structural element of mathematics as a deductive system. As a matter of fact, it shapes the way in which mathematics progresses when it is presented deductively, for after the definition is presented, theorems that give additional information about the concept are formulated and proved (Linchevsky, Vinner, and Karsenty 1992, p. 54).

For example, if a rhombus is defined as a quadrilateral with four congruent sides, then the fact that the diagonals are perpendicular bisectors of each other can be proved as a theorem. Conversely, if a rhombus is defined as a quadrilateral with diagonals that are perpendicular bisectors of each other, then the fact that the four sides are congruent becomes a theorem. Both definitions are economical insofar as the defining conditions contain no superfluous information. This goes back to the idea that definitions are arbitrary. Different sets of definitions produce different theorems within a system.

However, in a few instances in mathematics, definitions are not minimal. A familiar example is the way in which some textbooks define congruent triangles: "Congruent triangles are triangles that have all pairs of corresponding sides congruent and all corresponding angles congruent." We know that it is sufficient to require less than that for two triangles to be congruent, and the fact that less is required is expressed by each of the four postulates normally accepted in high school textbooks for the congruency of triangles.

Convenient Definitions

Obviously a good definition, as we have seen in the foregoing, avoids redundant information; it must be economical. But a good definition also has other characteristics. A good, or convenient, definition is one that also allows us to deduce the other properties of the concept easily; that is, it should be *deductive-economical.*

A valuable exercise for students is to have them compare different definitions according to this criterion. For example, the definition of a rhombus as a quadrilateral with two axes of symmetry through the opposite angles is more deductive-economical than the standard textbook definition of it as a quadrilateral with all sides congruent. For the former, the other properties (e.g., perpendicular, bisecting diagonals, all sides congruent, and so on) follow immediately from symmetry, whereas with the latter, we have to use congruency and somewhat longer arguments to deduce the other properties.

A Teaching Sequence

Young children most easily learn what a table or a chair is by seeing many different examples of those concepts, not by being provided with formal definitions of a table and a chair. (See also Battista [2009] for more about this phenomenon of concept formation.) Similarly, without starting with a formal definition, children can easily learn what a square, rectangle, or kite is.

In an interactive geometry environment, concepts such as the special quadrilaterals can be introduced in three stages. The first stage uses the software to help students learn what a specific shape is (a concept image). For example, a concept such as an isosceles trapezoid can easily be introduced as shown in figure 13.7 by first having students construct a line *AB,* then construct a line segment *CD,* and

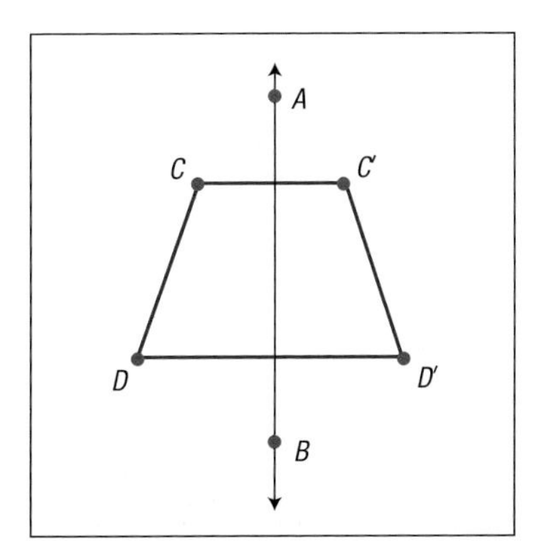

Fig. 13.7. Introducing an isosceles trapezoid by construction

subsequently reflect CD in the line AB to obtain $C'D'$. By connecting vertices, students can then be asked to explore the properties of quadrilateral $CDD'C'$, which they may be told is called an *isosceles trapezoid* (compare de Villiers [2003]).

By dragging the figure and measuring its attributes, students can develop a sound concept image of an isosceles trapezoid as a quadrilateral having many properties, including congruent diagonals, two pairs of congruent adjacent angles, at least one pair of opposite sides congruent, at least one pair of opposite sides parallel, and others. Moreover, students can discover that they can drag an isosceles trapezoid into the shape of a rectangle and square, but not a general rhombus, parallelogram, or kite, thus forming the foundation for a hierarchical view.

The second stage involves challenging students to write their own correct, economical definitions for an isosceles trapezoid, and then to test those definitions by means of construction and measurement. Equivalently, students can be challenged to devise different ways of constructing a dynamic isosceles trapezoid that always remains an isosceles trapezoid no matter how it is dragged. Such constructions, as reported in Smith (1940), help develop an understanding not only of the difference between a *premise* and a *conclusion* but also of the *causal* relationship between them, that the conditions force the result. For example, consider a circle with two parallel lines intersecting it (see fig. 13.8). Then the quadrilateral formed by the points where these two parallel lines intersect the circle must have congruent diagonals and congruent opposite sides and, hence, is an isosceles trapezoid. This result shows that the condition is sufficient to ensure an isosceles trapezoid, and that we could define an isosceles trapezoid as any cyclic quadrilateral with at least one pair of opposite sides parallel.

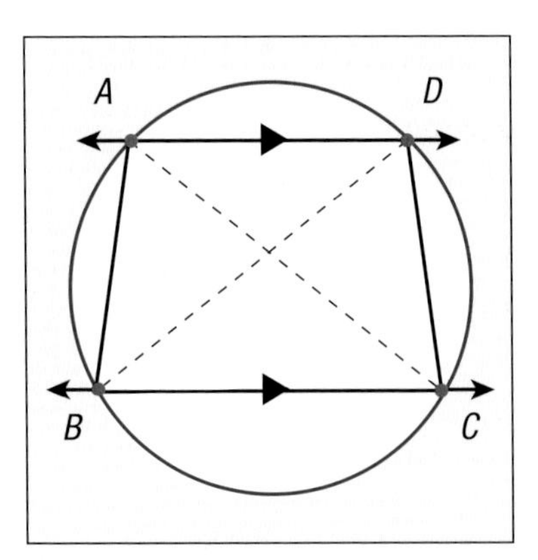

Fig. 13.8. An isosceles trapezoid constructed from two parallel lines intersecting a circle

The third stage involves the formal systematization of the properties of an isosceles trapezoid. For example, by starting from any given definition, students then have to deduce the other properties logically from it as theorems. A popular initial choice suggested by students, teachers, and some textbooks is the following: "An isosceles trapezoid is any quadrilateral in which at least one pair of sides are parallel and the other pair of opposite sides are congruent." However, this definition is *incorrect*. For example, although the conditions can produce an isosceles trapezoid, the conditions can also produce a parallelogram. Students need to realize that a general parallelogram cannot be considered an isosceles trapezoid, as parallelograms do not necessarily have all properties of an isosceles trapezoid (e.g., congruent diagonals, two pairs of adjacent angles congruent, cyclic, axis of symmetry, and so forth).

Although students usually attempt to improve this definition, after a while they find no satisfactory way of correcting it. If they formulate it in such a way as to exclude the parallelograms—for example "An isosceles trapezoid is any quadrilateral with one pair of opposite sides parallel, and another pair of opposite sides equal but not parallel"—they exclude not only the (general) parallelogram but also rectangles and squares.

After further discussion and critical comparison of various correct, economical definitions for an isosceles trapezoid, students usually settle with a convenient definition, such as "an isosceles trapezoid is any quadrilateral with an axis of symmetry through a pair of opposite sides." This definition is much easier to use to derive other properties. Contrast it, for example, with a definition such as "an isosceles trapezoid is any cyclic quadrilateral with at least one pair of opposite sides parallel."

Conclusion

Students should be given the opportunity to engage in the process of constructing definitions. Interactive geometry software is a tool that can help promote this goal (e.g., de Villiers [2003]). It is plausible to conjecture that, through experiences in which definitions are not supplied directly, students' understanding of geometric definitions, and of the concepts to which they relate, will increase. Furthermore, students are likely to develop a better understanding of the nature of definitions as well as skill in defining objects on their own. In particular, they may come to realize that definitions should be economical and that they are a matter of choice. Recent results from Govender and de Villiers (2002), de Villiers (1998, 2004), and Sáenz-Ludlow and Athanasopoulou (2007) do indicate some improvement and positive gains in students' understanding of the nature of definitions, as well as in their ability to define geometric concepts themselves.

REFERENCES

Battista, Michael T. "Highlights of Research on Learning School Geometry." In *Understanding Geometry for a Changing World,* Seventy-first Yearbook of the National Council of Teachers of Mathematics (NCTM), edited by Timothy V. Craine, pp. 91–108. Reston, Va.: NCTM, 2009.

Battista, Michael T., and Douglas H. Clements. "Students' Cognitive Construction of Squares and Rectangles in Logo Geometry." In *Proceedings of Sixteenth International Conference for the Psychology of Mathematics Education (PME)* (University of New Hampshire, Durham, New Hampshire, USA) vol. 1, edited by William Geeslin and Karen Graham, pp. 57–64. Cape Town, South Africa: PME, 1992.

Blair, Stephen, and Daniel Canada. "Using Circle and Square Intersections to Engage Students in the Process of Doing Geometry." In *Understanding Geometry for a Changing World,* Seventy-first Yearbook of the National Council of Teachers of Mathematics (NCTM), edited by Timothy V. Craine, pp. 283–96. Reston, Va.: NCTM, 2009.

Craine, Timothy V., and Rheta N. Rubenstein. "A Quadrilateral Hierarchy to Facilitate Learning in Geometry." *Mathematics Teacher* 86 (January 1993): 30–36.

de Villiers, Michael. *The Role of Axiomatization in Mathematics and Mathematics Teaching.* Stellenbosch, South Africa: Research Unit for Mathematics Education of the University of Stellenbosch, 1986. Available from mysite.mweb.co.za/residents/profmd/axiom.pdf .

———. "The Role and Function of a Hierarchical Classification of Quadrilaterals." *For the Learning of Mathematics* 14, no. 1 (1994): 11–18. Available from mysite.mweb.co.za/residents/profmd/classify.pdf.

———. "To Teach Definitions in Geometry or Teach to Define?" In *Proceedings of the Twenty-second International Conference for the Psychology of Mathematics Education (PME)*, vol. 2, edited by Alwyn Olivier and Karen Newstead, pp. 248–55. Cape Town, South Africa: PME, 1998. Available from mysite.mweb.co.za/residents/profmd/define.htm.

———. *Rethinking Proof with The Geometer's Sketchpad.* Emeryville, Calif.: Key Curriculum Press, 2003.

———. "Using Dynamic Geometry to Expand Mathematics Teachers' Understanding of Proof." *International Journal of Mathematical Education in Science and Technology* 35, no. 5 (2004): 703–24. Available from mysite.mweb.co.za/residents/profmd/vanhiele.pdf.

Freudenthal, Hans. *Mathematics as an Educational Task.* Dordrecht, Netherlands: Reidel, 1973.

Govender, Rajendran, and Michael de Villiers. "Constructive Evaluation of Definitions in a Sketchpad Context." Paper presented at the Association for Mathematics Education of South Africa National Conference, University of Natal, Durban, July 2002. Available from mysite.mweb.co.za/residents/profmd/rajen.pdf.

Griffiths, H. Brian, and A. Geoffrey Howson. *Mathematics: Society and Curricula*. Cambridge: Cambridge University Press, 1974.

Human, Piet. "Wiskundige Werkwyses in Wiskunde-Onderwys" [Mathematical processes in mathematics education]. D.Ed. diss., University of Stellenbosch, South Africa, 1978.

Linchevsky, Leonora, Shlomo Vinner, and Ronnie Karsenty. "To Be or Not to Be Minimal? Student Teachers' Views about Definitions in Geometry." In *Proceedings of the Sixteenth International Conference for the Psychology of Mathematics Education (PME)* (University of New Hampshire, Durham, New Hampshire, USA), vol. 2, edited by William Geeslin and Karen Graham, pp. 48–55. Cape Town, South Africa: PME, 1992.

Ohtani, Minoru. "Telling Definitions and Conditions: An Ethnomethodological Study of Sociomathematical Activity in Classroom Interaction." In *Proceedings of the Twentieth International Conference for the Psychology of Mathematics Education (PME)* (University of Valencia, Spain), vol. 1, edited by Angel Gutierrez and Luis Puig, pp. 75–82. Cape Town, South Africa: PME, 1996.

Pólya, George. *Mathematical Discovery*. Vol. 2. New York: John Wiley, 1981.

Sáenz-Ludlow, Adalira, and Anna Athanasopoulou. "Investigating Properties of Isosceles Trapezoids with the GSP: The Case of a Preservice Teacher." In *Proceedings of the Ninth International Conference: Mathematics Education in a Global Community*, edited by David Pugalee, Alan Rogerson, and Amelie Schinck, pp. 577–82, University of North Carolina at Charlotte, 2007.

Smith, Rolland R. "Three Major Difficulties in the Learning of Demonstrative Geometry." *Mathematics Teacher* 33 (March, April 1940): 99–134, 150–78.

Vinner, Shlomo. "The Role of Definition in the Teaching and Learning of Mathematics." In *Advanced Mathematical Thinking*, edited by David Tall, pp. 65–81. Dordrecht, Netherlands: Kluwer Academic Publishers, 1991.

Wittmann, Erich. "Teaching Units as the Integrating Core of Mathematics Education." *Educational Studies in Mathematics* 15 (1981): 25–31.

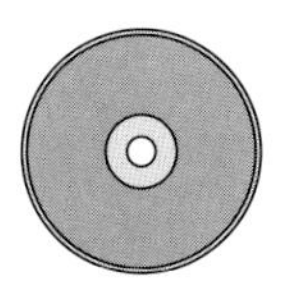

Geometer's Sketchpad files that support this article are found on the CD-ROM disk accompanying this Yearbook.

Advancing Elementary School Students' Understanding of Quadrilaterals

Tutita M. Casa
M. Katherine Gavin

Alex and Jessie, who are classmates and play on the same baseball team, are waiting to be picked up from practice. Jessie asks Alex, "Why do you think they call the infield a 'diamond?' I mean, shouldn't they call it a 'baseball square?'" Alex replies, "That's what everyone calls it, so it must be a diamond. Plus, it must be a diamond because it is tilted, and it can't be both a diamond and a square."

ALEX'S understanding of quadrilaterals is not unlike that of a typical elementary school student. Most students in these grades identify shapes according to their appearance, and not until the middle or high school years do they eventually come to understand that shapes are defined according to their properties and that relationships exist among those shapes (Clements 2003).

The work reported in this article was funded in part by the Jacob K. Javits Gifted and Talented Students Education Act, grant number S206A020006. The opinions, conclusions, and recommendations expressed in this article are those of the authors and do not necessarily reflect the position or policies of the U.S. Department of Education.

Fundamental Concepts and Inherent Challenges in the Study of Quadrilaterals

Van Hiele (1999) postulated that developing the levels of geometric thinking is primarily dependent on instructional experiences. Thus, what happens in the mathematics classroom during the elementary school years is of paramount importance in students' growing understanding of shapes. In an effort to move students beyond the visual level (level 1) of identifying shapes, elementary school teachers need to be aware of some important ideas that students will need for geometric reasoning in higher grades. First, shapes can have different names. Second, shapes typically tend to be referred to by their most specific name, such as *square,* rather than other correct names, such as *rectangle* or *parallelogram.* Third, shapes are named according to their attributes.

The study of quadrilaterals poses a fourth challenge, namely, that relationships among these particular shapes are hierarchical. (Further discussion of the importance of hierarchical classification can be found elsewhere in this volume in the article by de Villiers, Govender, and Patterson [2009].)

Figure 14.1 shows the relationships among quadrilaterals that were used in the project described subsequently. We should point out that *trapezoid* is a particularly challenging term because it has been defined differently by various authors. For example, Craine and Rubenstein (1993) define trapezoid as a quadrilateral with *at least* one pair of parallel sides. In this article, we define trapezoid as a quadrilateral with *exactly* one pair of parallel sides in accordance with the definition used by Gavin and her colleagues (2001) in *Navigating through Geometry in Grades 3–5*, part of the National Council of Teachers of Mathematics (NCTM) Navigations series.

The concepts foundational to the study of quadrilaterals were addressed in the geometry unit *Getting into Shapes* (Gavin et al. 2007) that resulted from Project M^3: Mentoring Mathematical Minds, a national research study devoted to creating high-level curriculum units for mathematically promising elementary school students. This article presents an overview of the strategies that teachers implemented to encourage fourth graders to study these concepts in depth. In the design of these materials, student activities were sequenced to increase in complexity and to support teachers in using mathematical language as a tool to promote students' development of important concepts. However, before teachers could implement these strategies effectively, they needed to be aware of the levels and progression of their students' understanding of these concepts. Thus, we begin with a discussion of the van Hiele (1999) framework for describing the development of students' understanding of properties of shapes and relationships among them.

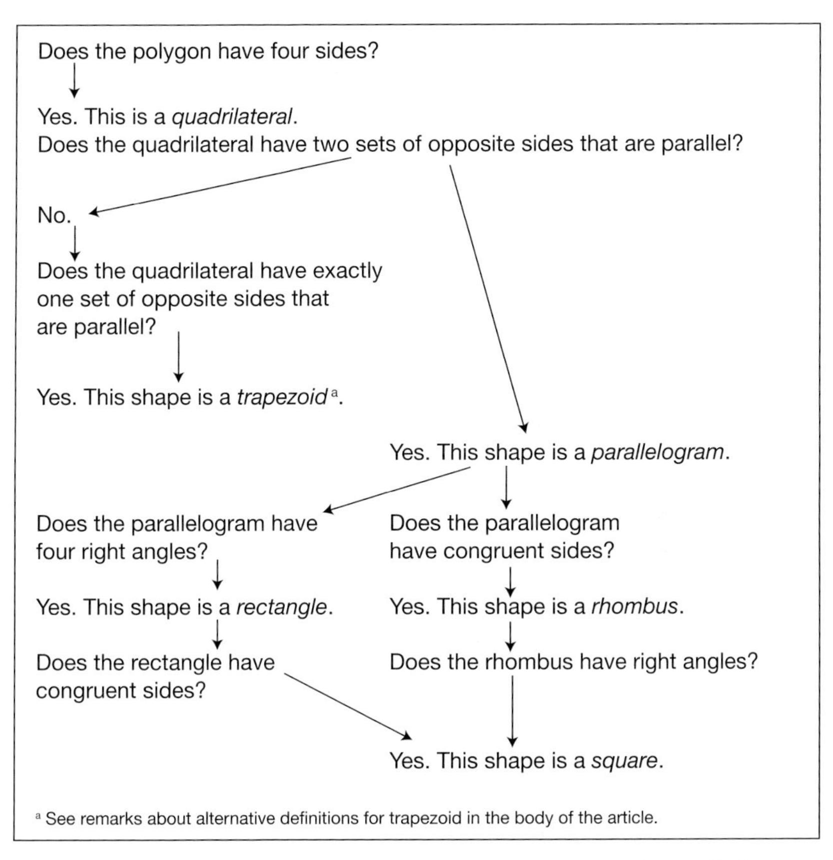

Fig. 14.1. Relationships among quadrilaterals that were studied

Progression of Students' Understanding

Developmentally, students begin their understanding of shapes at level 1, the *visual level* (van Hiele 1999), and from this point they build to reach subsequent levels. Because they have yet to begin to discern the properties of shapes, students operating at this level have only a tenuous visual sense of what categorizes a shape and may harbor misconceptions. One needs to look no further than the local toy store, the pages of children's books, or educational programs targeting primary-school-age children to see this influence. For example, a square rotated to "stand" on one of its vertices and other rhombi are often referred to as "diamonds." Also, although *square* is a more precise term, its name is introduced at the same time as a rectangle, which can lead children to assume that no relationship exists between squares and rectangles or even that these two shapes are mutually exclusive. This introduction may lead children to generalize incorrectly that all the sides of a rectangle cannot be congruent.

Students operating at level 2, the *descriptive level*, discern shapes' specific properties but do not yet logically order them (van Hiele 1999). Students see that some shapes have parallel sides, some have right angles, some have congruent sides, and other such properties.

The properties of shapes become logically ordered when students operate at level 3, the *informal deduction* (or *relational*) *level*. "Students use properties they already know to formulate definitions, for example, for squares [and] rectangles …, and use them to justify relationships, such as explaining why all squares are rectangles" (van Hiele 1999, p. 311). Research indicates that although U.S. students are not progressing through these levels, they can do so with quality curriculum and instruction (Clements 2003, p. 154): "Given that even young children refer to components and attributes of shapes, there is every reason to enrich multiple types of geometric thinking in students of all ages."

Instructional Strategies Used to Advance Students' Understanding of Quadrilaterals

We next discuss the strategies that our team (consisting of curriculum writers, professional development leaders, and classroom teachers) collectively implemented in twenty fourth-grade classrooms using the unit *Getting into Shapes* (Gavin et al. 2007). A primary goal of the unit was to help students move beyond van Hiele level 1, the visual level, and develop more advanced understanding of quadrilaterals. Specifically, the unit addressed the four fundamental concepts identified earlier in this article:

- shapes can have different names;
- shapes tend to be referred to by their most specific name;
- shapes are named according to their attributes; and
- quadrilaterals have hierarchical relationships.

Van Hiele asserted that "instruction intended to foster development from one [geometric] level to the next should include sequences of activities, beginning with an exploratory phase, gradually building concepts and related language, and culminating in summary activities that help students integrate what they have learned into what they already know" (1999, p. 311). Thus, we took two approaches to help students further their conceptual understanding. First, we designed strategies addressing all four of these concepts to be increasingly complex to coincide with students' developmental progress. Specifically, students engaged in activities to visualize different quadrilaterals and identify their properties. We then introduced students to a real-life analogy to help them realize that shapes can have different names and are usually referred to by their most specific one. Finally, students experienced multiple activities to investigate relationships among the shapes. Second, because language becomes a vehicle to describe shapes beginning at level 2 (van Hiele 1999), we also incorporated strategies focused on using mathematical language. The strategies encompassing both approaches are described next.

Conceptual Development Strategies of Increasing Complexity

Encourage Students to Discover Properties of Shapes

Before learning the specific names of shapes, students first participated in an activity to help them understand the concepts that define the shapes. The National Council of Teachers of Mathematics (NCTM 2000) recommends that students study both examples and nonexamples of two-dimensional objects to begin classifying shapes. The shapes included in the "You Either Have It or You Don't" activity encouraged students to examine properties of polygons with particular attention being given to quadrilaterals. For example, students were given the two groups of shapes shown in figure 14.2 and asked to identify the properties of rectangles. They then applied this understanding by adding their own drawings of a rectangle and a nonrectangle. Similar activities were done for other special quadrilaterals, such as rhombi and trapezoids. This activity illustrates the "reconstructive approach" to definitions (de Villiers, Govender, and Patterson 2009) and is far more challenging than giving students the definition of a shape and expecting them to memorize and then apply it. In this activity, students had to use high-level, analytical thinking (comparing, contrasting, synthesizing, and evaluating) to deduce characteristics of the shapes.

Present Real-Life Scenarios to Help Students Conceptualize the Naming of Shapes

The ideas that shapes can have different names, that shapes belong to classes or "families," and that shapes tend to be referred to by their most specific name can be confusing for many students, particularly ones who judge shapes solely by their appearance. To make these ideas more concrete for students, the lead author, who is of Puerto Rican descent, explained to students that a common practice in many Spanish cultures is for females to keep their maiden names, then add their husband's surname after they married. She shared her first name (Tutita), middle

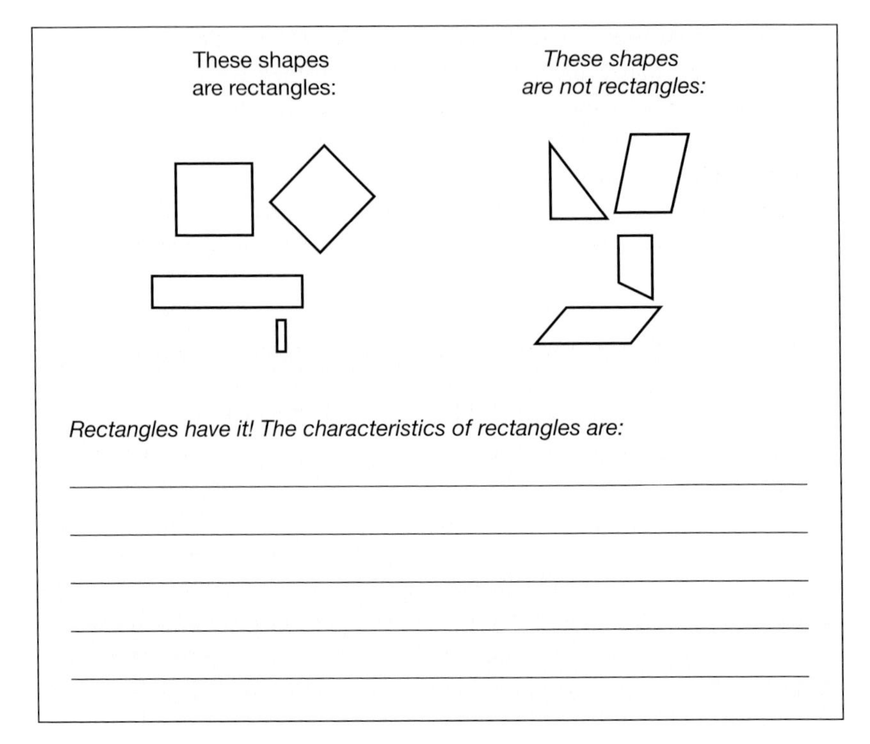

Fig. 14.2. Example of a "You Either Have It or You Don't" problem

name (Maria), married surname (Casa), and maiden name (Pérez) with classes. The students clearly understood that "Dr. Casa," "Tutita," "Mrs. Casa," "Miss Pérez," and "Tutita Maria" referred to the same person, just as several different names can refer to the same shape. They also saw that she belonged to more than one family (similar to some classes of shapes), including her immediate family, her husband's family, and the family in which she was raised. The students were able to relate to this idea as they discussed that they, too, had immediate families as well as belonged to extended families headed by their parents, stepparents, and grandparents. Students thought it also was rather comical for the lead author to be addressed at all times as "Tutita Maria Casa Pérez." Coupled with the fact they were usually addressed by their first names, students came to understand that shapes, too, tend to be referred to by their most specific name (e.g., square) even though they may have other names (e.g., rectangle, parallelogram).

Give Students Opportunities to Investigate Relationships among Quadrilaterals

In the elementary school years, "there should be growing emphasis on making conjectures about geometric properties and relationships and formulating mathematical arguments" (NCTM 2000, p. 9). Our team believed that multiple and frequent exposures to these fundamental concepts among quadrilaterals would be necessary for students to solidify these relationships and justify them. Students had the opportunity to work with these concepts in a variety of ways including the following activities: "All, Some, or None," "Three of These Things Belong Together," "Triple Play," and "Creating a Mobile" (Gavin et al. 2007).

In "All, Some, or None," students decided whether all, some, or none of a particular shape embodied a particular attribute and provided counterexamples for any statement that was false. For example, when asked to assess the statement "No rectangle is a rhombus," students listed the defining attributes of each shape. They were able to conclude that this statement was false because a square possesses all the attributes of a rectangle and rhombus. This insight represents quite a breakthrough because students previously thought of squares and rectangles as contrasting shapes.

In "Three of These Things Belong Together," students were given a set of four shapes and asked to identify which one did not belong and to explain why. They had to apply their developing understanding of the relationships among shapes to determine a single classification that could incorporate three of the shapes but not the fourth. For instance, when working with the problem shown in figure 14.3, students noted, "The trapezoid does not fit. It has only one pair of parallel sides and the other shapes all have two pairs" as one possible answer. (Note that other responses are possible and acceptable so long as they are justified.)

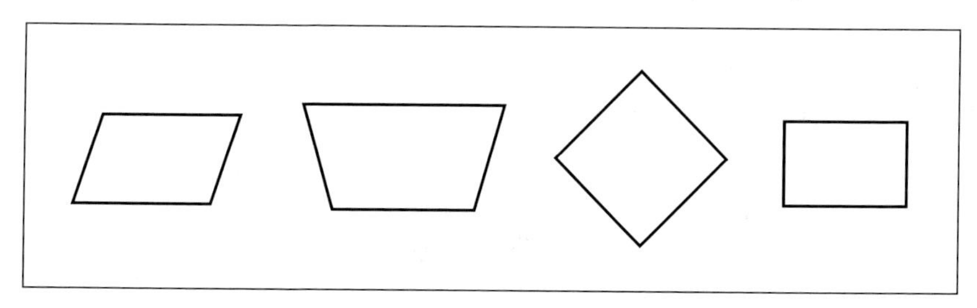

Fig. 14.3. Example of a "Three of These Things Belong Together" problem

Groups of four students played a motivational game called "Triple Play" to further develop their understanding of the relationships among shapes. In this game, each team of two students tried to make sets of three shapes in the same class using a deck of cards with pictures of quadrilaterals and other polygons. They were awarded points for each set of three: 10 points for sets of squares or trapezoids, 5 points for sets of rectangles or rhombi, and 3 points for sets of parallelograms or regular polygons (polygons with congruent sides and angles). For example, if a team had the six cards shown in figure 14.4, they could get a maximum of 8 points by creating two sets of three cards: 5 points for the three rhombi (cards B, C, and F) or any set of three rectangles (cards A, B, D, and F), and 3 points for three parallelograms (cards A–F).

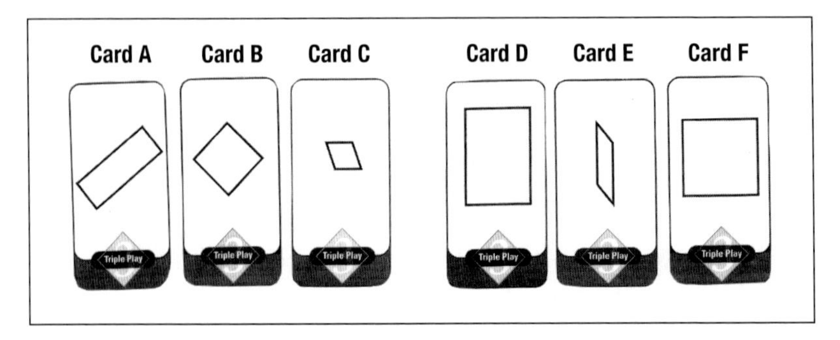

Fig. 14.4. A sample hand in the Triple Play game

Students quickly came to realize that getting sets of squares or trapezoids was more challenging (hence, the greater number of points that were awarded) because those shapes have more uncommon properties.

Relationships among shapes were also addressed by having students create a mobile. Students were asked to show the different levels of specificity with their mobile, with the bottommost shape—the square—being the most specific (see fig. 14.5). Simply to include the names of the shapes was not sufficient; they also had to list defining characteristics of each shape. This activity encouraged students to see that all rhombi, for example, are also parallelograms, but not all rhombi are squares.

Strategies to Develop Mathematical Language

Communicating orally and in writing were central practices in our classrooms. This type of environment encouraged students to appreciate the need for mathematically precise language, an important instructional goal (NCTM 2000). Although our students used informal language as they developed a conceptual understanding of the fundamental concepts, the introduction of formal mathematical language arose naturally. Once students had a basic notion of the concepts, a need for more accurate language presented itself as they discussed attributes and relationships among quadrilaterals. Teachers used a variety of strategies to help connect students' everyday language with more formal language.

Realizing that "imprecise language plagues student work in geometry" (Clements 2003, p. 154), we first describe the challenges with geometric language that we addressed in our program. First, students have limited exposure to

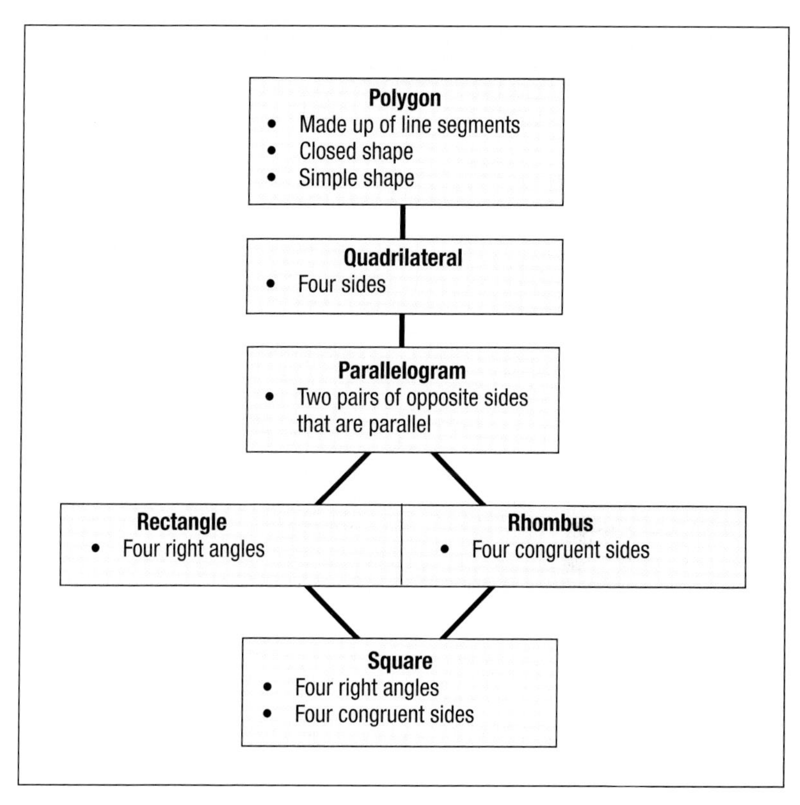

Fig. 14.5. A mobile illustrating relationships among shapes

mathematical vocabulary outside the classroom (Monroe and Orme 2002; Thompson and Rubenstein 2000). Second, some mathematical terms are used only within the context of mathematics (Thompson and Rubenstein 2000; Rubenstein and Thompson 2002), such as the word *rhombus,* so students are unable to draw from their everyday experiences to build meaning. Third, some vocabulary words are used in both mathematics and everyday language, but they have different meanings in those contexts (Monroe and Orme 2002; Thompson and Rubenstein 2000; Rubenstein and Thompson 2002). Further, students may be familiar with the everyday meaning of a word but not know its specific mathematical definition (Monroe and Panchyshyn 1995/1996) or even realize that such a distinction exists. The word *right* is an example. To say that a rectangle has four right angles means that all the angles measure 90 degrees, not that the angles are correct! Fourth, some mathematical vocabulary terms, such as *perpendicular* and *parallel,* are often taught together and students may not distinguish between their distinct meanings. Finally, students may use nonmathematical terms (Rubenstein and Thompson 2002), such as *corner* instead of *vertex,* to describe mathematical concepts. Table 14.1 summarizes these challenges and some examples related to the study of quadrilaterals.

Table 14.1
Challenges Inherent to the Learning of Vocabulary in the Study of Quadrilaterals

Language Challenges	Examples
Some words are used only in mathematics.	Congruent Quadrilateral Rhombus Trapezoid
Different meanings of the same word are used in mathematics and everyday language.	Closed, as in a closed shape Opposite, as in an opposite side Right, as in a right angle Simple, as in a simple shape
Some mathematical words are related but have distinct meanings.	Acute, right, and obtuse angles Horizontal and vertical Parallel and perpendicular
Some everyday words are used in lieu of mathematically precise ones.	Corner instead of vertex Diamond instead of rhombus

The following strategies had students build on their understanding of fundamental mathematical concepts while being engaged with and communicating using mathematical language in multiple ways (Krussel 1998; Leung 2005; Miller 1993; NCTM 2000).

Develop Student-Authored Definitions of Shapes Using Multiple Representations

Relying on a dictionary definition gives students only a surface understanding of mathematical concepts and is not enough for them to construct a robust meaning (Monroe and Orme 2002). Students need to own the meaning themselves so that the mathematical language makes sense to them (Miller 1993). One way to support this ownership is to have students address the concept using multiple representations, including writing and illustrations (Rubenstein and Thompson 2002). Consequently, our students wrote personal definitions for polygon, quadrilateral, parallelogram, rectangle, trapezoid, rhombus, and square in their own words; drew a real-life representation of each; added an example of the shape; and drew a nonexample that exemplified a common misconception about an attribute of the shape. Figure 14.6 shows how Lacy initially described a parallelogram in her own words.

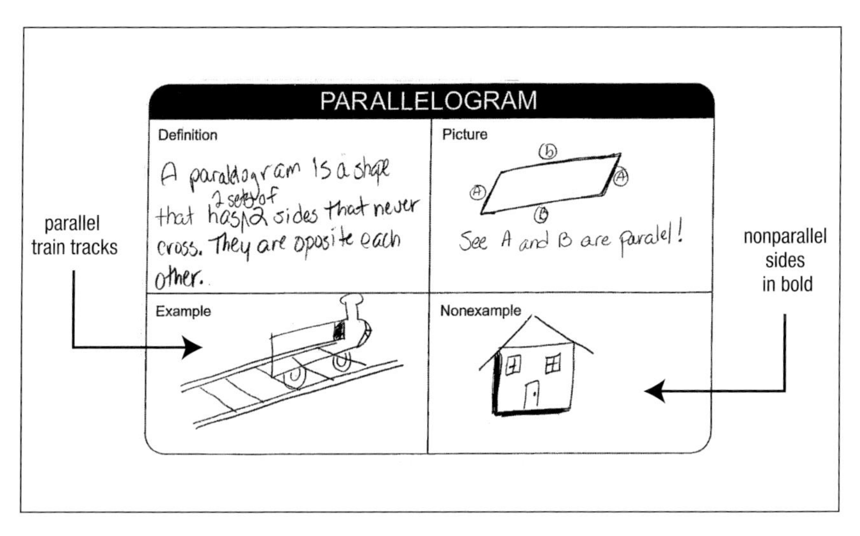

Fig. 14.6. An example of a student's description of a parallelogram

As a culminating experience, students came together as a class to form a consensus about an appropriate definition for each term. Although not all the students were ready to use formal language, Lacy decided to revise her definition of a parallelogram by substituting the word *parallel* to indicate this is what she meant by "sides that never cross." Because students need time to develop an understanding of these concepts, teachers had students continue to revisit and revise their definitions.

Encourage Students to Recognize the Need for Precise Mathematical Language

Our students regularly spoke in small groups and as a whole class. They also engaged in what Chapin, O'Connor, and Anderson (2003) refer to as "partner talk." After the teacher posed a question, students had a few minutes to transfer their ideas into words and address their closest neighbor. This approach allowed them to organize their thoughts and decide on the language they would use to express their ideas prior to sharing them with the class. Having multiple audiences, including their peers as partners and in whole-class discussion, encouraged students to be as clear as possible when expressing their ideas. These regular conversations, along with written work, addressed the need for students to use mathematical language intensely to develop fluency, experience ownership, and become comfortable using it over time (Thompson and Rubenstein 2000).

Model Appropriate Use of Mathematical Language

"Teachers should pay close attention to the vocabulary used, encouraging correct mathematical language so that geometric misconceptions are avoided" (Gavin et al. 2001, pp. 9–10). Our teachers modeled appropriate usage for students, a practice encouraged by others (e.g., Miller 1993; Monroe and Panchyshyn 1995/1996; Thompson and Rubenstein 2000). The teachers used precise mathematical language at an appropriate time when students had begun to become comfortable with the concepts. For example, when students described their ideas about "corners," "L-shaped angles," and sides being "the same," teachers shared with students that mathematicians refer to these ideas, respectively, as *vertices, right angles,* and *congruent sides*. One specific way they did so was to revoice (Chapin, O'Connor, and Anderson 2003) students' informal language using more precise vocabulary to demonstrate their meanings within the context. For instance, when a student said, "All parallelograms have opposite sides that will never touch," the teacher asked the student what she meant when she said "sides that will never touch." The student then went to the board to point out two opposite sides and motioned that if each side continued as a line, the lines would not cross. The teacher then first explained that in mathematics the word *parallel* represents the idea that the sides would never cross. She then restated the original idea by explaining, "So parallelograms have opposite sides that are *parallel*." Teachers also encouraged students to take on a similar role by having them rephrase (Chapin, O'Connor, and Anderson 2003) one another's ideas using appropriate language.

Use Mathematical Writing as a Tool to Further Develop Understanding

Our activities had students ponder questions and write specifically about fundamental mathematical concepts focused on the properties of shapes and the relationships among them. For instance, some questions asked students whether they thought a particular quadrilateral also belonged in another class, such as whether a rectangle is also a parallelogram. Others asked them about the design of the Triple Play game and to consider why squares and trapezoids were awarded the most points.

Students were also required to justify their reasons, not simply state their positions, further giving students an opportunity to make connections between everyday and formal mathematical language. Jaden's comment, quoted below, is an example. Jaden, operating between the descriptive and informal deductive levels (levels 2 and 3; van Hiele 1999), used precise terms to demonstrate his emerging understanding of the hierarchical relationship between shapes when he responded to the following:

> "Miranda made a discovery. She claims that all squares are rectangles! Do you agree or disagree? Explain your answer."

> I agree and disagree because a square meets most of the requirements of a rectangle and a square is consitered (sic) a rectangle but sometimes it is not consitered (sic) a rectangle because a rectangle does not have all congruent sides and a square does. But rectangles do have all 90° angles and so do squares [. A]lso a rectangle and a square both have at least two lines of semmetry (sic) and both shapes have two sets of parallel lines on opposite sides. But a square is also its own individual shape and it may have all of the requirements as a rectangle but it also has some additional thing to be a square. It has all congruent sides.

Tania, operating at level 3, the informal deductive level (van Hiele 1999), demonstrates how mathematical language became a tool for her to represent her conceptual understanding.

> I agree to Miranda's theory. I agree because a square has all the attributes of a rectangle. Those attributes are: 4 sides, 4 90° angles, and 2 sets of oppisite parellel (sic) and congruent lines. A square fits all those atributes (sic) but it also has 1 extra attribute. That all its sides are congruent. A square also has many other names. Those are: rectangle, parallelogram, rhombus, and quadrilateral. But its clearest name is square.

Encourage the Development of Precise Mathematical Vocabulary to Showcase Understanding

Our teachers were aware that they could not always infer students' meanings when they used informal language. That is, if students were imprecise in their verbal or written explanations, teachers did not assume they knew what the students meant. Teachers helped students develop the use of precise language to represent their concepts in several different ways. One way was to create an interactive word wall. As seen in figure 14.7, a word wall was placed in a prominent location in each room, in every student's line of vision so as to be easily read from the back of the room. The word wall had moveable parts. Individual terms were listed, and some word walls included definitions or drawings, or both. Teachers asked students to refer to the word wall whenever they rephrased another student's idea. They also grouped different terms (e.g., rhombus, trapezoid, and square) to have students determine what was similar about them (e.g., they are

Fig. 14.7. An interactive word wall

all quadrilaterals) and removed individual terms, definitions, or pictures to have students complete the missing pieces. Teachers added terms to the wall as they were used in class or prior to a lesson.

Final Remarks

Unfortunately, many students do not attain the level of understanding associated with the third van Hiele level until middle or secondary school years, if at all. We conjectured that those students may never have been given the opportunity to investigate these concepts in the earlier grades. We addressed this problem by providing fourth-grade teachers with instructional strategies and their students with activities to express their ideas through language with the aim of moving students' understanding beyond the visual level.

Understanding the fundamental concepts that describe quadrilaterals can be a complex endeavor for elementary school students. High-quality curriculum materials and teaching that supports concept and language development can help move students from visually identifying shapes in isolation to defining their attributes, classifying them, and recognizing relationships among them.

REFERENCES

Chapin, Suzanne H., Catherine O'Connor, and Nancy C. Anderson. *Classroom Discussions: Using Math Talk to Help Students Learn.* Sausalito, Calif.: Math Solutions Publications, 2003.

Clements, Douglas. "Teaching and Learning Geometry." In *A Research Companion to "Principles and Standards for School Mathematics,"* edited by Jeremy Kilpatrick,

W. Gary Martin, and Deborah Schifter, pp. 151–78. Reston, Va.: National Council of Teachers of Mathematics, 2003.

Craine, Timothy V., and Rheta N. Rubenstein. "A Quadrilateral Hierarchy to Facilitate Learning in Geometry." *Mathematics Teacher* 86 (January 1993): 30–36.

de Villiers, Michael, Rajendran Govender, and Nikita Patterson. "Defining in Geometry." In *Understanding Geometry for a Changing World,* Seventy-first Yearbook of the National Council of Teachers of Mathematics (NCTM), edited by Timothy V. Craine, pp. 189–204. Reston, Va.: NCTM, 2009.

Gavin, M. Katherine, Louise P. Belkin, Ann Marie Spinelli, and Judy St. Marie. *Navigating through Geometry in Grades 3–5. Principles and Standards for School Mathematics* Navigations Series. Reston, Va.: National Council of Teachers of Mathematics, 2001.

Gavin, M. Katherine, Judith Dailey, Suzanne H. Chapin, and Linda Jensen Sheffield. *Getting into Shapes*. Dubuque, Iowa: Kendall/Hunt Publishing, 2007.

Krussel, Libby. "Teaching the Language of Mathematics." *Mathematics Teacher* 91 (May 1998): 436–41.

Leung, Constant. "Mathematical Vocabulary: Fixers of Knowledge or Points of Exploration?" *Language and Education* 19, no. 2 (2005): 126–34.

Miller, L. Diane. "Making the Connection with Language." Special Issue, *Arithmetic Teacher* 40, no. 6 (February 1993): 311–16.

Monroe, Eula Ewing, and Michelle P. Orme. "Developing Mathematical Vocabulary." *Preventing School Failure* 46 (Spring 2002): 139–42.

Monroe, Eula Ewing, and Robert Panchyshyn. "Vocabulary Considerations for Teaching Mathematics." *Childhood Education* 72 (Winter 1995/1996): 80–83.

National Council of Teachers of Mathematics (NCTM). *Principles and Standards for School Mathematics.* Reston, Va.: NCTM, 2000.

Rubenstein, Rheta N., and Denisse R. Thompson. "Understanding and Supporting Children's Mathematical Language Development." *Teaching Children Mathematics* 9 (October 2002): 107–12.

Thompson, Denisse R., and Rheta N. Rubenstein. "Learning Mathematics Vocabulary: Potential Pitfalls and Instructional Strategies." *Mathematics Teacher* 93 (October 2000): 568–74.

van Hiele, Pierre M. "Developing Geometric Thinking through Activities That Begin with Play." *Teaching Children Mathematics* 5 (February 1999): 310–16.

Using Interactive Geometry Software to Teach Secondary School Geometry:

Implications from Research

Karen F. Hollebrands
Ryan C. Smith

WHO would have predicted that Apple's introduction of the Macintosh computer with a mouse-driven graphic-user interface in the 1980s would have such a profound effect on the computing world and consequently the learning and teaching of mathematics? This technology inspired the development of interactive geometry software, such as Cabri (Laborde and Bellemain 2005) and The Geometer's Sketchpad (Jackiw 2001), that allows students to manipulate geometric objects directly on the screen (Laborde and Laborde 1995).

Many researchers who investigate the teaching and learning of geometry with interactive software make subtle distinctions in the use of such words as *drawing*, *construction*, *diagram,* and *figure*. In this article, we refer to *drawing* as a process that involves the use of "freehand" tools to create a geometrical object. In a pencil-and-paper environment, a pencil is a freehand tool but a compass is not, because it is constrained to create only circles and arcs. With interactive geometry software, freehand drawing might involve the use of the segment tool to create an object that looks like a square without using construction-menu commands to create perpendicular lines or congruent segments. When a student views a visual representation of a geometrical object and focuses only on its perceptual characteristics and not its properties, then we refer to this process as reasoning about a *drawing*.

When a geometrical object is created using specific tools, such as a perpendicular line command to ensure that all angles are right angles, then this process is referred to as *constructing* and the final object is called a *construction*. When a student views a visual representation of a geometrical object and attends to its properties, then we state that the student is reasoning about a *figure*. In contrast, we use the term *diagram* when it is not evident whether the student is attending to the properties or the perceptual characteristics of the visual representation of a geometrical object.

Interactive geometry software enables users to create constructions by selecting a sequence of actions that are defined geometrically. When an element of such a construction is dragged, the geometrical object is modified while all the geometric relationships used in its construction are preserved. Thus, the construction relies on, and behaves according to, the geometrical relationships. This behavior is not true of drawings that can be distorted by students to meet their expectations. A question that naturally arises is "How do students' interactions with interactive geometry software influence their developing understandings of properties of geometrical figures and their views of proof?"

During the past twenty years, research on the use of interactive software has shed light on two particularly important aspects of geometrical thinking: (1) its effects on students' understandings of geometrical figures and properties and (2) its effects on students' deductive reasoning. This article offers a research-based discussion of these ideas.

Students' Understandings of Geometrical Figures and Properties

Students often create and reason about drawings in the computer environment when they are being asked and expected to reason about abstract geometrical objects (Hollebrands, Laborde, and Straesser 2008). For example, for students to operate on or manipulate points in the plane using The Geometer's Sketchpad, physical representations of the points must be created. Yet students are expected to reason about all points in the plane, not just those that are physically represented. As a result, students may focus on only those points that have been created and come to believe a plane contains only a finite number of points rather than an infinite number of points (Hollebrands 2007). For example, students were asked to determine which points are mapped to themselves under a reflection. Rather than state all points on the line of reflection, many students responded that only two points are mapped to themselves. This notion may be due to the fact that only two points are represented when a line is created with the technology.

Whether students focus on visual appearances or geometrical properties may be related to the ways in which students use the processes of drawing and constructing when working with a dynamic sketch. For example, a student may use the segment tool to draw segments *AB, BC, CD,* and *AD* and arrange them so they appear to form a square. As described previously, this process is referred to as *drawing.* That "square," however, is destroyed when the "drag test" is applied because the properties needed were not included in its production (see fig. 15.1).

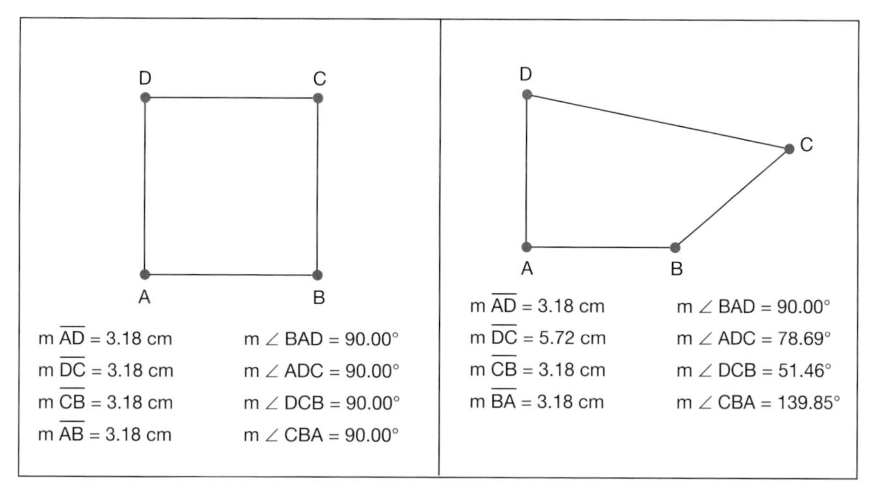

Fig. 15.1. A square drawn using measurements, shown before and after the "drag test" is applied

This method is contrasted with the process of constructing in which a student uses menu commands to assure that the properties of the object are maintained when any element of it is dragged. For example, when using The Geometer's Sketchpad a student may use the following procedure to construct square *ABCD.*

1. Use the segment tool to create segment *AB.*
2. Select points *A* and *B* and segment *AB,* and choose "Perpendicular Lines" from the Construct menu to construct perpendicular lines through points *A* and *B*.
3. Select point *A* and segment *AB* to construct a circle with center *A* and radius *AB*.
4. Construct one of the points of intersection of this circle with the perpendicular line passing through point *A* as the third point of the square, and name this point *D*.

5. Select point *D* and segment *AD,* and choose "Perpendicular Line" from the Construct menu to construct a perpendicular line through point *D* perpendicular to segment *AD*.
6. Construct the point of intersection of the perpendicular line through point *B* and the perpendicular line through point *D,* and label this fourth point of the square as point *C*.

To perform the latter construction, a student needs to understand geometric properties of a square and how to use the program's tools to construct an object that embodies those properties. The squares that result from the drawing process and the construction process may look the same, but the way in which they behave under the drag mode will differ. The use of the "drag test" may encourage students to reason using the properties of a square rather than rely only on its appearance. Researchers (Healy et al.1994; Hoyles and Noss 1994) challenged students to create objects on the screen that were not "messable" so that when any point of the object is dragged, the geometrical relationships remain invariant, thus requiring students to construct rather than simply draw the object. They suggest that the activity of constructing has the potential for moving students toward thinking about the theoretical geometric object rather than the characteristics of the drawing that appear on the screen.

Findings from research foster insights into ways in which students make links between drawings and theoretical geometrical objects. Laborde (1996) analyzed the relationships among drawings and theoretical geometrical objects in the processes students used as they completed various geometry tasks in paper-and-pencil and software environments. She found that with construction tasks and proof tasks, students initially appealed to perceptual features of the drawing but then moved to reason about theoretical properties.

The links that students make between drawings and theoretical objects may be related to the proactive and reactive strategies they employ when using interactive geometry software (Hollebrands 2007). Reactive strategies are those strategies in which a student responds to visual feedback as he or she manipulates dynamic drawings. Proactive strategies are those in which a student determines what actions to perform prior to performing them. Students using a tool in a reactive manner might not know what to expect prior to acting; students using a tool proactively have expectations of what they want to do with the technology, determine what actions will achieve the desired result, perform the actions, and finally reflect on the results. The proactive student appears to have in mind a plan about how to use the computer based on understandings of geometrical properties and relationships.

To illustrate the differences between reactive and proactive strategies, consider a task that requires the use of interactive geometry software to tessellate the plane using a scalene triangle (fig. 15.2). A student using the technology in a reactive manner might copy the scalene triangle, *ABC;* paste it; change the interior color from yellow to red; and then drag it so that it is next to triangle *ABC*. He or she may realize the need to rotate triangle *ABC* to fill in the space between the yellow and red triangles but may not be sure of the center of rotation. The student may place a point *D* somewhere in the sketch, mark it as the center, and then choose to rotate triangle *ABC* about point *D* through an angle of 180 degrees. The student may notice that it is not in the correct location and drag point *D* until the blue triangle fits between the other two triangles.

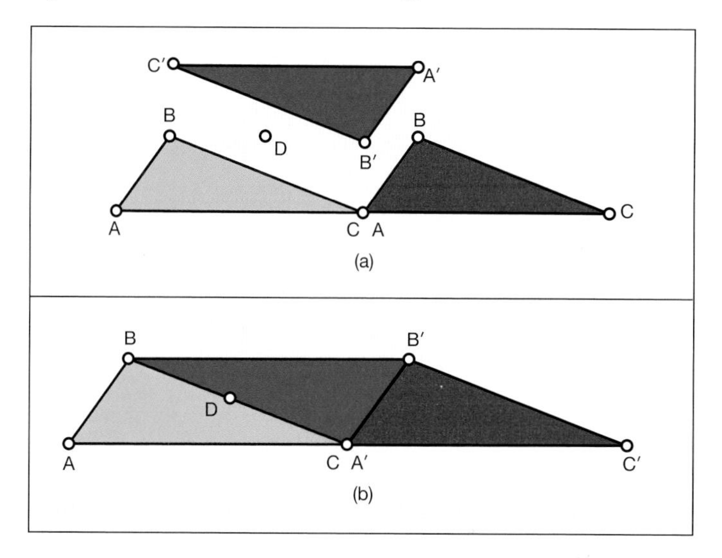

Fig. 15.2. An example of a reactive use (a) and a proactive use (b) of the tool to tessellate the plane with a scalene triangle

This approach is contrasted with a proactive strategy based on a student's understanding of transformations and tessellations. Triangle *ABC* is translated using vector *AC*. The student then reasons that at point *C* the measures of angles $B'A'C'$ and *BCA* sum to 180 minus the measure of angle *ABC* and determines that triangle *ABC* needs to be rotated. To rotate the triangle such that angle *ABC* is mapped to the appropriate location, a student may determine that the midpoint of side *BC* needs to be used as the center of rotation and 180 degrees as the angle of rotation.

Although both strategies produce a correct image, one strategy relies more heavily on visual appearances and feedback provided from the computer, whereas the other relies on mathematical properties and uses the computer to carry out a specific plan.

Empirical and Deductive Reasoning

Students use empirical evidence (e.g., measurements, diagrams) from the computer, or drawings produced on paper, in a variety of ways. This evidence may provide examples from which students inductively infer a conjecture and then prove it using deductive methods. Or students may view this evidence as sufficient proof that their conjecture is always true (Chazan 1993a; Edwards 1997). Because students need to appreciate the role of deduction and incorporate it into their ways of thinking, the teacher should know how to move students from merely describing drawings or examples toward deducing and proving geometrical properties and theorems. The role of empirical and deductive reasoning in students' conjecturing, exploring, and justifying is discussed in the following section.

Students' Abilities to Reason Deductively

Often students create conjectures based on the appearance of a drawing, but to prove the conjecture the student must attend to the theoretical properties. A student's ability to reason about the drawing or theoretical properties may be related to the proof scheme that the student brings to the situation. According to Harel and Sowder (1998), a proof scheme refers to what a student finds as convincing and persuading for herself or himself. Proof schemes can be organized in three general categories: external conviction proof schemes, empirical proof schemes, and analytical proof schemes. To illustrate the differences among the three different categories, consider the following problem: Given triangle *ABC* with midpoints *D, E,* and *F,* prove that the area of triangle *DEF* is one-fourth the area of triangle *ABC* (fig. 15.3).

A student exhibiting an external conviction proof scheme depends on an authority such as a teacher or book, the appearance of an argument, or symbols used in a meaningless way. For example, a student portraying an external conviction proof scheme might accept a proof found on the Internet on the basis of its two-column structure and the perceived authority of the Web site rather than the logical construction of the argument presented. Empirical proof schemes rely on evidence from examples or measurements or perceptions. A student exhibiting an empirical proof scheme might assume that triangle *ABC* in figure 13.3 is isosceles on the basis of the appearance of the drawing or measurements and might construct an argument regarding the area of triangle *DEF* based on measurements produced using interactive geometry software. Analytical proof schemes include deductive reasoning and reasoning about mathematics as an axiomatic system. A student exhibiting an analytical proof scheme might attempt to prove the four triangles congruent as a way to prove that the area of triangle *DEF* is one-fourth the area of triangle *ABC*.

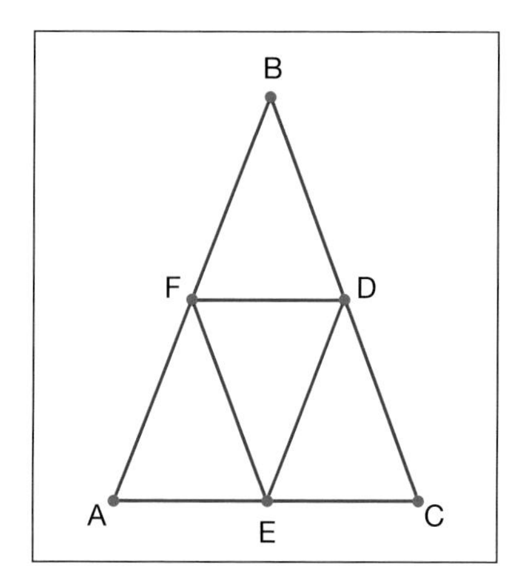

Fig. 15.3. Triangle *ABC* and midpoint triangle *DEF*

According to Harel and Sowder (1998), external conviction or empirical proof schemes may be valid for certain situations. External conviction proof schemes may be warranted for situations in which the content of an argument is outside the area of expertise of the individual. Empirical proof schemes can also be beneficial because the use of examples guides and enriches the students' knowledge of a topic. However, problems arise when a formal mathematical proof is required and the student believes that the validation of multiple examples is sufficient.

Several researchers have examined the effect of students' uses of interactive geometry software on their proof schemes (Galindo et al. 1997). Galindo and his colleagues studied high school students' predominant proof schemes when solving problems using The Geometer's Sketchpad in classes that used the software as a regular part of classroom instruction. When two students were completing the tasks at the beginning of the year, they showed a preference for working with paper and pencil, and they exhibited some external conviction proof schemes. At the end of the year, these students consistently exhibited empirical proof schemes. The authors suggest that the use of the software helped move students from a reliance on authoritarian proof schemes toward a reliance on measurements and examples. The study by Galindo and colleagues provides evidence of the difficulty in moving students from empirical proof schemes toward more deductive means of reasoning. This difficulty has also been noted by other researchers investigating the use of geometry software programs without dynamic capabilities (e.g., Geometric Supposer, Schwartz and Yerushalmy [1986]) on students' beliefs about evidence and proof (Chazan 1993a; Edwards 1997).

High school students' understandings of the differences between empirical evidence and deductive proof were investigated by Chazan (1993a). The students in this study were using an early version of the Geometric Supposer that did not have dynamic capabilities. Justifications that students in Chazan's study offered to support their conjectures suggested that they believed that empirical evidence is sufficient proof or that proof is simply evidence to support a particular case. On the basis of his findings, Chazan recommended that students should understand proof as a means to persuade others of the validity of a mathematical statement, that proofs should be valued for their explanatory power, and that they should not be considered an end in and of themselves. He also suggested that teachers should provide activities that require students to confront their beliefs about evidence and proof (Chazan 1993b).

Prior to writing a formal mathematical proof, students should engage in activities using interactive geometry software that include noticing patterns and invariants, describing those patterns informally and formally, conjecturing, checking examples or cases, and creating a deductive argument for a single case or example. Edwards (1997) refers to this collection of activities as part of the "territory before proof." Students in Edwards's study were asked to investigate geometric transformations using a microworld by participating in all the aforementioned activities except deductive reasoning. Edwards found that as students worked with the microworld on specific examples, they were able to notice patterns that they believed were generalizable. However, students seemed satisfied to test their conjecture with a single example, and so they did little inductive and even less deductive reasoning. Students were also less willing to reformulate a conjecture or to look for disconfirming evidence. Similar to the conclusions of Chazan, Edwards's findings suggested that teachers engage students in thinking about how certainty is established when technology is present. Teachers, researchers, and curriculum developers should be aware of those findings and consider them when developing lessons and activities whose purpose is to foster geometric reasoning using interactive geometry software.

Students' Understanding of the Role of Proof

Some researchers have focused on the development of instructional activities for the purpose of introducing or fostering deductive reasoning and proof (Sanchez and Sacristan 2003). These activities are designed to promote students' understandings of the need for, and roles of, proof. Building on students' transition from drawing to constructing, Jones (2000) described an instructional unit on the classification of quadrilaterals that was used to transition students to formal proofs. Students were challenged to reproduce a drawing that could not be "messed up" and, in some instances, that satisfied additional conditions. After completing the construction, students had to explain why they believed that the construction was created correctly. Explanation in these tasks foreshadows proof because a student's response must include a description of the conditions needed to ensure that the construction is the expected type of quadrilateral.

The use of interactive geometry software seems to have shifted the focus from examining students' ability to write formal proofs to a focus on students' understanding of the role and purpose of proofs (Hanna 1998). This view of proof as explaining empirical evidence as opposed to proof as resolving uncertainty was introduced and developed by de Villiers (1991). In his analysis of mathematicians' work, he found that in practice, mathematicians often do not begin the process of proving a statement until they have some inner conviction of the truth of the statement. At the secondary and collegiate levels, the use of interactive geometry software may be useful initially in providing a context for students to convince themselves of the validity of a mathematical statement. Then, by using axioms, postulates, theorems, and definitions, they can write a formal justification that helps explain the phenomena observed in the computer environment.

Some researchers have investigated how the use of features of a dynamic software program for geometry, such as adding only a few tools at a time, affects students' understandings of proof. Mariotti (2000) reports on a long-term teaching experiment in which the system of axioms and theorems is constructed by students themselves and in which the students do not at the beginning of the course have any commands available to them in the software program. Proof is used to justify the inclusion of a new command, and students must explain why it will do the job it was designed to do. In one example (see fig. 15.4), the students constructed the angle-bisector command by constructing a circle centered at the vertex, *A,* and then constructed congruent circles centered at *B* and *C,* the intersection points of circle *A* and the rays of angle *A*. The students verified that the ray constructed from *A* through *D,* the intersection of the congruent circles *B* and *C,* is the angle bisector by dragging parts of the angle to verify that the ratio between the measures of the large angle and each of the two smaller angles remained two to one. The construction of the commands of the system is similar to the construction

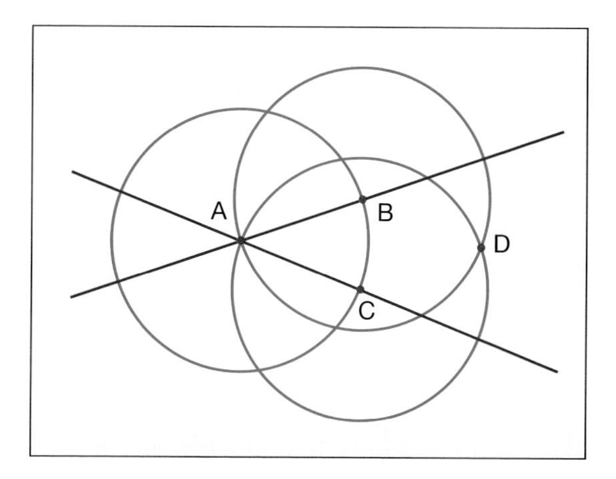

Fig. 15.4. A student's construction of an angle-bisector command (Mariotti 2002)

of theorems. The teacher facilitated the establishment of classroom norms that required every student to defend his or her construction in front of the class. Proof thus fulfils two roles: establishing the validity of a construction for each individual and convincing other students to accept the construction process.

Another method to help students learn about the important role of proof is to present them with tasks that may produce unexpected or surprising results that they then try to explain through proof. Hadas, Hershkowitz, and Schwarz (2000) carefully designed two sequences in which the order of the tasks led students to develop expectations that turned out to be incorrect when they attempted to validate them with interactive geometry software. This outcome created a cognitive conflict and generated students' curiosity about why this unexpected result was true. Although presentation of the same sequence of tasks to students who are not using the computer may seem possible, the unexpected result would not be possible without the software because the false conjectures came only in the context of other valid conjectures developed within the technological environment. In other instances the technological environment gave a counterexample for an expected result. This interplay of conjectures and checks, of certainty and uncertainty, was made possible by the exploration power and checking facilities unique to interactive geometry software.

Studies such as those of Hadas, Hershkowitz, and Schwarz (2000), Jones (2000), and Mariotti (2000) illustrate the usefulness of interactive geometry software in breaking down the traditional separation between actions on drawings and making deductions about properties of theoretical geometrical objects. Because the behavior of a construction created using the software is determined by the geometrical relationships used in its creation, students may abstract those relationships through their interactions with the technology. Those interactions provide students with feedback that can guide their understanding of properties and relationships of the geometrical objects. Each of the three studies showed improvements in students' abilities to construct and appreciate the role of proof. However, the aspect that made these results possible was not simply the addition of a computer. Rather, it was the use of the computer along with a sequence of carefully designed tasks implemented by a skilled teacher who was able to establish norms in the classroom that encourage students to construct justifications to account for empirical phenomena.

Conclusion

For more than twenty years, teachers have used interactive geometry software to facilitate students' understandings of abstract geometric objects and their associated properties as well as promote their development of deductive reasoning. Research in this field may inform us about the strategies students use, interpretations students make, and formal reasoning students engage in while using this technology to learn geometry. In particular, it sheds light on the ways in which students may use the tool proactively or reactively, how they reason about drawings and figures, and how interactive geometry software may influence students' proof schemes. Although these programs have been shown to have a positive impact on students' understanding of geometry and reasoning ability, the implementation of these tools and the nature of the tasks selected by teachers are crucial factors in determining their effects on students' learning.

REFERENCES

Chazan, Daniel. "High School Geometry Students' Justification for Their Views of Empirical Evidence and Mathematical Proof." *Educational Studies in Mathematics* 24 (December 1993a): 359–87.

———. "Instructional Implications of Students' Understandings of the Differences between Empirical Verification and Mathematical Proof." In *The Geometric Supposer: What Is It a Case Of?* edited by Judah Schwartz, Michal Yerushalmy, and Beth Wilson, pp. 107–16. Hillsdale, N.J.: Lawrence Erlbaum Associates, 1993b.

de Villiers, Michael. "Pupils' Needs for Conviction and Explanation within the Context of Geometry." In *Proceedings of the Fifteenth Conference of the International Group for the Psychology of Mathematics Education (PME),* vol.1, edited by Fulvia Furinghetti, pp. 255–62. Assisi, Italy: Program Committee of the Fifteenth PME Conference, 1991.

Edwards, Laurie D. "Exploring the Territory before Proof: Students' Generalizations in a Computer Microworld for Transformation Geometry." *International Journal of Computers for Mathematical Learning* 2 (October 1997): 187–215.

Galindo, Enrique, Gudmundur Birgisson, Jean-Marc Cenet, Norm Krumpe, and Mike Lutz. "The Development of Students' Notions of Proof in High School Classes Using Dynamic Geometry Software." In *Proceedings of the Eighteenth Annual Meeting, North American Chapter of the International Group for the Psychology of Mathematics Education (PME)* (Bloomington, Ill., October 1997), edited by John A. Dossey, Jane O. Swafford, Marilyn Parmantie, and Anne E. Dossey, pp. 207–14.Columbus, Ohio: ERIC/CSMEE, 1997.

Hadas, Nurit, Rina Hershkowitz, and Baruch B. Schwarz. "The Role of Contradiction and Uncertainty in Promoting the Need to Prove in Dynamic Geometry Environments." *Educational Studies in Mathematics* 44, nos. 1–2 (December 2000): 127–50.

Hanna, Gila. "Proof as Explanation in Geometry." *Focus on Learning Problems in Mathematics* 20 (Spring 1998): 4–13.

Harel, Guershon, and Larry Sowder. "Students' Proof Schemes: Results from Exploratory Studies." In *Research in Collegiate Mathematics Education III,* edited by Alan H. Schoenfeld, Jim Kaput, and Ed Dubinsky, pp. 234–83. Providence, R.I.: American Mathematical Society, 1998.

Healy, Lulu, Reinhard Holzl, Celia Hoyles, and Richard Noss. "Messing Up." *Micromath* 10, no. 1 (Spring 1994): 14–16.

Hollebrands, Karen F. "The Role of a Dynamic Software Program for Geometry in the Strategies High School Mathematics Students Employ." *Journal for Research in Mathematics Education* 38, no. 2 (March 2007): 164–92.

Hollebrands, Karen, Colette Laborde, and Rudolf Straesser. "Technology and the Learning of Geometry at the Secondary Level." In *Research on Technology and the Teaching and Learning of Mathematics: Syntheses, Cases, and Perspectives:* Vol. 1, *Research Syntheses,* edited by M. Kathleen Heid and Glendon W. Blume, pp. 155–205. Greenwich, Conn.: Information Age, 2008.

Hoyles, Celia, and Richard Noss. "Dynamic Geometry Environments: What's the Point?" *Mathematics Teacher* 87, no. 9 (December 1994): 716–17.

Jackiw, Nicholas. The Geometer's Sketchpad. Version 4.0. Software. Emeryville, Calif.: Key Curriculum Press, 2001.

Jones, Keith. "Providing a Foundation for Deductive Reasoning: Students' Interpretations When Using Dynamic Geometry Software and Their Evolving Mathematical Explanations." *Educational Studies in Mathematics* 44, nos. 1–2 (December 2000): 55–85.

Laborde, Collette. "The Computer as Part of the Learning Environment: The Case of Geometry." In *Learning from Computers: Mathematics Education and Technology,* edited by Christine Keitel and Kenneth Ruthven, pp. 48–67. Berlin: Springer-Verlag, 1993.

———. "A New Generation of Diagrams in Dynamic Geometry Software." Paper presented at the Eighteenth Conference of the North American Chapter of the International Group for the Psychology of Mathematics Education, Panama City, Florida, 1996.

Laborde, Collete, and Jean-Marie Laborde. "What about a Learning Environment Where Euclidean Concepts Are Manipulated with a Mouse?" In *Computers and Exploratory Learning,* edited by Andrea A. diSessa, Celia Hoyles, Richard Noss, and Laurie D. Edwards, pp. 241–62. Berlin: Springer-Verlag, 1995.

Laborde, Jean-Marie, and Franck Bellemain. Cabri 2. Software. Temple, Tex.: Texas Instruments, 2005.

Mariotti, Maria A. "Introduction to Proof: The Mediation of a Dynamic Software Environment." *Educational Studies in Mathematics* 44, nos. 1–2 (December 2000): 25–53.

Sanchez, Ernesto, and Ana I. Sacristan. "Influential Aspects of Dynamic Geometry Activities in the Construction of Proofs." In *Proceedings of the Joint Meeting of International Group for the Psychology of Mathematics Education and North American Chapter,* vol. 4, edited by Neil A. Pateman, Barbara J. Dougherty, and Joseph. T. Zilliox, pp. 111–18. Honolulu, Hawaii: Curriculum Research and Development Group, College of Education, University of Hawaii, 2003.

Schwartz, Judah L., and Michal Yerushalmy. The Geometric Supposer series. Software. Pleasantville, N.Y.: Sunburst Communications, 1986.

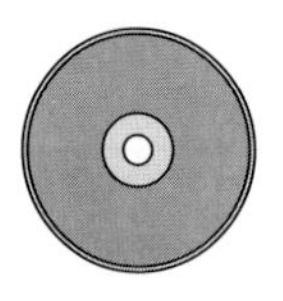

Video clips that support this article are found on the CD-ROM disk accompanying this Yearbook.

Representing, Modeling, and Solving Problems in Interactive Geometry Environments

José N. Contreras
Armando Martinez-Cruz

PAST and current calls for reform in mathematics education advocate that problem solving be an important, if not the central, goal of the mathematics curriculum (e.g., NCTM 1989, 2000). Unfortunately, achieving this goal has been like searching for the Lost Ark. The difficulty, in part, resides in the fact that problem solving is a complex process rather than a step-by-step procedure. Problem solving involves a task for which the solution is not immediately attainable. The problem solver needs to draw on his or her knowledge and problem-solving strategies as he or she struggles to find what sometimes seems the elusive solution. To facilitate the process of solving complex problems, one should use any available resource. Some of the powerful resources that we have now include interactive geometry software, such as The Geometer's Sketchpad (GSP) (Jackiw 2001) or Cabri Geometry II (Texas Instruments 1998). Interactive geometry software facilitates the construction of geometric configurations as well as the execution of computations and repetitive tasks. These features, along with dragging capabilities, support geometrical problem solving greatly. As a consequence, the technology promotes and facilitates conjecturing, gaining insight, discovering, exploring, and investigating.

This article illustrates how learners can use interactive geometry software to represent, model, and solve geometric problems dynamically. Doing so, in turn, enhances, fosters, motivates, and facilitates solving complex problems. We perform our classroom investigations with GSP using computers, but any other type of interactive geometry software including programs available on calculators can be used.

This article consists of two main sections. In the first section we describe some of the problem-solving situations that preservice and practicing secondary school mathematics teachers have investigated using interactive geometry software during regular classroom instruction and professional development seminars. These investigations include such classics as solving Viviani's and Heron's problems and finding the Fermat point for a triangle or quadrilateral. In the second section we describe some of the knowledge, behaviors, and dispositions we have observed that students may develop as a result of performing these problem-solving investigations with interactive geometry software.

A Sample of Problem-Solving Activities Performed in Interactive Geometry Environments

Problem-Solving Activity 1: Gamow's Hidden Treasure Problem

Our first example illustrates an interesting problem that preservice and practicing teachers have found very enjoyable. This problem is related in Gamow (1947). Our version follows:

> A young and adventurous man finds some directions to locate a hidden treasure in a remote, isolated, and deserted island. The directions are as follows:
>
> In the center of the island, close to the dry lake, there is a lonely oak, a lonely pine, and an old gallows. Walk from the gallows to the oak and count the steps. At the oak, make a 90° turn to the right. Walk the same number of steps and put a spike in the ground. Return to the gallows. Walk towards the pine counting the steps. At the pine, make a 90° left turn and take the same number of steps. Put a second spike in the ground. Behold! The treasure is halfway between the spikes.
>
> The man found the island but to his dismay the gallows was gone. Frantically, he began digging everywhere randomly but without any luck. Finally, he gave up and returned home with empty hands, so the treasure may still be there.

Before reading on, we encourage the reader to solve this problem using any type of interactive geometry software. The solution is a delightful surprise! Figure 16.1 displays the geometric configuration for this problem. Notice that points A, B, and C are draggable, whereas the point representing the treasure cannot be dragged. Since the positions of the pine and oak are fixed although the position of the gallows is unknown, we drag the point that represents the gallows. As we drag point A, we discover an amazing result: *No matter where the gallows was located, the location of the treasure is invariant*. Changing the position of points B and C further confirms our conjecture about the independence of the location of the treasure with respect to the position of the gallows. Here is a coordinate proof of this fact:

> Without loss of generality, we can take line BC as the x-axis and take the midpoint of segment BC as the origin as indicated in figure 16.2. If $A = (x, y)$ and $B = (a, 0)$, then $C = (-a, 0)$. Using congruence of triangles, we deduce that $D = (y - a, -x - a)$ and $E = (-y + a, x - a)$. Using the midpoint formula, we conclude that the coordinates of F are $(0, -a)$, thereby showing that the coordinates of F are independent of the coordinates of A.

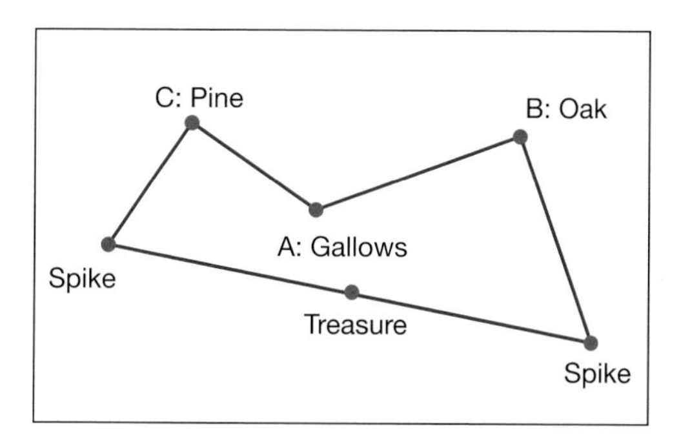

Fig. 16.1. Treasure problem

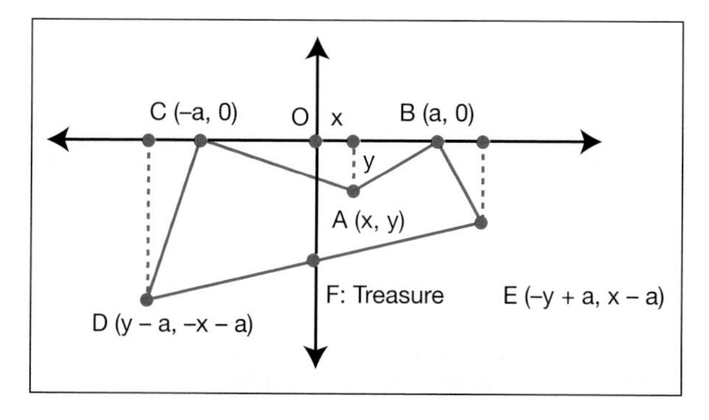

Fig. 16.2. Coordinatization of treasure problem

Problem-Solving Activity 2: Fermat's Problem for Triangles

Fermat's problem for triangles can be paraphrased using an applied context as follows:

> Three towns—Armon (*A*), Betania (*B*), and Calista (*C*)—are planning to build an airport to serve the three cities. To keep costs at a minimum, the airport will be constructed at a place where the sum of its distances to each of the cities is minimal. Describe the minimum distance point, called the *Fermat point,* for the location of the airport.

Figure 16.3 displays a configuration for Fermat's problem for triangles. In this figure all points *A, B, C,* and *D* can be dragged. As we drag point *D* to find an optimal point, we get the configuration displayed in figure 16.4. Point *D* seems to be a point that minimizes the sum of its distances to each of the vertices of the given triangle. A cursory examination of the diagram may not provide a clue as to how to characterize this optimal point; however, we can use the software to measure angles *ADB, BDC,* and *ADC* (fig. 16.4). We then make an amazing discovery: the Fermat point seems to be the point that is the vertex of three congruent adjacent angles whose sides contain the vertices of the original triangle. This point is also called the equiangular point of the triangle.

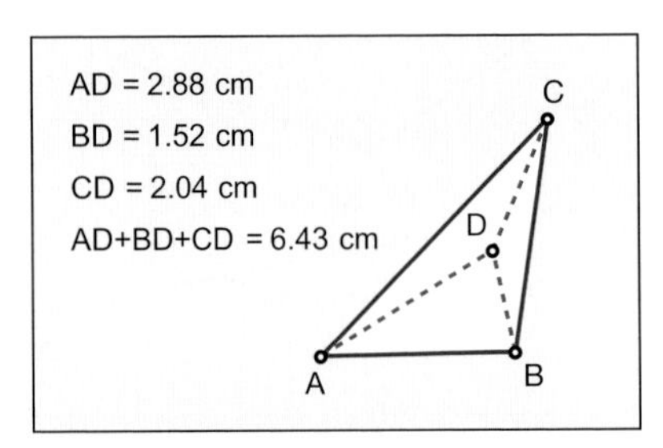

Fig. 16.3. Searching for the Fermat point of a triangle

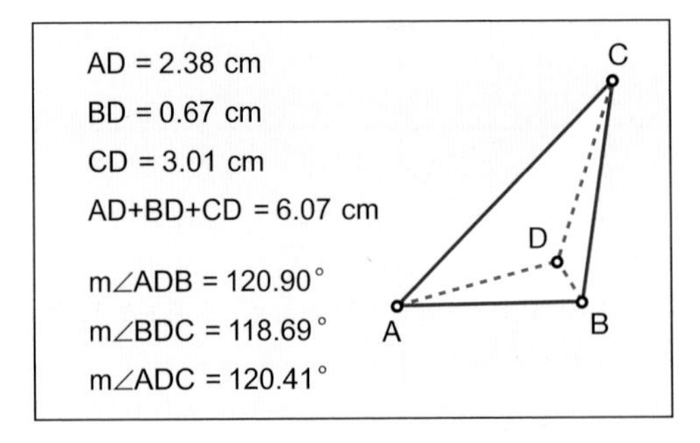

Fig. 16.4. Measuring angles for the Fermat point of a triangle

To verify our conjecture for additional triangles, we need to construct the equiangular point of the given triangle. Because a way to construct such a point is not evident, we provide students with a procedure to construct it (fig. 16.5). First we construct an outward equilateral triangle on each side of the given triangle. We next construct segments from a vertex of the original triangle to the remote vertex of the equilateral triangle constructed on the opposite side. The custom tools or macros of interactive geometry software facilitate this construction tremendously. These three segments are concurrent, and their point of intersection H is the equiangular point of the given triangle. After constructing the equiangular point of the triangle, we can verify that this is the optimal point by dragging D and showing that $AD + BD + CD$ is always greater than or equal to $AH + BH + CH$ for any point D (fig. 16.5). Further dragging of points A, B, and C supports our conjecture for additional triangles. However, as we keep testing our conjecture for other triangles, we may realize that the equiangular point seems to disappear for some triangles (fig. 16.6). Again, the dragging and measuring capabilities of the software allow us to discover that the optimal point for triangles with an interior angle measuring 120 degrees or more is the vertex of such an angle (figs. 16.7 and 16.8).

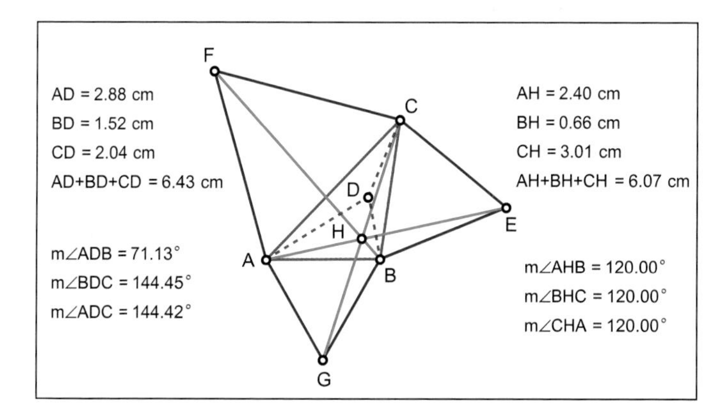

Fig. 16.5. Constructing the Fermat point of a triangle

In summary, interactive geometry software helped us find a tentative solution to a complex problem: the Fermat point of a triangle with no interior angle measuring 120 degrees or more is the equiangular point. The Fermat point of a triangle with an interior angle measuring 120 degrees or more is the vertex of that angle. A beautiful proof of the first result is due to J. E. Hofmann (Coxeter 1969). The proof is along the following lines:

> Let F be any arbitrary point with distances to the vertices of a triangle ABC as indicated in figure 16.9. Let F' and C' be the images of points

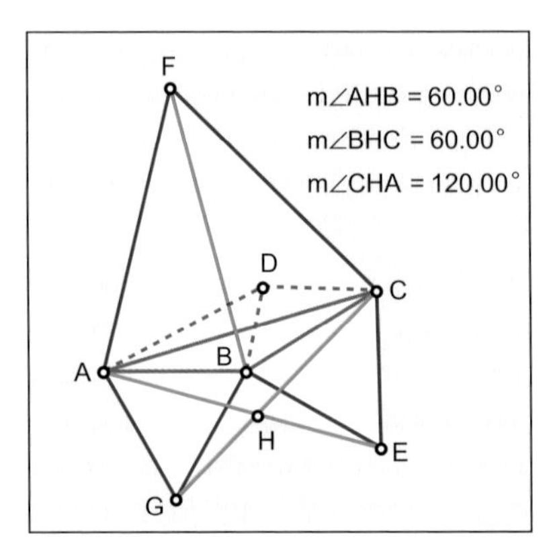

Fig. 16.6. The equiangular point does not exist for some triangles.

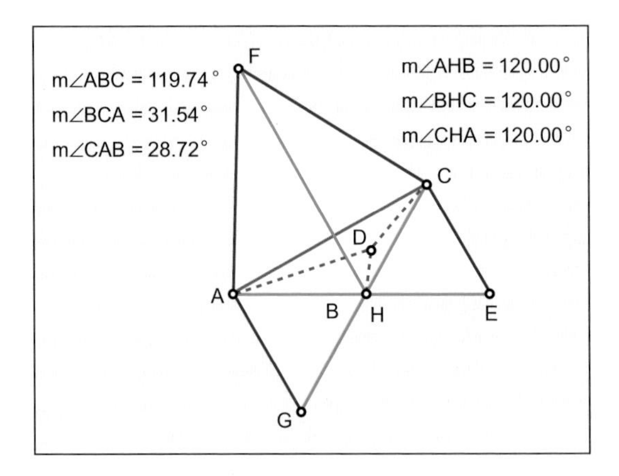

Fig. 16.7. The Fermat point of a triangle with an angle measuring 120 degrees

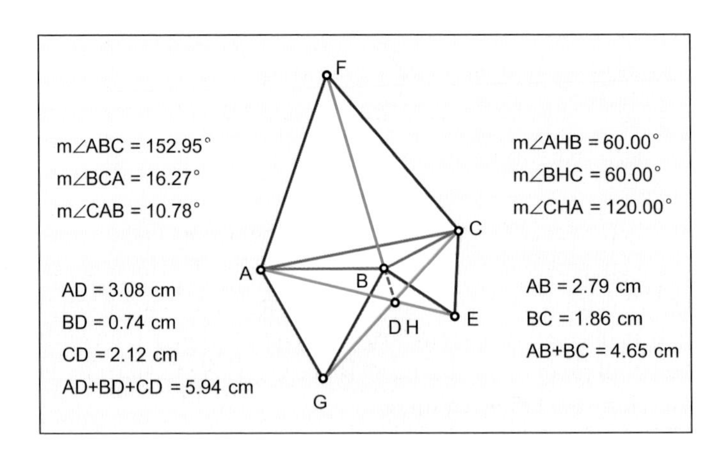

Fig. 16.8. The Fermat point of a triangle with an angle measuring more than 120 degrees

F and C when these points are rotated 60 degrees clockwise around point B. Hence triangles BFF' and BCC′ are equilateral. Therefore $AF + BF + CF = AF + FF' + F'C'$ because $BF = FF'$ and $CF = C'F'$. Minimizing $AF + BF + CF$ is equivalent to minimizing $AF + FF' + F'C'$. But the path $AF + FF' + F'C'$ is minimal when it is a straight line (fig. 16.10). Thus $m\angle AFB = 120$ degrees because $m\angle BFF' =$ 60 degrees. Similarly, $m\angle C'F'B = 120$ degrees. Therefore, $m\angle CFB =$ 120 degrees because $\triangle C'F'B \cong \triangle CFB$. As a consequence, $m\angle AFC = 120$ degrees. This construction is valid for any triangle with no interior angle greater than 120 degrees.

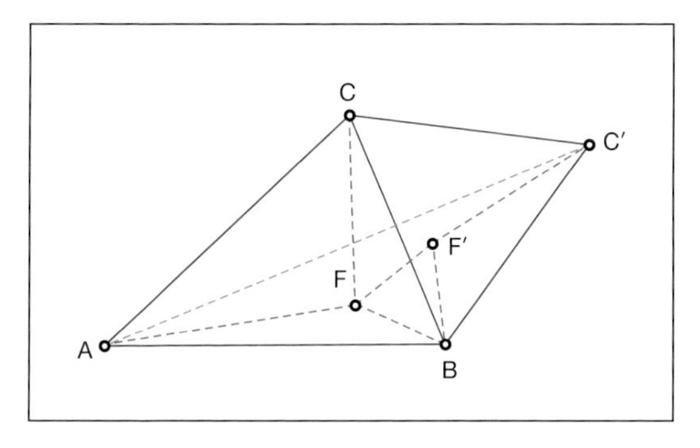

Fig. 16.9. Searching for the optimal solution

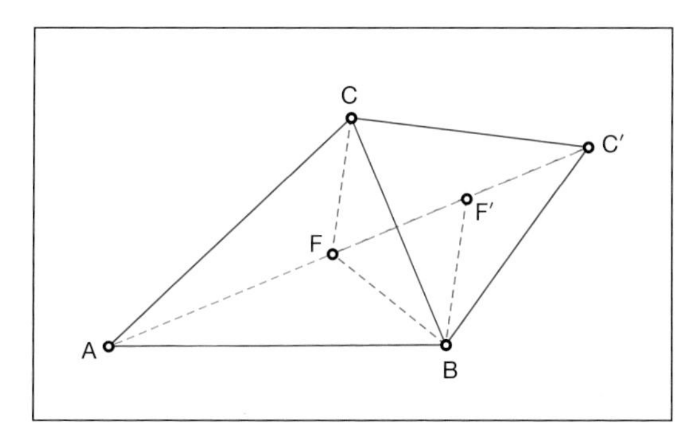

Fig. 16.10. The optimal solution

Hofmann's argument can be modified for the cases when one of the interior angles measures 120 degrees or more. We leave the proof to the reader.

Figure 16.10 can be used to justify the construction we made in figure 16.5. Because F lies on the segment AC', whose length equals the minimum distance, we need to construct point C', the remote vertex of equilateral triangle BCC'. Likewise F lies in each of the other two segments having as endpoints a vertex of the original triangle and the remote vertex of the equilateral triangle constructed on the opposite side (F is point H in fig. 16.5). Thus the three segments AE, BF, and CG in figure 16.5 are concurrent at H, and *because their lengths equal the minimal distance, they are congruent as well.*

Figure 16.10 also suggests an alternative construction of the equiangular point F. Because angles CFB and $CC'B$ are supplementary, points B, C', C, and F are *concyclic;* that is, F lies on the *circumcircle* of triangle $BC'C$. Point F can then be constructed as the intersection of segment AC' and the circumcircle of triangle $BC'C$. Alternatively, F can also be constructed as the point of intersection of the circumcircles of two of the outward equilateral triangles constructed on two of the sides of triangle ABC. In addition, segments AE, BF, and CG (fig. 16.5) are congruent; a proof is based on the congruence of appropriate triangles in figure 16.5. For example, $\triangle BCF \cong \triangle ECA$ by the side-angle-side congruence criterion.

Problem-Solving Activity 3: Fermat's Problem for Quadrilaterals

> Fermat's problem for quadrilaterals can be stated as follows:
>
> Four towns—Athenea (A), Belico (B), Casat (C), and Delore (D)—want to build an electric-power-generating plant. The four towns are coplanar and are the vertices of a quadrilateral. To keep the costs of the power cables at a minimum, the plant will be constructed at a place where the sum of its distances to each of the four towns is minimal. Describe the optimal place for the location of the plant.

Figure 16.11 displays the configuration for Fermat's problem for quadrilaterals. Notice that all points (A, B, C, D, and E) can be dragged. As we drag point E to locate its optimal position, we get the configuration shown in figure 16.12.

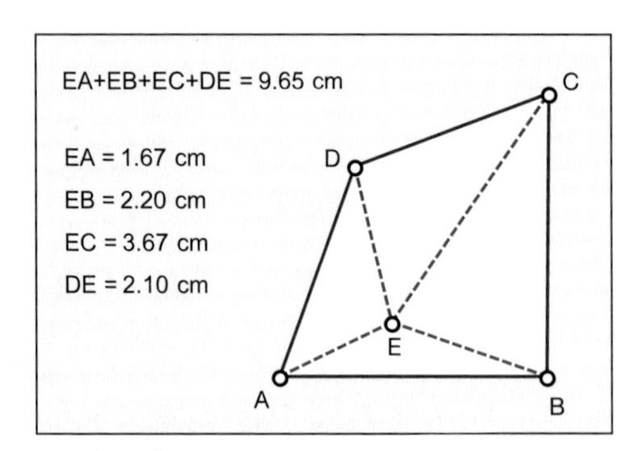

Fig. 16.11. Searching for the Fermat point of a quadrilateral

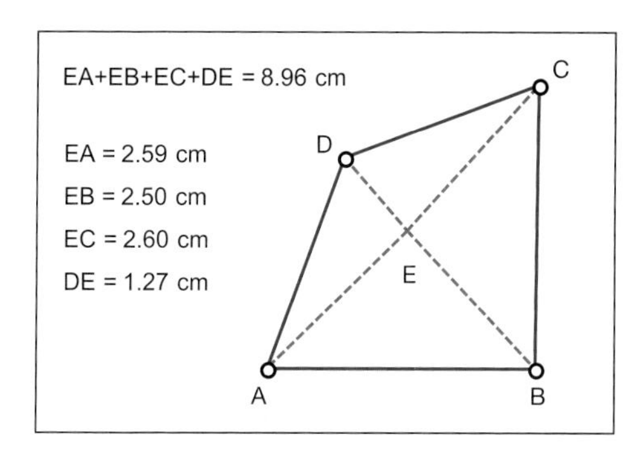

Fig. 16.12. The Fermat point of a convex quadrilateral

As figure 16.12 suggests, the optimal point seems to be the point of intersection of the diagonals. To check our conjecture for other quadrilaterals, we need to construct Fermat's point, *F*, as the point of intersection of the diagonals and drag point *E* (fig. 16.13). Additional dragging of points *C* (or any other vertex) and *E* provide more experimental evidence that Fermat's point is the intersection of the diagonals of a quadrilateral. However, as we keep dragging one of the vertices of the quadrilateral—say, *C*—we realize that the point of intersection of the diagonals disappears for nonconvex quadrilaterals (fig. 16.14). When we perform this investigation with our students, we ask them to characterize for what types of quadrilaterals their diagonals do not intersect.

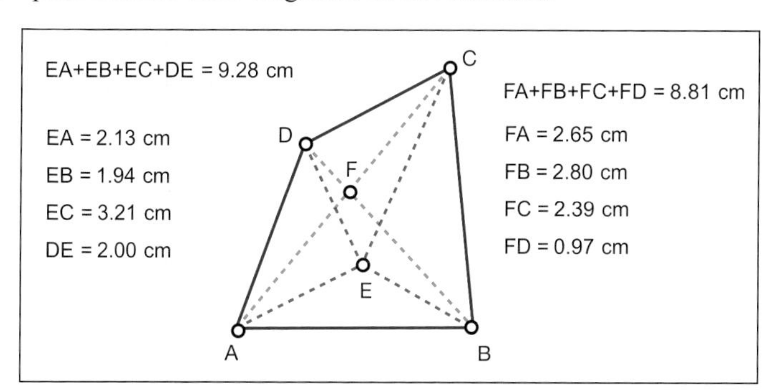

Fig. 16.13. Verifying that *F* is the Fermat point of a convex quadrilateral

Again, the location of Fermat's point for nonconvex quadrilaterals can be investigated using interactive geometry software. As we drag point *E*, we discover that Fermat's point is the vertex of the quadrilateral with a reflex interior angle (fig. 16.15).

Here is a proof of these results:

Let *ABCD* be a convex quadrilateral, *E* an arbitrary point, and *F* the intersection of the diagonals (fig. 16.13). $EA + EC > AC$ and $EB + ED >$

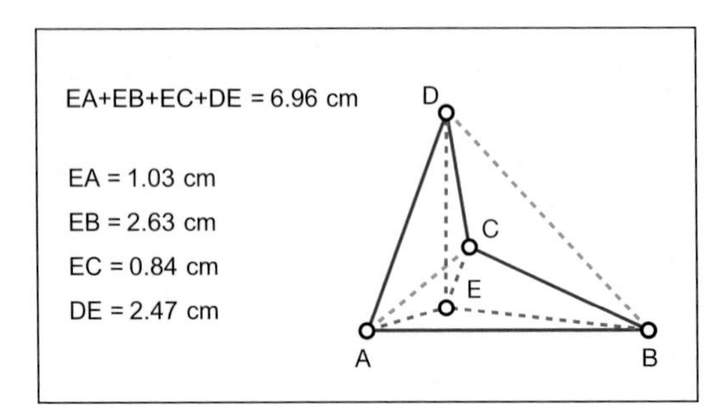

Fig. 16.14. The diagonals of a nonconvex quadrilateral do not Intersect.

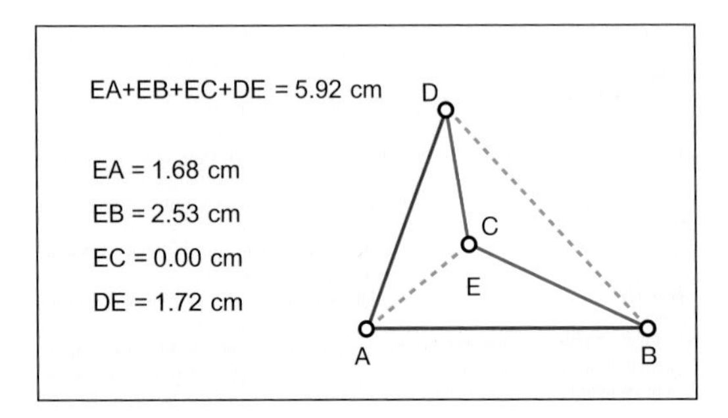

Fig. 16.15. The Fermat point of a nonconvex quadrilateral

BD by the triangle inequality. Adding both inequalities, we obtain $EA + EB + EC + ED > AC + BD$, which is what we wanted to prove. For the nonconvex case, let F be the point of intersection of lines BC and DE (fig. 16.16). Applying the triangle inequality twice, we get $BE + DE > BF + FD > BC + CD$. Furthermore, $AE + CE > AC$. These two inequalities imply that $AE + BE + CE + DE > BC + DC + AC$, showing that C is the Fermat point. This argument can be adapted for different positions of point E.

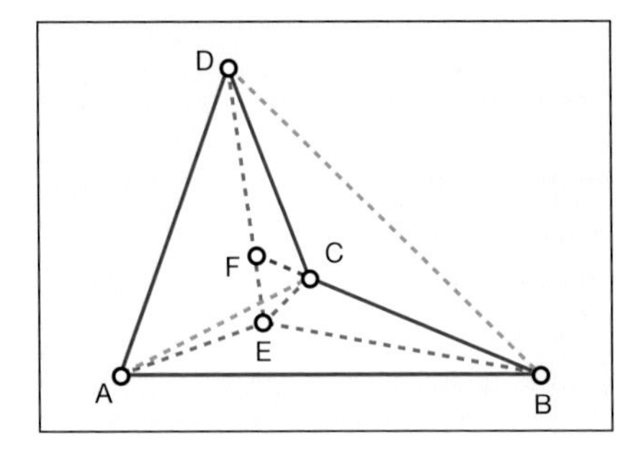

Fig. 16.16. Proving that *C* is the Fermat point for a nonconvex quadrilateral

Problem-Solving Activity 4: A System of Roads for a Square

Our next problem is a variation of Fermat's problem. It also illustrates the power of interactive geometry software to investigate the solution.

> There are four main towns in Lateralia. We will call them *A, B, C,* and *D*. They lie at the corners of a ten-mile square. In order to improve communication between the towns, the Lateralian Department of Transport decided to build a new road linking all four towns. Because the towns have very little money, it was decided that the new road system should be as short as possible and still allow access from any one town to any other (www.geom.umn.edu/~lori/mathed/problems/sloanA515.html). Propose several designs, and recommend the optimal one.

A different version of this problem can also be found in Coxeter (1969, p. 23).

Figure 16.17 displays two of several possible designs produced by our students. They are models constructed to scale (1 in : 10 miles). The first design requires 30 miles, whereas the second one requires about 28.3 miles ($20\sqrt{2}$).

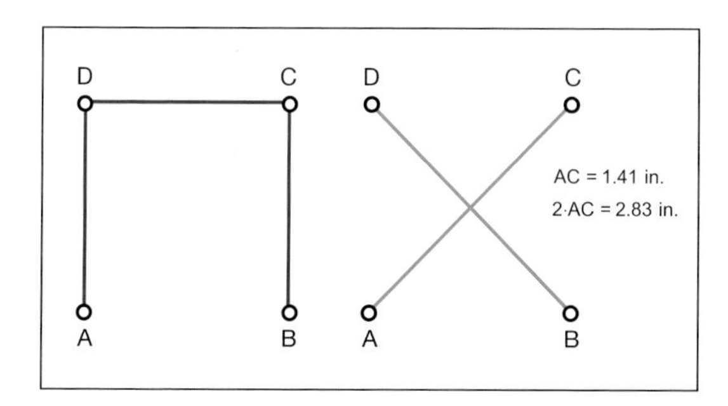

Fig. 16.17. Possible systems of roads for a square

Students who previously studied Fermat's problem for quadrilaterals often offer the construction shown in figure 16.18 to argue that the second design in figure 16.17 is optimal. We then tell students that when the engineers of the transportation department proposed that design, the queen, who was a mathematician, accused them of extravagance and proposed a better, and optimal, solution. The task for students to accomplish is to find the queen's road-system design.

Because a way to find the optimal design is not obvious, we give students a hint: What if the road system has more than one vertex in the square's interior? By dragging points *E* and *F* in figure 16.19, after some trial and error, most students conjecture that a road system with two internal vertices produces a better solution. We then analyze cases in which the road system consists of more than

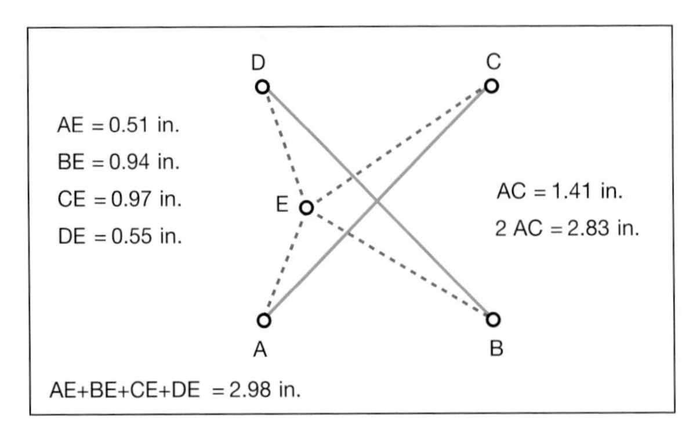

Fig. 16.18. Searching for a possible optimal road system for a square

two internal vertices. The triangle inequality, however, shows that a road system with two vertices is better than a road system with more than two vertices.

Our next task is to construct the solution and to prove that the road system depicted in figure 16.19 is, indeed, a better solution. The symmetry of the solution leads us to conjecture that *E* and *F* belong to the perpendicular bisector of side *AD*. To locate *E* and *F* on the perpendicular bisector of *AD,* we need more information. Again, interactive geometry software comes to our rescue: we can measure angles *ADE* and *BCF* (fig. 16.19). On aesthetic grounds, we conjecture that angles *ADE* and *BCF* each measure 30 degrees. With this information we proceed to construct the solution. To locate *E* and *F,* we need to construct 30-degree angles. First, we construct an equilateral triangle with base *AD* and then the angle bisectors of the 60-degree angles with vertices *A* and *D.* We repeat the process to locate point *F.* We can now prove that this solution with two vertices is better than the one involving the diagonals (fig. 16.18). Trigonometric analysis shows that the length of the network with two vertices is $10 + 10\sqrt{3}$, or approximately 27.3 miles, as opposed to $20\sqrt{2}$, or approximately 28.3 miles. A good

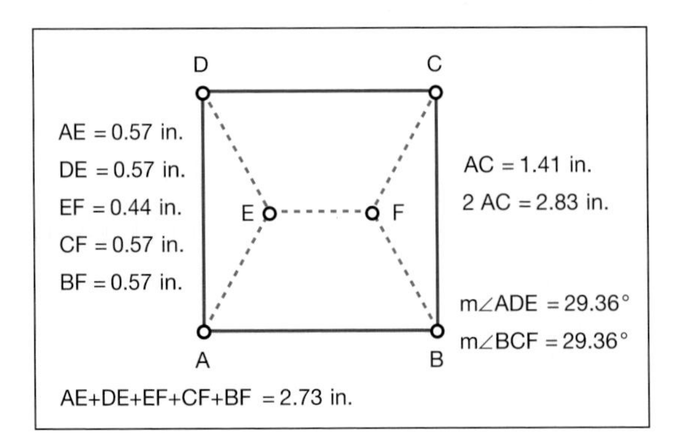

Fig. 16.19. A possible optimal road system for a square

exercise for students is to prove that $10 + 10\sqrt{3} < 20\sqrt{2}$, without using approximate numerical values.

A proof that $\left(1+\sqrt{3}\right)AD$ is the length of the optimal road network can be developed using Hofmann's argument along the following lines:

> In figure 16.20 we want to minimize path $AE + DE + EF + BF + CF$. Rotate points C, F, D, and E 60 degrees about points A and B as indicated in the figure. Because rotations preserve congruence, $DE = D'E'$ and $CF = C'F'$. Because triangles AEE' and BFF' are equilateral, $AE = EE'$ and $BF = FF'$. Minimizing path $AE + DE + EF + BF + CF$ is equivalent to minimizing $D'E' + E'E + EF + FF' + F'C'$, which is a path connecting point C' and D'. Thus, the optimal road network is achieved when points D', E', E, F, F', and C' are collinear (fig. 16.21), from which it can be shown that each of the angles AED, DEF, EFC, and BFC measures 120 degrees. Finally, trigonometric analysis shows that
>
> $$D'C' = \left(\frac{\sqrt{3}}{2} + 1 + \frac{\sqrt{3}}{2}\right)AD = \left(1+\sqrt{3}\right)AD.$$

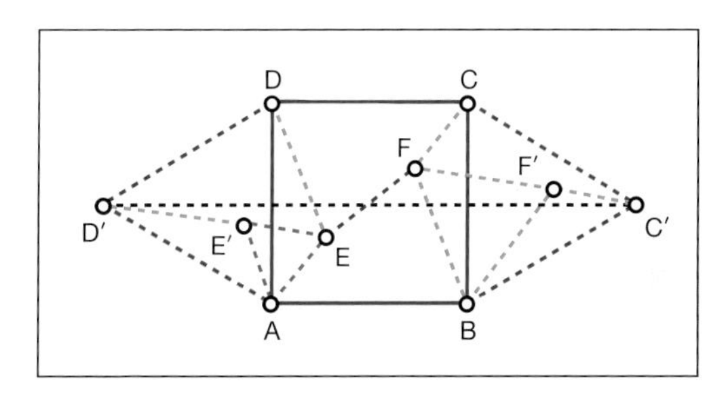

Fig. 16.20. Investigating the optimal solution of the road system for a square

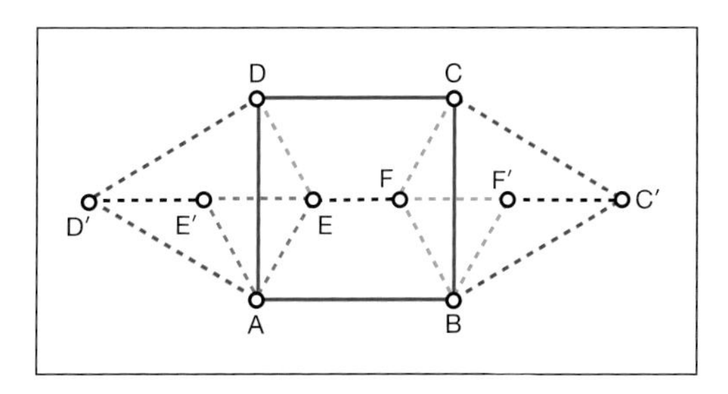

Fig. 16.21. The optimal solution of the road system for a square

Additional information regarding optimal network problems or minimum spanning trees can be found in Bern and Graham (1989), Skiena (1990), and Stewart (1991).

Problem-Solving Activity 5: Viviani's Problem

Viviani's problem can be stated as follows: Locate the set of points for which the sum of their distances to the sides of an equilateral triangle is minimum.

An example of an application of Viviani's problem is the following:

> Three towns are the vertices of an equilateral triangle. The sides of the triangle are the roads that connect the towns. A picnic area will be constructed such that the sum of its distances to the roads is as small as possible.
>
> 1. What are all the possible locations for the picnic area?
> 2. For practical reasons, what is the best location for the picnic area? Justify your response.
> 3. Give a geometric interpretation for the sum of the distances of the optimal point to the sides of the triangle.

Figure 16.22 displays the configuration for Viviani's problem. In this figure points *A, B,* and *P* can be dragged. As we drag point *P* to locate the optimal place, we discover a striking finding: Any point of the interior of the triangle seems to satisfy the criterion (fig. 16.23). Additional dragging confirms our conjecture. At this point a question arises: Are the interior points of the triangle the only points that satisfy the minimum sum of the distances? When *P* is on the sides of the triangle, we realize that also any point of the triangle seems to satisfy the required condition (fig. 16.24).

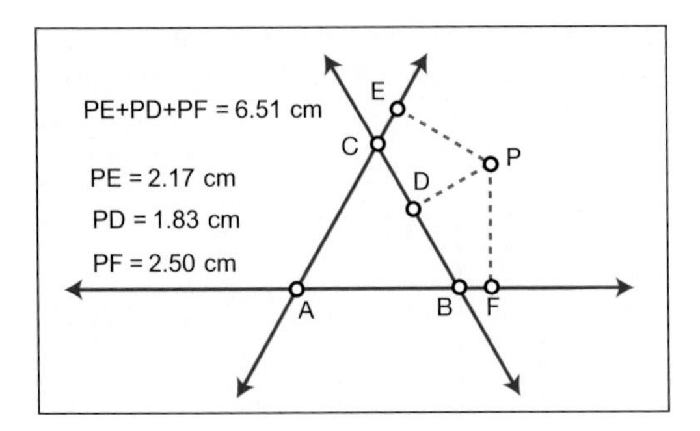

Fig. 16.22. Searching for the optimal location of the picnic area

We can now respond to the first question: Any point in the interior of the triangle or on its sides (the highways) is a potential place to construct the picnic

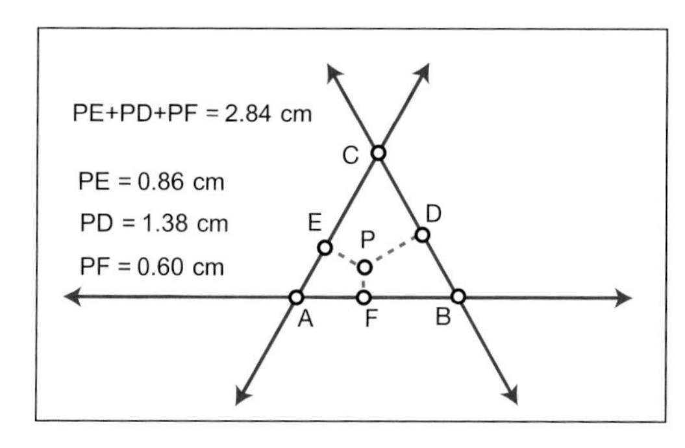

Fig. 16.23. A Possible location of the picnic area

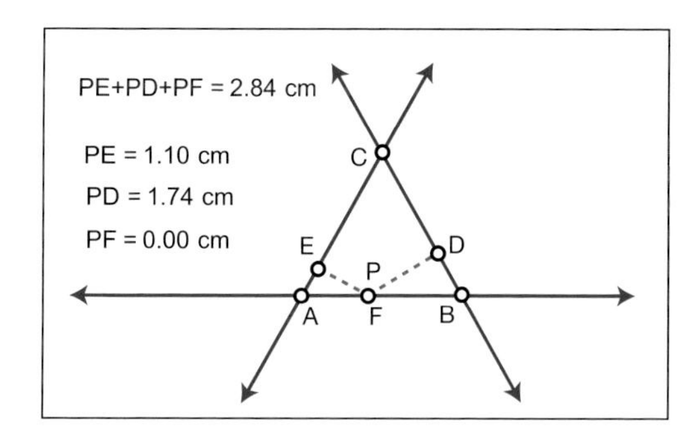

Fig. 16.24. Another possible location of the picnic area

area. However, for practical reasons, we should select points in the interior of the equilateral triangle. Probably the best place would be the center of the equilateral triangle. Because this point is equidistant from the highways and from the towns, the picnic area would be at a central location. To answer the last question, we can also rely on the power of interactive geometry software. As we drag point P close to any of the vertices, we discover that the sum of the distances from any interior point or any point on an equilateral triangle to the sides of the triangle is the height of the triangle (fig. 16.25). Although interactive geometry software does not give us much insight into how to construct a proof, doing so is not difficult, as the following argument shows:

> As indicated in figure 16.23, Area($\triangle ABC$) = Area($\triangle APB$) + Area($\triangle BPC$) + Area($\triangle CPA$) = $AB*PF/2 + BC*PD/2 + CA*PE/2$ = $AB(PF + PD + PE)/2$ because $\triangle ABC$ is equilateral; but Area($\triangle ABC$) = $AB*h/2$, where h is the height of $\triangle ABC$. So we have that $AB*h/2$ = $AB(PF + PD + PE)/2$, which leads us to conclude that $h = PF + PD + PE$.

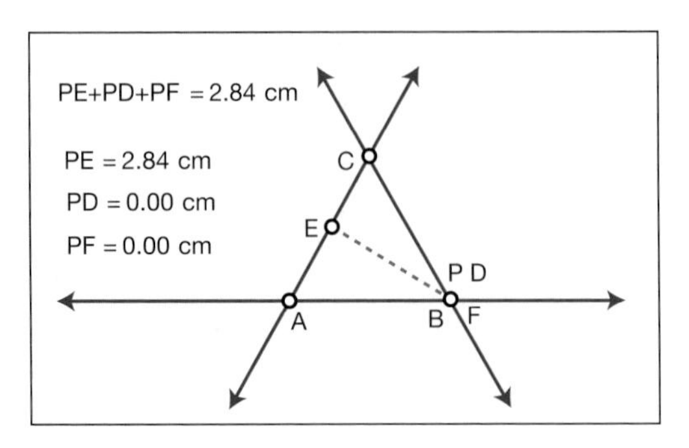

Fig. 16.25. Geometric interpretation of the sum of the distances of the optimal location to the sides of the triangle

We still need to prove that $PD + PE + PF > h$ when P lies on the exterior of $\triangle ABC$ (fig. 16.22). P can be on any of the six exterior regions determined by the extended sides of $\triangle ABC$. For cases when P lies as shown in figure 16.22, we have that $AB*h/2$ = Area($\triangle ABC$) = Area($\triangle ABP$) + Area($\triangle ACP$) – Area($\triangle BCP$) = $AB(PF + PE - PD)/2$ or $h = PF + PE - PD$. Therefore, $PD + PE + PF > h$. Similar arguments can be developed for the other cases.

Problem-Solving Activity 6: Heron's Problem

We state Heron's problem in an applied context:

> Let A and B be two towns on the same side of a railroad track. A railroad platform will be built so that the sum of the distances from it to the two towns is minimal. Where must the railroad platform be built?

Figure 16.26 shows the geometric configuration for this problem. Notice that in this figure all points (A, B, C, D, and E) can be dragged. As we drag point C along the railroad track, we find the location of the optimal point (fig. 16.27). However, a question remains: How do we construct the optimal point such that when we change the location of points A and B, we still have the optimal point? Using the geometry software's measurement tool, students generally conclude that C is the point where angles ACD and BCE are congruent (fig. 16.27). Some students then reason that we need to construct point C such that angles ACD and DCF are congruent, as shown in figure 16.28. Because one method of constructing congruent angles involves constructing congruent triangles, it is natural for some students to construct the perpendicular to DE that goes through A. Because $\triangle AHC \cong \triangle GHC$ by angle-side-angle congruence, we can conclude that $AH = HG$. This conclusion allows determining point C as follows: construct point G on the perpendicular AH such that $AH = HG$ (or equivalently, construct the symmetric point of A with respect to line DE). C is the point of intersection of segment BG and line DE. Our next task is to provide a proof for the conjecture that C is the optimal point.

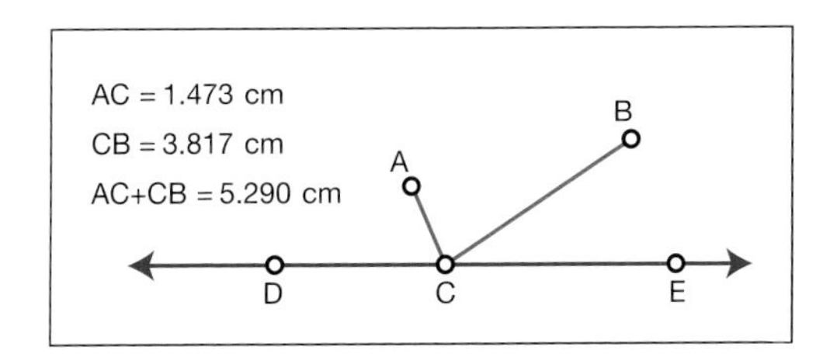

Fig. 16.26. Heron's problem

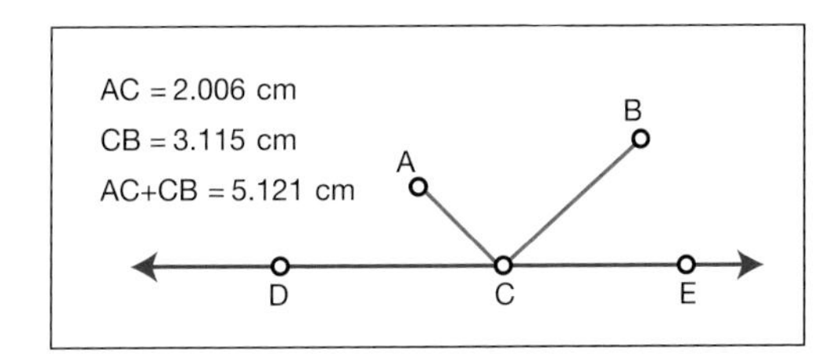

Fig. 16.27. Searching for the optimal location of the railroad platform

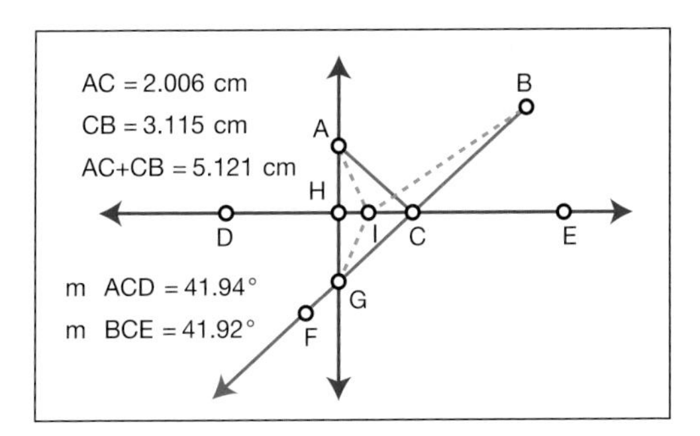

Fig. 16.28. Proving that C is the optimal location of the railroad platform

Let I be another point on DE. $AI = GI$ because $\triangle AHI \cong \triangle GHI$ by side-angle-side congruence. Similarly, $AC = GC$. Hence, $AI + IB = GI + IB > GB$ by the triangle inequality. Using the fact that $GB = GC + CB = AC + CB$, we conclude that $AI + IB > AC + CB$.

Students' Learning in Problem-Solving, Interactive Geometry Environments

In this section we discuss some of the knowledge, behaviors, and dispositions that we have observed that students may develop or continue to develop as a result of representing, modeling, and solving problems in interactive geometry environments. In our experience, we have observed that solving problems with geometry software may enable students to do the following.

- *Use representations to model physical and mathematical relationships.* In using interactive geometry software to model a mathematical relationship embedded in a problem, learners need to translate a verbal statement into an interactive representation (diagram).
- *Apply their knowledge to construct interactive diagrams that model problem situations.* Whereas drawing a static diagram requires understanding the structure of a problem, constructing an interactive diagram requires a deeper understanding of the relationships among the geometric objects involved in the problem situation, as well as knowing how to construct those geometric objects. Thus, interactive geometry diagrams give students opportunities to use or apply their knowledge.
- *Generate, justify, and refine mathematical conjectures related to the solution of the problem.* Once students have represented a problem situation with a dynamic diagram, they can use the software's capabilities (precision, measuring, dragging, and construction) to engage in conjecturing. The activity of conjecturing is not usually a linear process but a cycle of generation, justification, and refinement of a plausible solution.
- *Develop such habits of mind as motivation, disposition, and persistence to solve complex geometric problems.* We have noticed that students are more confident and enthusiastic in solving geometric problems when they are allowed to use interactive geometry software. Constructing a dynamic diagram allows students to explore and get acquainted with the problem, and knowing that the software frees us from time-consuming tasks seems to motivate students to work on a problem for longer periods of time.
- *Model and solve a variety of complex geometric problems.* Representing, modeling, and solving mathematical problems involve understanding the problem; devising a plan or deciding on a method, strategy, or tool; carrying out the plan; and looking back (Pólya 1945). Representing the structure of a problem with a dynamic diagram allows students

to assess for themselves whether they understand the problem. Systematically dragging dynamic vertices leads to plausible conjectures about a solution to the problem. Once students know what the solution seems to be, then they may be better equipped to devise a strategy to prove that a conjecture is indeed correct.

- *Experience the delight of discovering plausible solutions to problems instead of just verifying or proving that a given object is the solution to a problem with a well-defined goal.* Traditional textbook problems are presented to students with a well-defined goal (e.g., show that the point of intersection of the diagonals of a convex quadrilateral is the point that minimizes the sum of its distances to the four vertices of the quadrilateral). The availability of interactive geometry software allows us to modify such problems and reformulate them in a more open-ended way so that students and we, teachers and teacher educators, experience the joy of discovering the solutions to the problems.

When representing, modeling, and solving geometry problems in a dynamic environment, our students, working alone or in pairs in the computer lab, are actively involved in all the stages of the problem-solving cycle. This cycle can be described as follows. First, students are asked to solve a problem and are given enough time to translate the verbal statement into an interactive representation. After most students have constructed the interactive diagram, the instructor, guided by students, represents the problem with a computer connected to a projector. At this point, most students verify the correctness or appropriateness of their diagrams. Second, after students have produced an interactive representation, they use that representation to model the problem to find a plausible solution. Next, they refine and justify their conjectures experimentally by constructing, measuring, and dragging as appropriate to the problem situation. Students validate or refute their own conjectures and thoughts during whole-class discussions. In instances when all students are pursuing an incorrect or incomplete conjecture, the instructor poses leading questions to provoke a cognitive conflict or induce awareness of the faulty path (e.g., are you sure the diagonal point is the optimal point for *all* possible types of quadrilaterals? Has somebody tested the conjecture for *all* possible kinds of quadrilaterals?). Usually students themselves resolve their inconsistencies or faulty conjectures at this stage of the problem-solving cycle. Finally, the most challenging task is to provide a mathematical argument to prove that our plausible conjecture is, indeed, the solution to the problem. Again, the instructor allows students to go as far as they can on their own by giving hints or posing leading questions. Even in instances when the instructor guides students to prove an elusive conjecture, students are expected to justify the arguments. The instructor often allows students to pursue their own thoughts and ideas, even if they lead to a dead end.

Conclusion

We have illustrated that working in interactive geometry environments may allow students and teachers to go beyond the memorization and execution of formulas to solve routine geometric problems. Interactive representations are used to model and solve complex problems, that is, problems requiring non-algorithmic approaches. Teaching students and teachers to become better problem solvers is a challenging task, but interactive geometry software may facilitate this task enormously. The technology may also enable more students and teachers to understand, approach, and solve a wider variety of problems than would be possible without it.

We encourage other teachers, teacher educators, and curriculum developers to find or develop problems whose solution can be facilitated by the appropriate use of interactive geometry software. We believe that the technology has helped us become better problem solvers and that it can help all learners, including students, teachers, and teacher educators, do the same. To emphasize, the use of interactive geometry software allows students to "experience the tension and enjoy the triumph of discovery" (Pólya 1980, p. 2).

REFERENCES

Bern, Marshall W., and Ronald Graham. "The Shortest-Network Problem." *Scientific American* 260 (1989): 66–71.

Coxeter, Harold S. M. *Introduction to Geometry*. New York: John Wiley & Sons, 1969.

Gamow, George. *One, Two, Three, … Infinity: Facts and Speculations of Science*. New York: Viking Press, 1947.

Jackiw, Nicholas. The Geometer's Sketchpad. Version 4.0. Software. Berkeley, Calif.: Key Curriculum Press, 2001.

National Council of Teachers of Mathematics (NCTM). *Curriculum and Evaluation Standards for School Mathematics*. Reston, Va.: NCTM, 1989.

———. *Principles and Standards for School Mathematics*. Reston, Va.: NCTM, 2000.

Pólya, George. *How to Solve It*. Princeton, N.J.: Princeton University Press, 1945.

———. *Mathematical Discovery*. 2 vols. New York: John Wiley & Sons, 1962, 1965.

———. "On Solving Mathematical Problems in High School." In *Problem Solving in School Mathematics,* 1980 Yearbook of the National Council of Teachers of Mathematics (NCTM), edited by Stephen Krulik, pp. 1–2. Reston, Va.: NCTM, 1980.

Skiena, Steven. "Minimum Spanning Tree." In *Implementing Discrete Mathematics: Combinatorics and Graph Theory with Mathematica* by Steven Skiena, pp. 232–36. Reading, Mass.: Addison-Wesley Publishing Co., 1990.

Stewart, Ian. "Trees, Telephones, and Tiles." *New Scientist,* November 16, 1991, pp. 26–29.

Texas Instruments. Cabri Geometry II. Software. Dallas, Tex.: Texas Instruments, 1998.

Inventing a Geometry Theorem

Armando Martínez-Cruz
José N. Contreras

> [The student] should get the thrill of formulating new theorems, even if they are new only to [the student]. . . .
>
> —*James M. Moser*

PROBLEMS have been described as the pumping heart of mathematics (Halmos 1980). Mathematics is alive, for as soon as a problem is posed, it becomes the source of many more problems, even when it is not completely solved. This never-ending story has made mathematics grow significantly in different directions. Mathematicians are not the only ones expected to formulate new problems. *Principles and Standards for School Mathematics* (NCTM 2000) recommends that students be given opportunities to formulate their own problems. This process of generating new problems and reformulating old problems is referred to as *problem posing,* and it can occur prior to, during, or after problem solving (Silver 1994). In this article, we describe a mathematics content course for current and future secondary school teachers in which problem posing plays a central role and that culminates in a project ("Inventing a New Theorem") that students are required to complete and present with the aid of interactive geometry software.

We start with a discussion of how students can learn to pose problems. Then we illustrate two sample activities presented to students in our course and discuss how the problem-posing process takes place. We conclude the article with a sample from students' reactions and some of the work that has emanated from students' projects.

Learning to Pose Problems

Learning to pose problems is a challenging activity for which students need some systematic and useful strategies. Brown and Walter (1990) propose the "what if" strategy as a generic means for generating problems. In particular, they suggest listing the attributes of a problem and changing them systematically to generate new problems. These changes lead to the creation of converse problems, special problems, general problems, and extended problems. In our personal and classroom experiences, we have found that proving, reversing, specializing, generalizing, and extending are very fruitful problem-posing strategies. As an example, consider the following problem (fig. 17.1):

> Let *ABCD* be a parallelogram. Let *E, F, G,* and *H* be the centroids of triangles *ABC, BCD, CDA,* and *DAB,* respectively. What type of quadrilateral is *EFGH?*

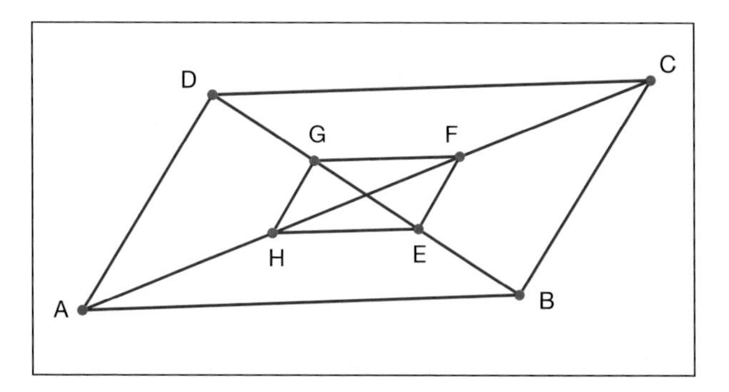

Fig. 17.1. In parallelogram *ABCD,* construct a diagonal. Find the centroids of the resulting triangles. Do the same for the second diagonal. What kind of quadrilateral is formed with the centroids of triangles *ABC, BCD, CDA,* and *DAB?*

As a first step, we can list or underline some of the attributes that can be modified to generate new problems: for example, parallelogram, centroids, triangle *ABC.* Of these attributes, consider *parallelogram* as an attribute that can be modified to create special, general, and extended problems. We use the hierarchy of quadrilaterals displayed in figure 17.2.

As figure 17.2 indicates, a rectangle is a special case of a parallelogram, a trapezoid provides a general case, and a kite provides an extended case of a

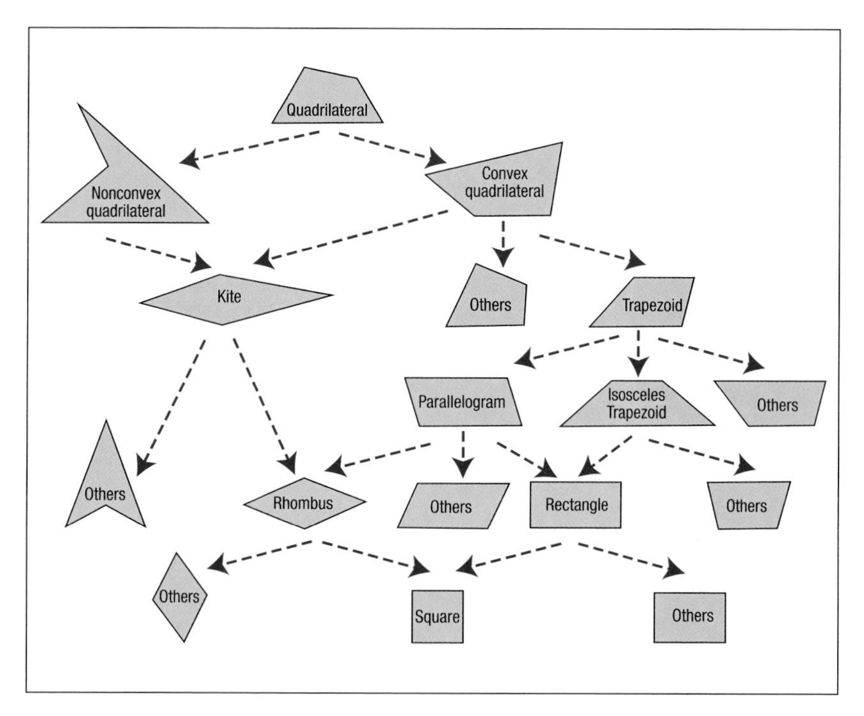

Fig. 17.2. A hierarchy of quadrilaterals

parallelogram. It is also possible to pose a converse problem. Following are sample corresponding problems some of which require proofs:

> Let *ABCD* be a rectangle. Prove that the centroids of triangles *ABC, BCD, CDA,* and *DAB* are the vertices of a rectangle. (Special case of the original problem)
>
> Let *ABCD* be a trapezoid. Prove that the centroids of triangles *ABC, BCD, CDA,* and *DAB* are the vertices of a trapezoid. (Generalization of the original problem)
>
> Let *ABCD* be a kite. Construct the centroids *E, F, G,* and *H* of the triangles *ABC, BCD, CDA,* and *DAB.* What type of quadrilateral is *EFGH?* (Extended problem)
>
> Let *ABCD* be a quadrilateral. If *EFGH* is a parallelogram whose vertices are the centroids of triangles *ABC, BCD, CDA,* and *DAB,* respectively, prove that *ABCD* is a parallelogram. (Converse problem)

Of course, we can further extend the problem by modifying the attribute *centroids* and consider other types of distinguished points of a triangle (e.g., incenters, circumcenters, and orthocenters), for example:

> Let $ABCD$ be a parallelogram. If E, F, G, and H are the orthocenters of triangles ABC, BCD, CDA, and DAB, what type of quadrilateral is $EFGH$?

We can also vary the types of triangles. Instead of considering triangles whose vertices are consecutive vertices of a quadrilateral, we can consider triangles whose vertices are the point of intersection (I) of the diagonals and consecutive vertices (i.e., ABI, BCI, CDI, and DAI, above). As we can see, problem posing can become an ongoing journey of discovery and delight as new problems become the source of additional problems. Readers can find additional help in learning to pose problems in Contreras (2003).

Problem posing and conjecturing are intrinsically related to problem solving. First, in a given problem we can find some *known* information, some *unknown* information, and often implicit or explicit *restrictions* (Moses, Bjork, and Goldenberg 1990). Once these features are identified, it is possible to pose new problems as suggested above. Second, Brown and Walter (1990) mention that even to understand a problem to be solved, we have to reformulate the problem or pose a simpler or similar problem. They note that we often do not "appreciate the significance of an alleged situation without generating and analyzing problems or questions" (p. 118).

The Geometry Course

In our course for current and future secondary school teachers, in addition to learning geometric content, students use interactive geometry software intensively for problem solving, constructions, and conjecture making. Students learn the software's capabilities and commands from the first day. As the semester unfolds, more commands are introduced as necessary. Connections among geometric ideas add excitement to the class.

For instance, students learn to construct a golden rectangle after studying the golden ratio,

$$\phi = \frac{1+\sqrt{5}}{2} \approx 1.618\ldots.$$

Next, when studying the pentagram and a regular pentagon, we discover that

$$\frac{\text{length of a diagonal of a regular pentagon}}{\text{length of a side of a regular pentagon}} = \phi.$$

This relationship allows us to use interactive geometry software to construct a regular pentagon using a golden rectangle. Later, when we discuss Euclid's work, we present Proposition 10 in Book XIII of his *Elements* (1956), which enables us to integrate several ideas related to the pentagon and the golden ratio. Here is our statement of this proposition.

> If a regular pentagon with side *EB,* a regular hexagon with side *AB,* and a regular decagon with side *FB* are inscribed in the same circle, then $AB^2 + BF^2 = EB^2$. In other words, *AB, FB,* and *EB* are the lengths of the sides of a right triangle (fig. 17.3).

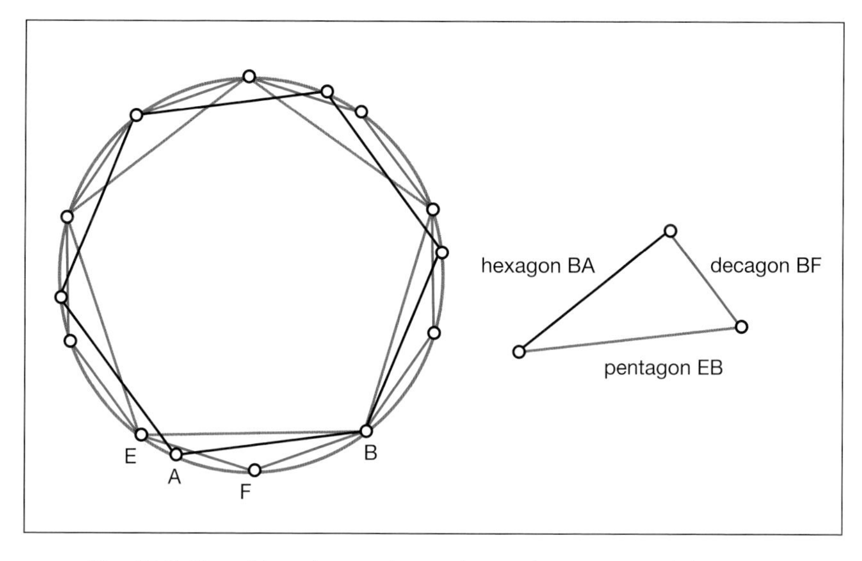

Fig. 17.3. The sides of a regular pentagon, hexagon, and decagon inscribed in the same circle form a right triangle.

This result is investigated with interactive geometry software prior to stating it. The instructor asks the students to construct a regular pentagon, circumscribe a circle, and inscribe a regular hexagon and a regular decagon. Next, students are asked to construct a triangle using one side of each of these polygons. Rarely has the instructor had to ask for the kind of triangle that is obtained. Students go about measuring the largest angle to test their observation. Once we conjecture that it is a right triangle, we work on a proof. Below is an algebraic proof provided by one student after the instructor suggested assuming that the length of the side of the hexagon is 1 unit. Notice that this student was already familiar with relationships among the golden ratio, the 36-72-72-degree triangle, and trigonometric ratios for 36 and 72 degrees.

> Proof. Let *C* be the center of the circle, and let us assume that the radius of the circle, *CB,* is 1. Then the side of the hexagon, *AB,* is 1 because $\triangle ABC$ is an equilateral triangle. To compute the side of the decagon *FB,* we look at $\triangle FBC$. This triangle is an isosceles triangle

with $\angle BCF = 36$ degrees, and $\angle CFB \cong \angle CBF = 72$ degrees. This is a golden triangle, and

$$\frac{\text{longest side}}{\text{shortest side}} = \frac{CB}{FB} = \phi,$$

the golden ratio. Therefore,

$$FB = \frac{1}{\phi}.$$

Finally, to determine the side of the pentagon, $\overline{EB}$, we look at ΔECB. In this triangle, $\angle ECB = 72^{\circ}$. Hence,

$$EB = 2 \bullet \sin\left(36^{\circ}\right) = \frac{\sqrt{2(5-\sqrt{5})}}{2}.$$

To summarize,

$$AB = 1,\ FB = \frac{1}{\phi},\ \text{and}\ EB = \frac{\sqrt{2\left(5-\sqrt{5}\right)}}{2}.$$

We now use the converse of the Pythagorean theorem, that is, we show that $AB^2 + FB^2 = \mathrm{EB}^2$ to prove that sides $\overline{AB}$, $\overline{FB}$, and $\overline{EB}$ form a right triangle:

$$AB^2 + FB^2 = 1 + \frac{1}{\phi^2} = 1 + \frac{4\left(6-2\sqrt{5}\right)}{16} = 1 + \frac{3-\sqrt{5}}{2} = \frac{5-\sqrt{5}}{2} = EB^2.$$

This equality proves the result.

The Interplay of Problem Solving and Problem Posing: Course Activities

Problem solving and problem posing are naturally intertwined in our geometry course. Students start with a problem, model it with interactive geometry software, investigate, and then solve or prove it. Next, they modify its conditions, using ideas from Brown and Walter (1990) and Contreras (2003); work to obtain a new problem or conjecture; and continue to investigate with interactive geometry software. The interplay between problem solving and problem posing is represented in figure 17.4. We include interactive geometry software explorations in

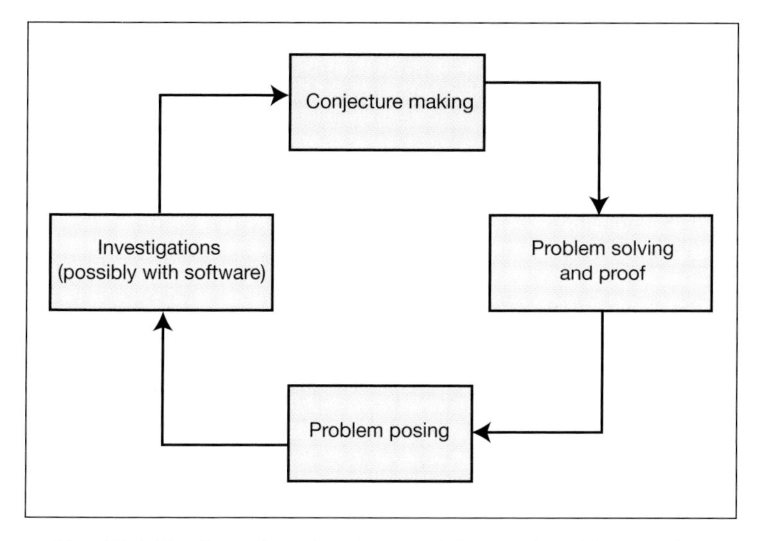

Fig. 17.4. The interplay of problem solving and problem posing

"Investigations" in the figure. This process is a cycle; each conjecture becomes a new base problem and the investigation continues.

Two Samples

After studying the Pythagorean theorem, students are shown the work of a former student, referred to as "Ron's theorem" (Martínez-Cruz, McAlister, and Gannon 2004). Squares are constructed on each side of a right triangle. Next, triangles are formed from adjacent sides of the squares. Students observe that the resulting triangles have the same area (fig. 17.5).

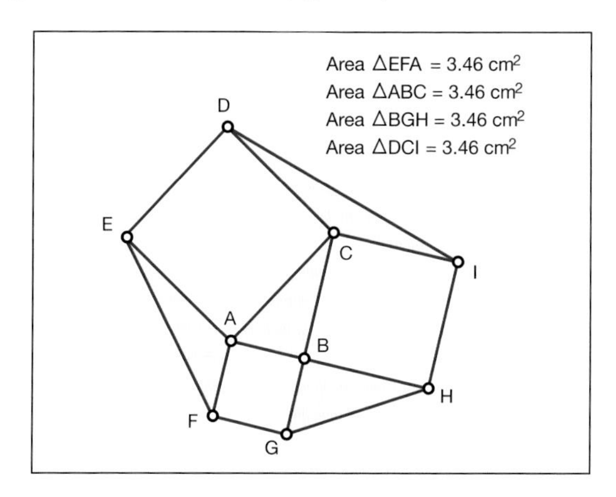

Fig. 17.5. Squares are constructed on each side of a right triangle. Triangles are formed from adjacent sides of the squares. The resulting triangles have the same area.

Here is a proof written by one of our students on the basis of the construction shown in figure 17.6. *IC* and *FA* are extended into the interior of square *ACDE*, and perpendiculars to these lines are dropped from *D* and *E* to meet at *J* and *K*, respectively.

> PROOF. $\triangle ABC \cong \triangle GBH$ by side-angle-side congruence, so these two triangles have the same area. Because $\overline{DC} \cong \overline{CA}$, the acute angles at *C* are congruent, and both triangles have right angles, $\triangle DJC \cong \triangle ABC$, by angle-angle-side congruence. Finally, notice that $\triangle DJC$ and $\triangle DCI$ have the same base and the same height, and therefore the same area. Hence, $\triangle DCI$ and $\triangle ACB$ have the same area. We can prove in a similar way (but using $\triangle EAK$) that $\triangle EAF$ and $\triangle CAB$ have the same area.

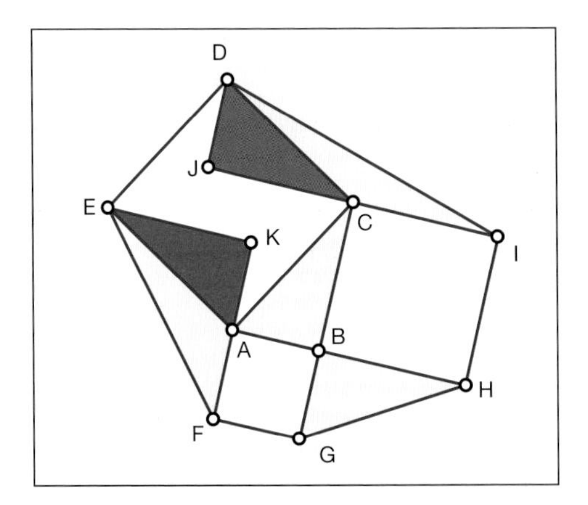

Fig. 17.6. Diagram for a proof of statement given in figure 17.5

When we study the golden rectangle, we revisit Ron's theorem. This time, as a way to modify one condition instead of using squares on each side of the triangle, we construct outward golden rectangles with widths from each side of right triangle *ABC* (fig 17.7). Students are surprised that the result is not quite the same. Although the three new triangles have the same area, the original right triangle has a different (and smaller) area. When they look at the ratio, say,

$$\frac{\text{area of } \triangle IDC}{\text{area of } \triangle ABC},$$

not everyone recognizes a familiar number right away: the golden ratio squared, ϕ^2! This result also serves to illustrate a context where ϕ^2 appears. This modification is a nice extension of Ron's result, and its proof is similar to the proof of Ron's theorem.

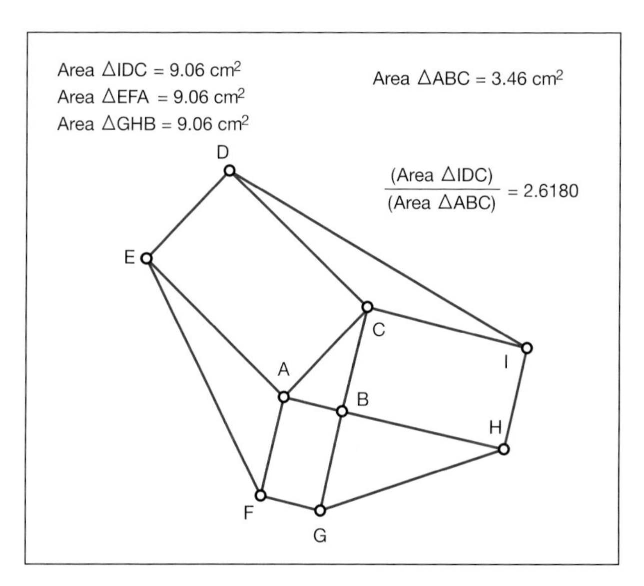

Fig. 17.7. Golden rectangles are formed with widths from each side of a right triangle. Triangles are formed from adjacent sides of rectangles. The resulting triangles have the same area.

Our second sample deals with centroids of triangles formed by three consecutive vertices in a quadrilateral (see fig. 17.8.). We mentioned this problem earlier to illustrate a systematic way to pose problems, and we have discussed this problem elsewhere (Contreras, Erickson, and Martínez-Cruz 2005). We describe here how we work this problem in class.

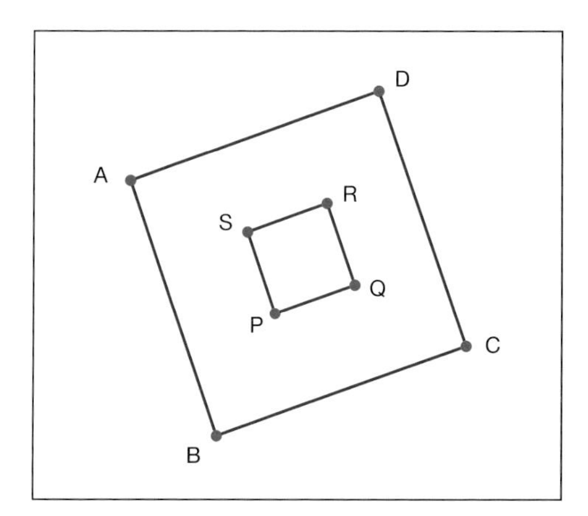

Fig. 17.8. In square *ABCD,* construct a diagonal. Find the centroids of the resulting triangles. Do the same for the second diagonal. The four centroids obtained form a square.

> Let *ABCD* be a square. Let *P, Q, R*, and *S* be the centroids of triangles *ABC, BCD, CDA,* and *DAB,* respectively. What type of quadrilateral is *PQRS* (fig. 17.8)?

First, students use interactive geometry software to produce and drag this construction. Next, they conjecture that *PQRS* is also a square. Even before we ask students to generate a new problem, some of them are already investigating two natural variations using rectangles and parallelograms (fig. 17.9). Again, students observe that the quadrilateral formed with the centroids appears to be the same shape as the original quadrilateral. We have observed that students do not naturally look at a *general* convex or nonconvex quadrilateral, which is a generalization of the problem (fig. 17.10). These two cases are necessary to investigate, first, because generalizing is an important activity in mathematics, and second, because both cases provide students with information that was hidden in the previous cases: the resulting quadrilateral, although similar, is *not* oriented in the same way! This observation coincides with Brown and Walter's (1990) conclusion that we gain a better understanding of one problem after we formulate other related problems. Our investigation concludes with an examination of cyclic quadrilaterals (fig. 17.11). Students use interactive geometry software at each stage to make and drag the constructions. As an aside, we note that one benefit of using interactive geometry software is that it often leads students to explore content that is otherwise rarely visited (e.g., nonconvex quadrilaterals, as shown in fig. 17.10).

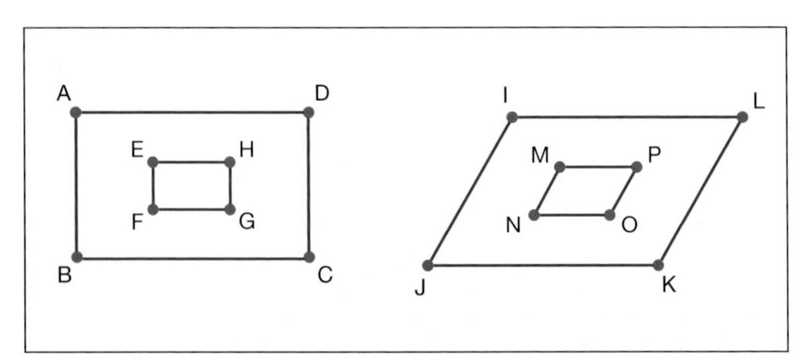

Fig. 17.9. Two variations of a problem: Rectangles and Parallelograms.
In a given rectangle (parallelogram), construct a diagonal.
Find the centroids of the resulting triangles.
Do the same for the second diagonal.
The four centroids seem to form a rectangle (parallelogram).

The cases examined are complementary. First, figures 17.10 and 17.11 help us understand the orientation of the resulting inside quadrilateral. Figures 17.8 and 17.9, however, help us conjecture relationships between the areas and perim-

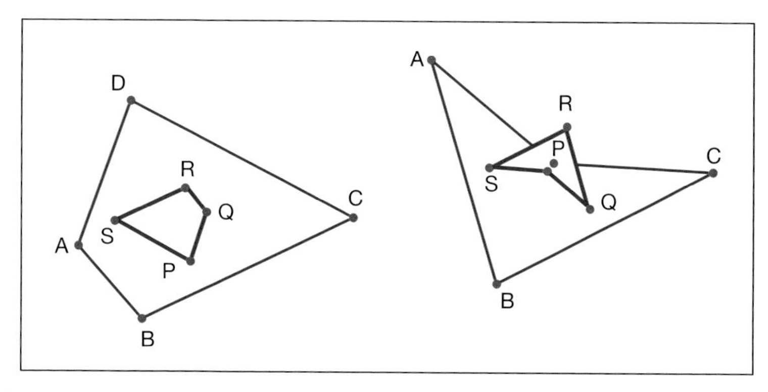

Fig. 17.10. A generalization of the problem. In any quadrilateral, including nonconvex quadrilaterals, construct each diagonal and the centroids of the two resulting triangles. Find the centroids of the four triangles. The centroids form a quadrilateral similar to the original quadrilateral.

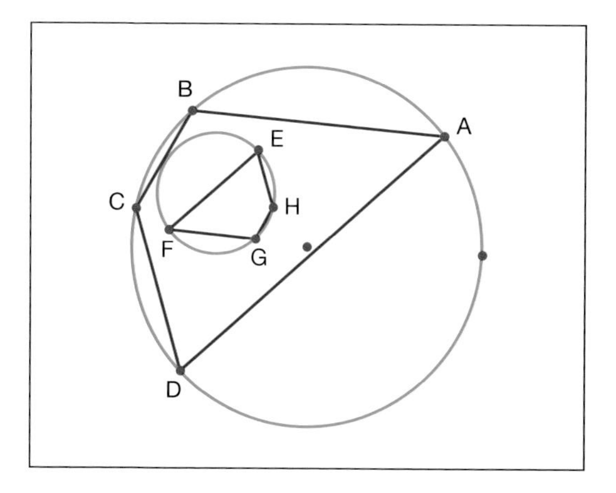

Fig. 17.11. A conjecture with a special case of the problem. In a cyclic quadrilateral, construct the diagonals. Find the centroids of the resulting triangles. The four centroids obtained in this way seem to form a quadrilateral similar to the original, but with a different orientation. Aha!

eters of the inside and outside quadrilaterals. These figures suggest that the ratio of the areas is 1:9. This conjecture is not derived easily from figures 17.10 and 17.11. Most students are able to provide the following proof of this result. Prior work with properties of the centroid and proofs of these properties fosters the necessary skills.

> Proof. In quadrilateral *ABCD, R* and *S* are the centroids of triangles *CDA* and *DAB,* respectively (fig. 17.12). We want to prove that *SR* is parallel to *CB*. Let *E* be the midpoint of *DA*. Then *EC* and *EB* are

medians of triangles CDA and DAB, respectively. Because R and S are centroids of these triangles, $3ER = EC$ and $3ES = EB$. Also $\triangle ESR$ and $\triangle EBC$ share angle E. Hence, $\triangle ESR \sim \triangle EBC$, $\overline{SR} \parallel \overline{BC}$, and $3SR = CB$. Using a similar argument, we obtain $\overline{RQ} \parallel \overline{AB}$. Because both pairs of corresponding sides are parallel, $\angle QRS \cong \angle ABC$. In the same way, the other corresponding angles are congruent and the sides are proportional. The two quadrilaterals are similar. Because the sides are in the ratio 1:3, their perimeters and areas are proportional, in the ratio 1:3 and 1:9, respectively.

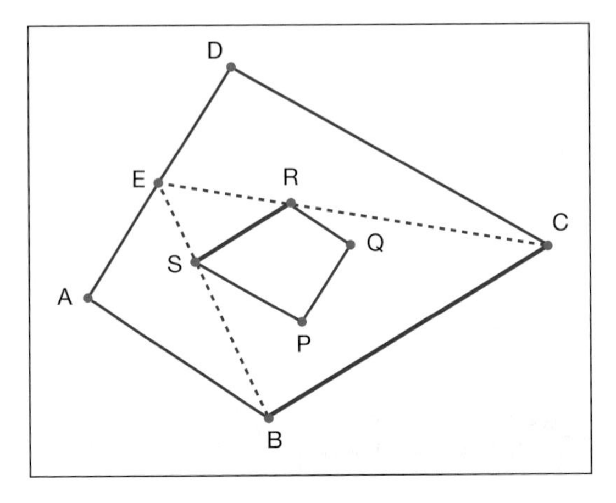

Fig. 17.12. In quadrilateral *ABCD, E* is the midpoint of *DA*. *R* and *S* are the centroids of triangles *DAC* and *DAB,* respectively. Then *RS* || *CB* and 3*RS* = *CB*.

Inventing a New Theorem: A Final Project in a Geometry Course

In this course, we ask students to complete a final project that requires them to “invent a new geometry theorem.” Below is the description of this final project as it appears in the course syllabus. Students are presented with this description the first day of class.

> The use of [interactive] software permits learners to investigate relationships dynamically so that they can see changes in geometric figures as they manipulate them, analyze their measurements, and communicate their findings to others. By using dynamic software participants will engage in conjecture making, which in turn gives the opportunity to experience what the mathematics community is about (Chazan and Houde 1989). Participants are expected to “discover a new theorem” using [interactive] geometry software. “New” should be understood as a result that you did not see in class, read in a book, or learn from a classmate. Do not panic! In class we will model various tech-

> niques to pose new problems. Your report should include all the explorations related to it, findings, and conjectures. Attempts to prove the result should also be included. This project is individual, but you may work in groups of 2 or 3 and individually submit variations of one single discovery. Presentations of research are scheduled for the last two weeks of the semester.

Certainly many students feel uneasy about this assignment, even when the word "original" has been explained—the idea of coming up with a conjecture on their own seems unthinkable. To clarify, we have found it beneficial not to require a proof. Under these circumstances, students are more willing to take chances while posing problems that they otherwise would not. We report with satisfaction that many do pursue a proof of their result, and a majority of these students are able to write good ones.

Students' Reactions to the Final Project

When students present their original theorem in class, they are required to submit files of their work and efforts to produce a proof. Additionally, they are required to include a paragraph with their reactions to the final project. During the presentation students explain their thinking about the problem, and most of them also comment that they thought the project was "impossible" at the beginning of the course. Below, we include one student's reaction, which is representative of the responses of most students. Notice how this student gained insight from the process of investigating.

> Working on the geometry theorem gave an opportunity to expand on what was done in class. There were many explorations that could be done with the software. It was difficult getting a starting point as to what to explore. When exploring my theorem, I began using the centroids and the incenters on a square, but the incenter did not yield any interesting results. From here, I decided to discard the incenter and just work with the centroid, but with other quadrilaterals. I decided to go on and try to prove my theorem and was successful. Once I saw the hidden lines on the construction, I was able to see why the result was as it was for the parallelogram and saw that it was rather intuitive.

Next Steps

The education of mathematics teachers is more than simply preparing them for their role in the classroom. It is about making them part of a larger community—the mathematics teaching profession. Having this outcome in mind, we have as our goal that teachers also disseminate their mathematical results and their experience from the class. One might think that once they came up with a result, students would be eager to present it to a larger audience. Unfortunately, they do not often choose to do so. However, we are pleased that some students have published their results (Martínez-Cruz, McAlister, and Gannon 2004), others have submitted their results for publication, some have made presentations along with us at local and national conferences, and still others have conducted workshops for mathematics teachers with us. Most report that they feel capable of contributing to mathematics. This outcome is the most gratifying part of the course. We have empowered the students through problem posing and problem solving. As Moser (1974) stated, they experienced the thrill of inventing a new geometry theorem.

REFERENCES

Brown, Stephen I., and Marion I. Walter. *The Art of Problem Posing*. 2nd ed. Hillsdale, N.J.: Lawrence Erlbaum Associates, 1990.

Chazan, Daniel, and Richard Houde. *How to Use Conjecturing and Microcomputers to Teach Geometry.* Reston, Va.: National Council of Teachers of Mathematics, 1989.

Contreras, José. "A Problem-Posing Approach to Specializing, Generalizing, and Extending Problems with Interactive Geometry Software." *Mathematics Teacher* 96, no. 4 (April 2003): 270–76.

Contreras, José, David Erickson, and Armando M. Martínez-Cruz. "On Creating Polygons with the Centroid of a Triangle: An Investigation with The Geometer's Sketchpad." *Ohio Journal of School Mathematics* 52 (Autumn 2005): 12–16.

Euclid. *The Thirteen Books of the Elements.* 3 vols. Translated by Thomas L. Heath. New York: Dover Publications, 1956. Translated from Johan Ludvig Heiberg, *Euclidis Elementa* (Leipzig, Germany: Teubner, 1883–88).

Halmos, Paul. "The Heart of Mathematics." *American Mathematical Monthly* 87, no. 7 (August–September 1980): 519–24.

Martínez-Cruz, Armando M., Ron McAlister, and Jerry Gannon. "Ron's Theorem and Beyond: A True Mathematician and GSP in Action." *Mathematics Teacher* 97, no. 2 (February 2004): 148–51.

Moser, James M. *Modern Elementary Geometry*. Englewood Cliffs, N.J.: Prentice-Hall, 1971.

Moses, Barbara, Elizabeth Bjork, and E. Paul Goldenberg. "Beyond Problem Solving: Problem Posing." In *Teaching and Learning Mathematics in the 1990s*, 1990 Yearbook of the National Council of Teachers of Mathematics (NCTM), edited by Thomas J. Cooney, pp. 82–91. Reston, Va.: NCTM, 1990.

National Council of Teachers of Mathematics (NCTM). *Principles and Standards for School Mathematics.* Reston, Va.: NCTM, 2000.

Silver, Edward A. "On Mathematical Problem Posing. *For the Learning of Mathematics* 14 (1994): 19–28.

Theorems Discovered by Students Inspire Teachers' Development

Antonio Quesada

THE INTRODUCTION during the mid-1990s of interactive geometry software, such as The Geometer's Sketchpad (Jackiw and Finzer 1993) and Cabri Geometry (Laborde and Bellemain 1994), has made it possible to change drastically the way we teach and learn geometry. Students using this software may not only enhance their geometric intuition by constructing elementary figures but also test or uncover figures' defining properties. Moreover, students can deform or transform their constructions and observe which properties remain invariant. This capability allows them to explore, discover, test, and conjecture new properties and relationships, anticipating the need for formal proofs.

Interactive geometry software gives instructors the ability to introduce, through discovery, geometrical concepts and properties ranging from elementary to advanced. The interactive nature of this software makes experimentation and discovery possible from the very beginning. From my experience with Cabri, with the aid of the publisher's manual (Texas Instruments 1996) students learn enough in two 75-minute labs to begin to do independent work. As they become comfortable with the software, students are motivated to seek proofs as a way to establish the truth of generalizations they conjecture.

In this article I address two aspects of interactive geometry software. First

I review some new discoveries by secondary school students and comment on what we, as teachers, can learn from them. Then I describe how the software has affected the preparation of preservice teachers.

Fostering New Mathematical Findings by Secondary School Students

In the 1990s I observed an increase in the number of mathematical findings published by secondary school students. These works illustrate how technology has been empowering students to do mathematics as never before, to explore and test ideas graphically or numerically, to find mathematical relationships, and to conjecture results. Because these findings captured the attention of mathematics educators around the world, I decided to publish compilations of these students' discoveries (Quesada 2001a, 2001b).

Many colleagues commented that it would be nice to have this kind of student in their classrooms. Eventually, I asked myself, "Are there common factors underlying these students' discoveries? And, if so, am I promoting these factors in my own classes?" After rereading the students' work and conferring with a teacher of one of the students involved, I found two common threads supporting these students' discoveries. The first thread is that these students used interactive geometry software and graphing calculators. Clearly, these technologies allow students to navigate cumbersome calculations, and, when properly used, facilitate an inquiry-based approach that promotes exploration and discovery. The second thread, not surprisingly, is that most students have been challenged by their teachers. To be sure that we are challenging our students, teachers should be asking themselves the following questions:

1. Do we ask our students to try to generalize their solutions?
2. How do we encourage our students to raise their own questions, and try to answer them?
3. How do we create extensions—for example, more challenging questions—to the activities students do?
4. Do we dare to ask our students truly challenging questions, questions for which we ourselves may not have an answer?
5. How do we reward students who accept these challenges?

If we do not challenge our students or risk posing problems whose solutions we do not know, we will be hiding the true nature of mathematics, and we will miss the opportunity to be rewarded with their discoveries!

Samples of Secondary School Students' Discoveries

The following is a selection of problems solved by secondary school students. Some of these solutions represent new mathematical findings; others are simply new to the learner. In some instances, clues to the proofs are given; for the others, the reader is encouraged to find a proof. The persistence and insight displayed by several of the young authors show unusual mathematical maturity.

Example 1: An Extension of Marion Walter's Theorem

The cover of the February 1992 *Mathematics Teacher* (Johnston 1992) showed that if selected trisection points of an equilateral triangle are joined to the opposite vertices, the area of the resulting central triangle is one-seventh the area of the original triangle (see fig. 18.1). This graphic prompted the publication in the *Mathematics Teacher* of a related theorem by Marion Walter (Cuoco, Goldenberg, and Mark 1993): If all trisection points of the sides of any triangle are connected to the opposite vertices, the resulting hexagon has area one-tenth the area of the original triangle (see fig. 18.2).

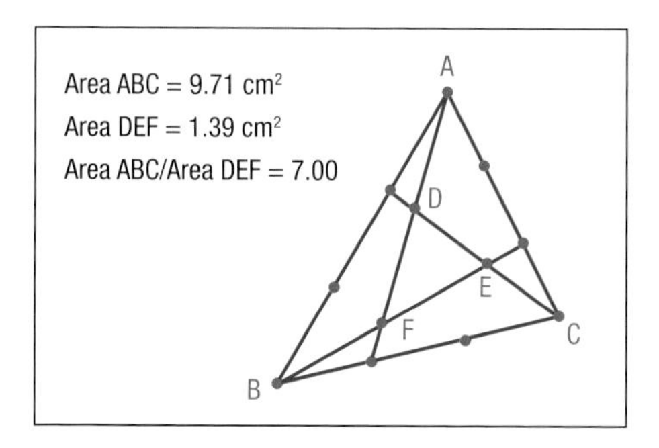

Fig. 18.1. Area of $\triangle DEF$ = 1/7 area of $\triangle ABC$.

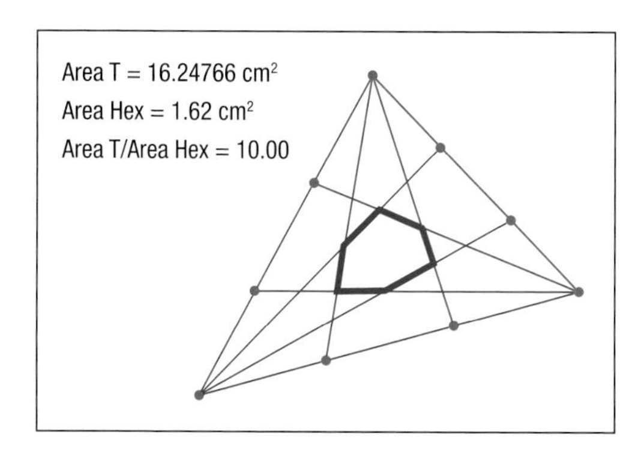

Fig. 18.2. Area of hexagon formed = 1/10 area of triangle.

In the fall of 1993, Frank D. Nowosielski, a teacher at Patapsco High School, Baltimore County, Maryland, proposed that his ninth-grade geometry class "rediscover" Marion Walter's theorem using The Geometer's Sketchpad. One of the students in this class, Ryan Morgan, after verifying the theorem, became interested in determining what would happen if the sides of the triangle were partitioned into more than three congruent segments. He and his teacher later termed this process *n-secting* (Watanabe, Hanson, and Nowosielski 1996).

Ryan first noticed that to have two "central" points, the side of the triangle had to be subdivided into an odd number of parts. Then, using The Geometer's Sketchpad, Ryan found that for each odd n greater than 3, the ratio of the area of any triangle to its central hexagon appeared to be constant.

Figures 18.3 and 18.4 show the value of this constant for $n = 5$ and $n = 11$. Grabbing one of the vertices of any of these triangles and deforming it in any direction shows that, for a given n, the ratio found remains constant. Table 18.1 contains the ratios determined for the first six values of n.

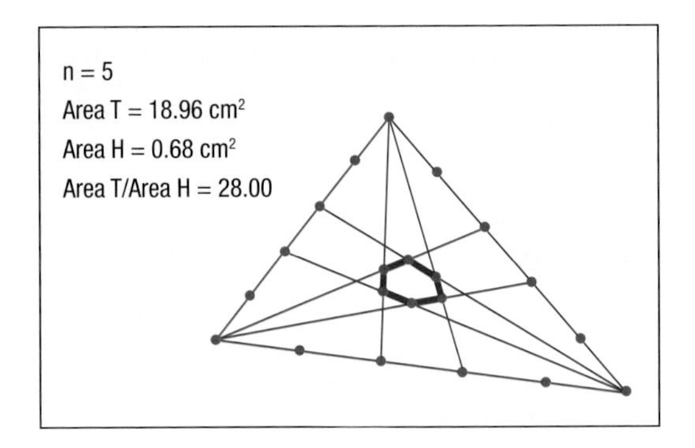

Fig. 18.3. Ratio of the areas of triangle to hexagon for $n = 5$

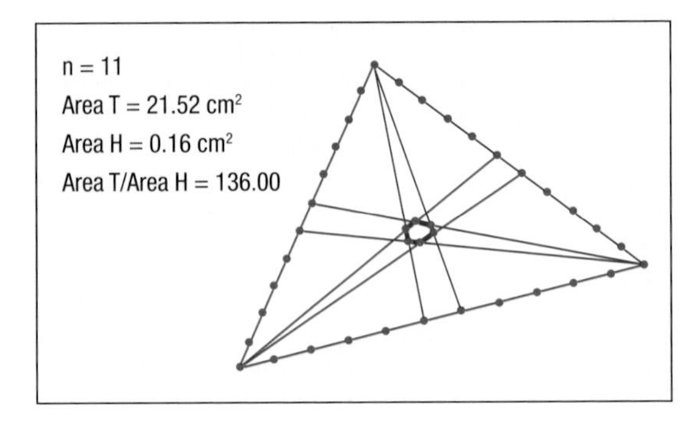

Fig. 18.4. Ratio of the areas of triangle to hexagon for $n = 11$

Table 18.1

Ratios of Areas of Triangles and Interior Hexagons Formed for Different n-Sections

n-sections	3	5	7	9	11	13
Ratio of areas	10:1	28:1	55:1	91:1	136:1	190:1

After plotting these points on a graphing calculator, Ryan noticed the parabolic tendency and decided to fit the data with a quadratic regression, obtaining a perfect fit. Doing so allowed him to forecast the ratio for a larger value of n and confirm it using interactive geometry software. Then Ryan formulated his conjecture:

> For n odd, if the central n-section points of the sides of any triangle are connected to the opposite vertices, the ratio of the area of the original triangle to the area of the resulting hexagon is $(9n^2 - 1)/8$ to 1.

After his discovery, Ryan presented his result at Towson State University. As one member of the faculty noticed, the proof of his finding can be obtained (see fig. 18.5) using the following theorem due to Jacob Steiner (Todd 2006–2007):

> If the sides *AB, BC,* and *CA* of triangle *ABC* are divided at *P, Q,* and *R* in the respective ratios $x:1, y:1,$ and $z:1$, the ratio of the area of the triangle formed by intersecting the segments *AP, BQ,* and *CR* to the area of triangle *ABC* is $k = (xyz - 1)^2 / [(xy + x + 1)(yz + y + 1)(zx + z + 1)]$.

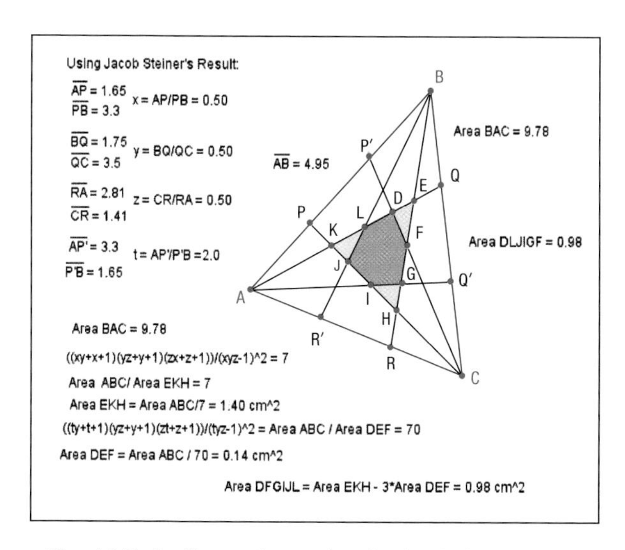

Fig. 18.5. An illustration, using Steiner's formula, of the area of the hexagon as the difference of areas of triangles

Ryan rejected some hints on how to prove his result and later submitted his own proof (Watanabe, Hanson, and Nowosielski 1996, p. 423). To this day we have been unable to find in the literature any reference to a result similar to Ryan's. Ryan became a software engineer, having studied computer science at the University of Maryland and Johns Hopkins University.

Ryan's original conjecture clearly would have been very difficult without a tool to draw polygonal figures and obtain areas quickly and accurately, and to deform the figures interactively so that invariant properties might be observed.

Example 2: Minimizing Distances from a Point in a Triangle

Arne Engebretsen, a teacher at Greendale High School, Greendale, Wisconsin, proposed to his freshman geometry class the following problem:

> Given a triangle ABC, find a point F such that the sum of the distances from the three vertices, $FA + FB + FC$, is minimized.

The account that follows can be found in vonder Embse and Engebretsen (1996, p. ii).

> Two of the students, Bridget and Connie,[1] proceeded the first day, by using trial and error, to place a point F inside of the triangle ABC and measure the distances FA, FB, and FC. Then they calculated the sum $S = FA + FB + FC$ and moved F around to find a location for F that will minimize S. Then, the second day, they looked for a systematic way for determining the point. However, five minutes before the end of the class, they had not found one. They had constructed equilateral triangles with their centroids over the sides of the given triangle.... Partly out of desperation, Connie suggested reflecting vertex B over the line joining the closest centroids of the equilateral triangles, $C1$ and $C2$ in figure [18.6]. The point seemed to coincide with the one obtained by trial and error! The third day, they eagerly continued their investigation, confirming that their method worked for all three vertices of every acute triangle they had time to explore. When it came time to present their findings, they reported their discovery with pride and confidence.

The point the students were seeking is called the Fermat point of a triangle.[2] Figure 18.7 shows a well-known solution for obtaining the Fermat point as the intersection of the circumcircles of the equilateral triangles constructed over the sides of the original triangle.

1. The teacher, Arne Engebretsen, did not include his students' last names, and sadly, he has since passed away.

2. For another approach to the Fermat point, see Contreras (2009).

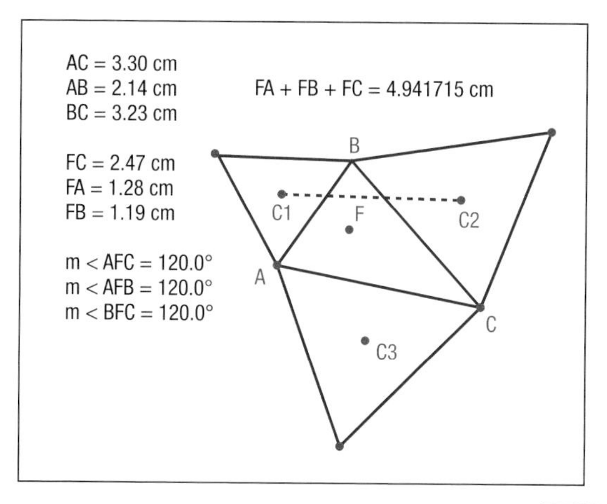

Fig. 18.6. Fermat's point as a reflection of B on $\overline{C_1C_2}$

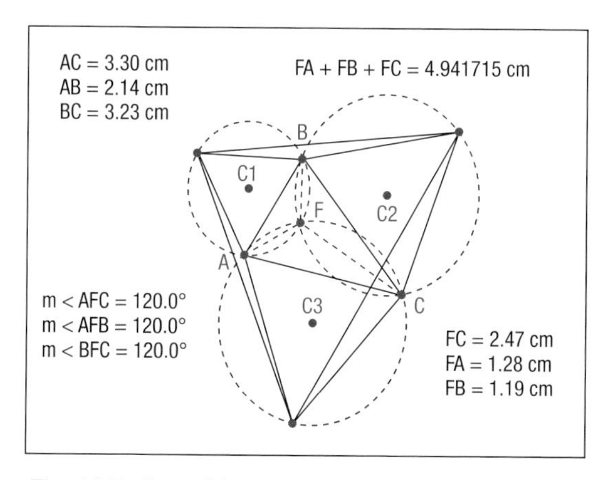

Fig. 18.7. A traditional approach to obtaining the Fermat point of a triangle

In presenting the students' work, vonder Embse and Engebretsen appropriately quoted Pólya (1954, p. vi), who points to mathematical exploration, investigation, and conjecturing as crucial first steps in the learning process that concludes with the rigor of proof.

> You have to guess a mathematical theorem before you prove it; you have to guess the idea of the proof before you carry through the details. You have to combine observations and follow analogies; you have to try and try again. The result of the mathematician's creative work is demonstrative reasoning, a proof; but the proof is discovered by plausible reasoning, by guessing. If the learning of math reflects at any degree the invention of math, it must have a place for guessing, for plausible inference.

Example 3: The GLaD Construction

Named after the initials of the students who found this construction and their teacher (**G**oldenheim, **L**itchfield, **a**nd **D**ietrich), this example became known worldwide after Leslie Chess Feller wrote about it in a *New York Times* article, "The Eternal Challenge of Euclid's Geometry," on March 7, 1999.

In June 1995, Charles H. Dietrich, a teacher at Greens Farms Academy in Westport, Connecticut, posed to his ninth-grade students the well-known problem of how to subdivide a given segment into n equal segments. However, Dietrich added the condition of doing the subdivision without using a compass (Dietrich, Goldenheim, and Litchfield 1997).

In a few hours Dave Goldenheim and Dan Litchfield produced two constructions, one for n odd, and another for n even, which can be combined as illustrated in figure 18.8 (Wilson 1998). They started with the segment to be divided, $\overline{AB}$, and constructed rectangle $ABCD$. They Let $P_1 = B$, and obtained P_{i+1} as the perpendicular projection of $\overline{DP_i} \cap \overline{AC}$ on $\overline{AB}$. This process yields the squence of points $P_2, P_3, P_4, \ldots$ so that $AP_n = AB/n$.

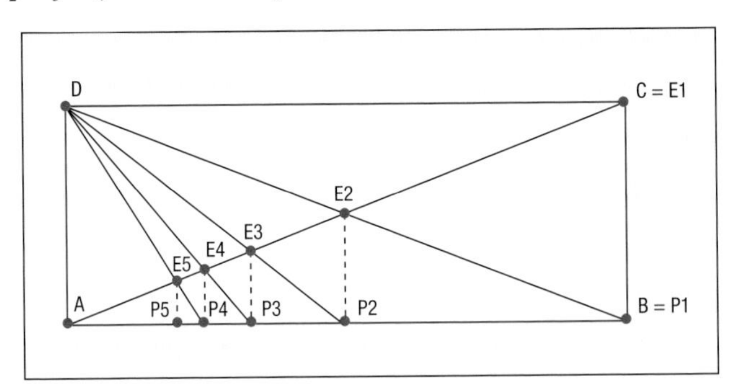

Fig. 18.8. GLaD Construction. Given the segment $\overline{AB}$, the rectangle $ABCD$ is constructed. Let $B = P_1$ and $C = E_1$. For $i > 0$, let E_{i+1} be $\overline{DP_i} \cap \overline{AC}$ and P_{i+1} be the perpendicular projection of E_{i+1} on $\overline{AB}$. Then $AP_i = AB/i$.

Next we sketch the first steps of a proof that $AP_i = AB/i$, based on two sequences of similar triangles:

(1) $$\Delta ACB \sim \Delta AE_2P_2 \sim \Delta AE_3P_3 \ldots$$

and

(2) $$\Delta AE_3D \sim \Delta E_2E_3P_2,\ \Delta AE_4D \sim \Delta E_3E_4P_3 \ldots.$$

As shown in figure 18.8, P_{i+1} is obtained as the perpendicular projection of $\overline{DP_i} \cap \overline{AC}$ on $\overline{AB}$. Because E_2 is the intersection of the diagonals and $\Delta ACB \sim \Delta AE_2P_2$ from (1), P_2 is the midpoint of $\overline{AB}$ and

$$E_2P_2 = \frac{1}{2}CP_1.$$

Therefore from (2), $\Delta AE_3D \sim \Delta E_2E_3P_2$ with similarity ratio $r = 2$. Hence $AE_3 = 2E_2E_3$, and it follows that

$$AE_3 = \frac{2}{3}AE_2.$$

Next, from (1), $AE_2P_2 \sim \Delta AE_3P_3$, so

$$AP_3 = \frac{2}{3}AP_2 = \frac{2}{3} \bullet \frac{1}{2}AP_1 = \frac{1}{3}AB.$$

The process above can be used as part of an induction proof of the general result.

The two students also found the subdivision shown in figure 18.9, which, with the help of their teacher, they recognized as having ratios that follow the Fibonacci sequence. Again let $ABCD$ be any rectangle constructed with base $\overline{AB}$, and let $\overline{MN}$ be its midsegment. Let $P_1 = C$, $Q_2 = T_2 = \overline{AP_1} \cap \overline{MN}$, and P_2 be the perpendicular projection of $Q_2 = T_2$ on $\overline{DP_1}$. For $i \geq 2$, define $T_{g(i)} = \overline{AP_{f(i)}} \cap \overline{MN}$, $Q_{f(i+1)} = \overline{AP_{f(i)}} \cap \overline{DT_{g(i-1)}}$, and $P_{f(i+1)}$ as the perpendicular projection of $Q_{f(i+1)}$ on $\overline{DC}$, where the points are indexed using the Fibonacci sequence $f = \{1, 2, 3, 5, 8, \ldots\}$ and a sequence with the same values doubled, $g = \{2, 4, 6, 10, 16, \ldots\}$.

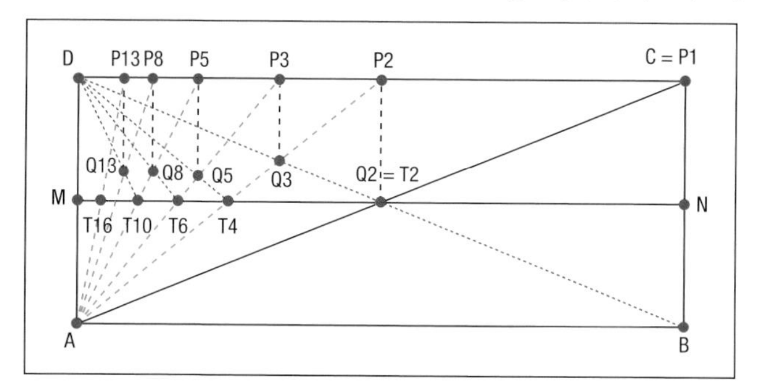

Fig. 18.9. Using the rectangle $ABCD$ with midsegment MN, Goldenheim and Litchfield also found the sequence $C = P_1, P_2, P_3, P_5, P_8, \ldots$ such that DC/DP_i generates the Fibonacci sequence.

It is claimed that $DP_{f(i)} = AB/f(i)$.[3] As in the previous case, the result can be established using sequences of similar triangles.

3. Jeff Levine, a student in my senior course for preservice secondary school teachers, developed the initial idea for this proof, which we worked on together. Jeff told me that our collaboration was a great experience for him, and I must say that was also true for me.

(1) $$\triangle AP_1P_2 \sim \triangle AT_2T_4,\ \triangle AP_2P_3 \sim \triangle AT_4T_6,\ \ldots,\ \triangle AP_{f(i)}P_{f(i+1)} \sim \triangle AT_{g(i)}T_{g(i+1)},\ \ldots$$

(2) $$\triangle DQ_3P_2 \sim \triangle T_2Q_3T_4,\ \triangle DQ_5P_3 \sim \triangle T_4Q_5T_6,\ \ldots,\ \triangle DQ_{f(i+1)}P_{f(i)} \sim \triangle T_{g(i-1)}Q_{f(i+1)}T_{g(i)},\ldots$$

(3) $$\triangle DP_2T_2 \sim \triangle DP_3Q_3,\ \triangle DP_4T_4 \sim \triangle DP_5Q_5,\ \ldots,\ \triangle DP_{g(i-1)}T_{g(i-1)} \sim \triangle DP_{f(i+1)}Q_{f(i+1)},\ \ldots$$

Without loss of generality assume that $DC = 1$. Our goal is to show that the points P_2, P_3, P_5, P_8, …, obtained by projecting the corresponding Q_2, Q_3, Q_5, Q_8, …, determine distances $DP_{f(i)}$, which are the reciprocals of the Fibonacci numbers. Clearly $DP_l = 1$ and

$$DP_2 = \frac{1}{2}.$$

Since the ratio of similarity of all the pairs of triangles in the first sequence is consistently 2, we see that

$$T_4T_2 = \frac{1}{2}P_2P_1 = \frac{1}{4} = MT_4.$$

Hence, the perpendicular projection of T_4 on CD yields P_4 with

$$DP_4 = \frac{1}{4}.$$

For the sake of clarity, that is, to display on CD only $P_{f(i)}$, we decided not to include the points $P_{g(i)}$ obtained by projecting $T_{g(i)}$ in figure 18.9.

From (2) we get that $\triangle DQ_3P_2 \sim \triangle T_2Q_3T_4$, with ratio 2. Hence $DQ_3 = 2Q_3T_2$, so

$$DQ_3 = \frac{2}{3}DT_2.$$

But from (3), $\triangle DP_2T_2 \sim \triangle DP_3Q_3$, so it follows that

$$DP_3 = \frac{2}{3}DP_2 = \frac{1}{3}.$$

Next we need to show that

$$DP_5 = \frac{1}{5}.$$

By (1), $\triangle AP_2P_3 \sim \triangle AT_4T_6$, and because $DP_3 = 1/3$, we have

$$MT_6 = \frac{1}{6},$$

from which it follows that $T_4T_6 = MT_4 - MT_6 = 1/12$. From (2) we know that $\triangle DQ_5P_3 \sim \triangle T_4Q_5T_6$ with similarity ratio

$$\frac{DP_3}{T_4T_6} = \frac{1}{3} \div \frac{1}{12} = 4.$$

Therefore $DQ_5 = 4Q_5T_4$. From (3), $\triangle DP_4T_4 \sim \triangle DP_5Q_5$, from which it follows that $DP_5 = 4P_5P_4$ and

$$DP_5 = \frac{4}{5} DP_4 = \frac{1}{5}.$$

The conclusion can be obtained by generalizing this process through induction.

This approach to the GLaD problem does not appear in most textbooks; however, a similar solution had been previously published in a little-known book (Leslie 1811). The construction that these students found is nevertheless an outstanding accomplishment.

Example 4: Graphical Representation of a Quadratic's Imaginary Solutions in R × R.

Consider the quadratic equation $a(x - h)^2 + k = 0$, where a, $k > 0$, whose imaginary roots are

$$x = h \pm i\sqrt{k/a}.$$

How can we find the two points on the real plane whose coordinates are the real and imaginary parts of these solutions?

Shaun Pieper (1997), a student at Saint Paul's School in Concord, New Hampshire, solved this problem using a graph similar to the one depicted in figure 18.10, where $y = (x - 7)^2 + 5$. He observed that, in general, by reflecting the graph of $y = a(x - h)^2 + k$ over $y = k$, $k > 0$, we obtain the graph of $y = -a(x - h)^2 + k$ whose x-intercepts are

$$h \pm \sqrt{k/a}.$$

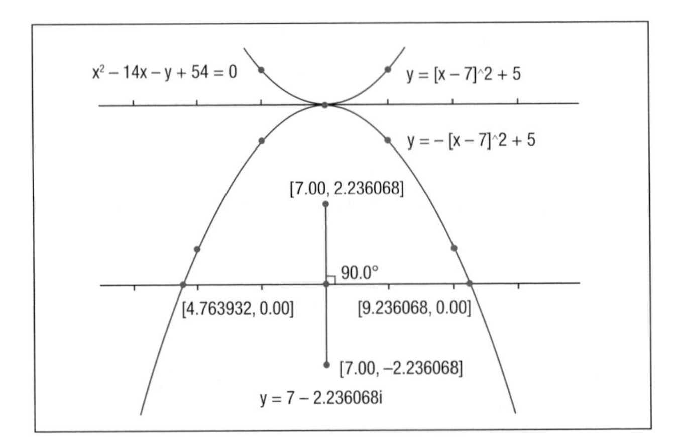

Fig. 18.10. Points on the real plane whose coordinates are the real and imaginary parts of a quadratic equation's imaginary solutions

By rotating 90 degrees about $(h, 0)$ the segment defined by these intercepts, we obtain a segment whose endpoints have for coordinates

$$\left(h, \pm\sqrt{k/a}\right),$$

the real and imaginary parts of the solutions.

Example 5. On Reflecting the Circumcenter of a Cyclic Quadrilateral over Its Sides

What type of quadrilateral is formed by joining the points obtained by reflecting the circumcenter of a cyclic quadrilateral over each of its four sides? This problem, posed by Michael de Villiers, was solved by four students. Figure 18.11 shows the solution found by the winner, Lori Sommars, a tenth grader at Wheaton North High School, Wheaton, Illinois, which appeared in *Sketchpad Puzzler* in the fall of 1999. As shown in figure 18.11 she began with circle A and

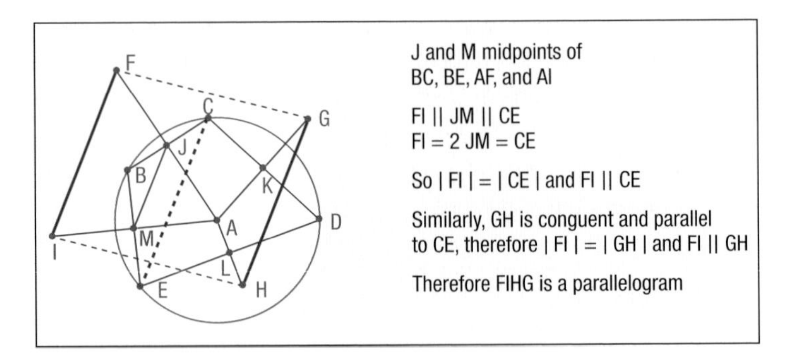

Fig. 18.11. Solution found by Lori Sommars

cyclic quadrilateral *BCDE*. She then reflected *A* over each of the four sides of the quadrilateral to obtain *FGHI*. She constructed $\overline{JM}$ as the midpoint connector for both triangles $\triangle AFI$ and $\triangle BEC$; consequently $\overline{JM}$ is parallel to their bases and one-half their lengths. It then followed that $\overline{FI}$ and $\overline{CE}$ are parallel and congruent. Similarly $\overline{GH}$ is congruent and parallel to $\overline{CE}$, and by transitivity so are $\overline{FI}$ and $\overline{GH}$. A comparable argument establishes the analogous result for $\overline{FG}$ and $\overline{IH}$. Therefore, *FGHI* is a parallelogram.

Influences of Interactive Geometry Software in Our Concepts in Geometry Course

If secondary school students are to invent and prove theorems, as the previous examples suggest, then their teachers need to be prepared with tools and strategies for fostering these activities. This and the following section describe the work that colleagues and I are doing in preservice education in that vein.

Formerly the Concepts in Geometry course for secondary school teachers had been a traditional survey of geometry. However, the diverse, fragile geometric background of many of our students, as well as the didactical possibilities provided by interactive geometry software, motivated instructors to transform this course to address what we consider the current needs of a secondary school teacher. The more recent approach is based on our belief that geometry is better learned with activities moving from the concrete to the abstract; by making lessons as visual and hands-on as possible, with students raising questions to explore and discover principles; by daring to venture into unknown territory; and by convincing ourselves and others of particular results through well-written logical arguments. Proofs should follow and validate discoveries.

The course follows a Socratic approach for doing proofs in class. We pose questions, provide hints when needed, and encourage students to think individually and with classmates. Using Cabri, students work in teams, mostly outside of class, to solve a set of inquiry-based activities in traditional, modern Euclidean, and in non-Euclidean geometries.

Additional exploration and discovery are encouraged and rewarded in each activity. Students also do a final project that includes creating new inquiry-based activities, a poster session, and an in-class presentation.

In this course students use interactive geometry software to—

1. discover important results from modern Euclidean (e.g., nine-point circle, Fermat's point, Simpson line, and so on) and non-Euclidean geometries;
2. solve challenging problems involving lesser-known results;

3. gain familiarity with interactive geometry software as a tool to support learning by discovery in algebra and calculus as well as geometry; and
4. develop the ability to raise their own questions and then try to solve them.

We reward students both for consistently raising questions about the problems they are working with and for making progress toward solutions, thus motivating them to start developing their disposition toward inquiry. Our initial guidance in this process seems to be very helpful. For example, when we present the task of constructing the perpendicular bisectors of the sides of a triangle to discover that they meet in a point equidistant from the vertices, we encourage students to ask these questions. Do the perpendicular bisectors always meet in a point? Is that point always equidistant from the vertices? Can this point lie inside, outside, or on a side of the triangle? What kind of triangles will have this point outside? Inside? On a side? Do quadrilaterals always have a similar point, that is, a point equidistant from the four vertices? If we draw a triangle inscribed in a circle and then move one of the vertices along the circle, what is the locus of the circumcenters?

As a result, interactive geometry software contributes to—

- increasing the depth and breadth of our course,
- engaging students as more active participants in the learning process,
- aiding students in developing more confidence in their own ability to solve problems, and
- initiating students into mathematics research.

Struggles of Inquiry Teaching

Most preservice secondary school mathematics teachers have learned, or at least heard about, an inquiry-based approach, but they have had very little experience, if any, with this kind of learning in their discipline, and even less experience developing inquiry-based activities. On several occasions, the day after a class has completed an inquiry-based activity, I have given an unannounced quiz on the important ideas addressed by the activity. Invariably, a number of students have failed to grasp the mathematical ideas involved. This outcome indicates that, after completing an inquiry-based activity, many students do not stop to review and retain the concepts and properties discovered. Hence, in addition to the group debriefing, asking the students at the end of each inquiry-based activity to write down the important definitions and properties discovered *in their own words* is helpful to facilitate transmission of the primary concepts from short-term into long-term memory.

Writing a good inquiry-based activity is time-consuming and at times frustrating. It is an excellent learning experience, however, because it requires not only mastering the concepts involved, breaking those concepts down to accessible parts and organizing them, and clearly establishing preexisting knowledge but also thinking continuously in the Socratic way. Many students think of problems merely in terms of "book exercises"; few students have done problem solving outside that context. Writing inquiry-based activities engages preservice teachers in raising their own questions to decompose a larger task into smaller more accessible subtasks.

Some preservice teachers do not value the inquiry-based approach, especially when exposed to it for only one semester. They are not used to "exploring and discovering," so they lack confidence in their abilities and in the conclusions they reach. Some students feel uneasy about, or question the time required for, this approach. Thinking of mathematics problems as akin to a set of cookbook recipes, they argue, "Why discover a known result when it can be read and memorized?" Sometimes teamwork is misinterpreted as a subdivision of goals without the appropriate planning and ongoing exchange of ideas.

Despite these challenges, we persist in teaching with inquiry and expecting future teachers to explore its use for themselves. Among other benefits, the use of inquiry-based activities, with the exploration and discovery that interactive geometry software facilitates, seems to have increased the self-confidence of many students in solving real mathematics problems.

Conclusion

The changes to the content of our geometry course for secondary school teachers and the number and depth of problems solved by secondary students give one an idea of the potential for mathematical thinking that technology in general, and interactive geometry software in particular, can bring to the classroom. The following statement by J. J. Lagowski (NASA Education Partnerships Forum 2006) is particularly appropriate at a time when new technologies with improved capabilities are having a dramatic impact on the way we learn and teach:

> We are attempting to educate students today so that they will be ready to solve future problems that have not yet been identified, using technologies not yet invented, based on scientific knowledge not yet discovered.

REFERENCES

Contreras, José N., and Armando Martínez-Cruz. "Representing, Modeling, and Solving Problems in Interactive Geometry Environments." In *Understanding Geometry for a Changing World,* Seventy-first Yearbook of the National Council of Teachers of Mathematics (NCTM), edited by Timothy V. Craine, pp. 233–52. Reston, Va.: NCTM, 2009.

Cuoco, Al, Paul Goldenberg, and June Mark. "Reader Reflections: Marion's Theorem." *Mathematics Teacher* 86, no. 8 (November 1993): 619.

Dietrich, Charles H., Dave Goldenheim, and Dan Litchfield. "Euclid, Fibonacci, and Sketchpad." *Mathematics Teacher* 90, no. 1 (January 1997): 8–12.

Feller, Leslie Chess. "The Eternal Challenge of Euclid's Geometry." *New York Times,* March 7, 1999.

Jackiw, Nicholas, and William Finzer. "Programming by Geometry: The Geometer's Sketchpad." In *Watch What I Do: Programming by Demonstration,* edited by Allen Cypher, pp. 293–308, Cambridge, Mass.: MIT Press, 1993.

Johnston, William. *Mathematics Teacher* 85, no. 2 (February 1992): cover.

Laborde, Jean-Marie, and Franck Bellemain. Cabri Geometry II. Software. Dallas, Tex.: Texas Instruments, 1994.

Leslie, John. *Elements of Geometry, Geometrical Analysis, and Plane Geometry.* Edinburgh, U.K.: Ballantyne, 1811.

NASA Education Partnerships Forum. *Summary Minutes* (September 2006): 13. education.nasa.gov/pdf/163824main_Forum Minutes.pdf.

Pieper, Shaun. "Visualizing the Complex Roots of Quadratics." *College Mathematics Journal* 28, no. 5 (November 1997): 359.

Pólya, George. *Mathematics and Plausible Reasoning, I.* Princeton, N.J.: Princeton University Press, 1954.

Quesada, Antonio. "New Students' Findings." In *Proceedings of the Thirteenth Annual International Conference on Technology in Collegiate Mathematics,* edited by Gail Goodell, pp. 312–16. Reading, Mass.: Addison Wesley Longman, 2001a.

———. "New Mathematical Findings by Secondary Students." *Universitas Scientiarum* (Javeriana, Bogota, Colombia) 6, no. 2 (2001b): 11–16. Available at www.javeriana .edu.co/universitas_scientiarum/vol6n2/ART1.htm.

Texas Instruments. *Getting Started with Cabri Geometry II for McIntosh, Windows, and MS-DOS:* Dallas, Tex.: Texas Instruments, 1996.

Todd, Philip. "Feynman's and Steiner's Triangle." *Journal of Symbolic Geometry* 1 (2006–2007): 85–90.

Vonder Embse, Charles, and Arne Engebretsen. *Geometric Investigations for the Classroom Using the TI-92.* Dallas, Tex.: Texas Instruments, 1996.

Watanabe, Tad, Robert Hanson, and Frank D. Nowosielski. "Morgan's Theorem." *Mathematics Teacher* 89, no. 5 (May 1996): 420–23.

Wilson, James. "Comments on the GLAD construction." University of Georgia, Department of Mathematics Education, 1998. Available at jwilson.coe.uga.edu/ Texts.Folder/GLaD/GLaD.Comments.html.

Using Circle-and-Square Intersections to Engage Students in the Process of Doing Geometry

Stephen Blair
Daniel Canada

WHAT does it mean to "do geometry"? Geometry is often presented to students as a finished product; they learn a preset system of definitions, theorems, constructions, and proofs. Although students need to learn about geometric knowledge that has been developed over the course of the human journey, they also need to learn that "geometry is not so much a branch of mathematics as a way of thinking" (Atiyah 1982). In essence, for students to really do geometry, they must develop and use geometric habits of mind (Goldenberg, Cuoco, and Mark 1998). (See also Driscoll et al. [2009] regarding habits of mind.) That is, they need to engage actively in open-ended problems that allow them to define mathematical objects and discover their properties. (See also de Villiers, Govender, and Patterson [2009] regarding definition.) Furthermore, they need to explore and make conjectures about relationships, develop explanatory proofs, and create classification schemes to systematize their understanding. Indeed, by highlighting the importance of both content and process standards, the National Council of Teachers of Mathematics (NCTM) has actively promoted the engagement of students in geometric thinking in the broad sense described above (NCTM 2000).

Given the need for students to engage actively in doing geometry, the challenge falls on teachers to develop, implement, and refine worthwhile tasks to use with students (NCTM 1991). An example of such a rich, open-ended task is given in this article. The story presented is a reflection of the authors' orchestration of this task in our own classes with high school students, preservice elementary school teachers, and in-service middle and high school teachers in Washington, Oregon, and Michigan over the past few years. In our article in the December 2006–January 2007 *Mathematics Teacher,* we introduced this task to readers and discussed how it could be used to promote meaningful discussion and group work (Canada and Blair 2006/2007). In this article, we focus on how we used this task to engage students in doing geometry in the spirit of authentic mathematical research, namely, as a cyclic process of conjecture, refutation, definition, and proof (Lakatos 1975). We offer our ideas and reflections not as an activity to be used in any particular classroom but as an example of a task that mathematics educators at all levels can use to consider some important issues in trying to engage students in doing geometry.

How Many Points of Intersection Are Possible between a Circle and a Square?

Our investigation with students always began with this simple, open-ended question, which we encouraged students to consider in small groups using pencil and paper. Students at both the high school and college levels generally began exploring the situation and made freehand drawings that convinced most of them that zero, one, two, four, and eight points of intersection are possible for a circle and square lying in the same plane. Some students produced drawings with five, six, or seven intersection points, but they were often unable to convince their peers because of the imprecision of their drawings. Indeed, some students became convinced that certain numbers of points were not possible, concluding, for example, "You can't get five unless you change the square to a rectangle or the circle to an oval" (Canada and Blair 2006/2007). After about half an hour of small-group exploration, we facilitated a whole-group discussion in which students presented and compared ideas based on their freehand sketches. The goal was for students to begin answering the question while also deciding as a group that they needed more-precise representational tools to explore the existence of certain possibilities. Another point highlighted during the discussion was the fact that the lack of a counterexample does not guarantee the truth of a conjecture.

The fact that their drawings were imprecise helped them realize that they needed to justify why a given situation was impossible.

Once students agreed on the need for more-precise representational tools, they were provided with graph paper, compasses, and rulers. Some classes were also able to use The Geometer's Sketchpad (Jackiw 2001) to explore the situation. These tools enabled students to make more-precise drawings, as illustrated in one preservice teacher's work (fig. 19.1), and they were eventually able to use these tools to convince one another that all the numbers of intersection points from zero to eight were possible.

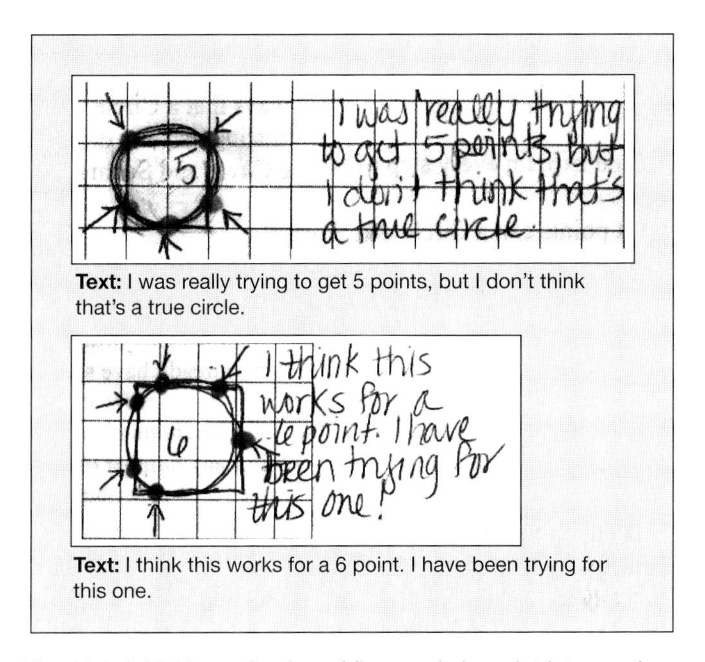

Fig. 19.1. Initial investigation of five- and six-point intersections

The level of discourse during this part of the investigation was kept informal for the task to be engaging for all students. Students were encouraged to refer to their existing knowledge when explaining their ideas but were not forced to present ideas in a formal way. We facilitated students' work at this stage mainly by asking clarifying and probing questions and by refraining from judging students' work, deferring questions such as "Is this correct?" to the small groups. We took this perspective so that students would take ownership of the activity and see themselves as geometers. We also wanted them to see an increasing role for rigor as their exploration of the material matured.

In What Different Ways Can a Circle and a Square Intersect?

The richness of this investigation lies not only in determining that each number of intersection points from zero to eight is possible but also in classifying the different ways in which each number of points can be attained. As students explored the initial question concerning the number of intersection points, they noticed that a configuration for a given number of intersection points could be done in multiple ways. When we asked students to describe the different ways they found, the focus of the investigation shifted to the geometric process of defining. Both high school and college students believed that the type of intersection point was rele-vant to whether two ways were the same, and most groups began to distinguish among intersections at a vertex, tangent to a side, and crossing through a side. Some groups of students also believed that the relative size and position of the circle to the square was also important, and they generally added descriptions to distinguish whether one was inside or "mostly" inside the other, as illustrated in figure 19.2.

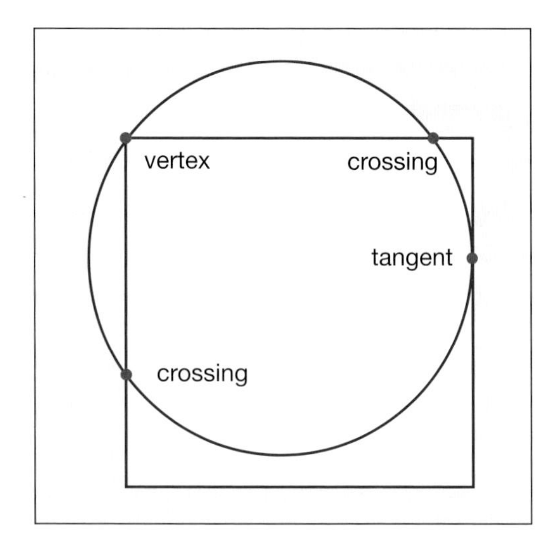

Fig. 19.2. An example of a vertex, a tangent, and two crossing intersection points with the circle "mostly" inside the square

An important aspect to note is that as students decided how they would describe different configurations, they also began to classify the different possibilities. Thus, the geometric processes of defining and classifying emerged together. As they worked, students soon realized that some of their methods were imprecise and also that the more complex their descriptions, the more difficult was the task of organizing all the possible configurations. For this reason, the more experienced groups of students generally decided not to include the relative position descriptions, and limited their system to the number and type of intersection points. Eventually, the students decided that different solutions should be based solely on the defining set of properties and no others, which is the essence of a precise mathematical definition. Thus, for example, all three of the configurations in figure 19.3 are different if relative position is included in the definition and the same if it is not included.

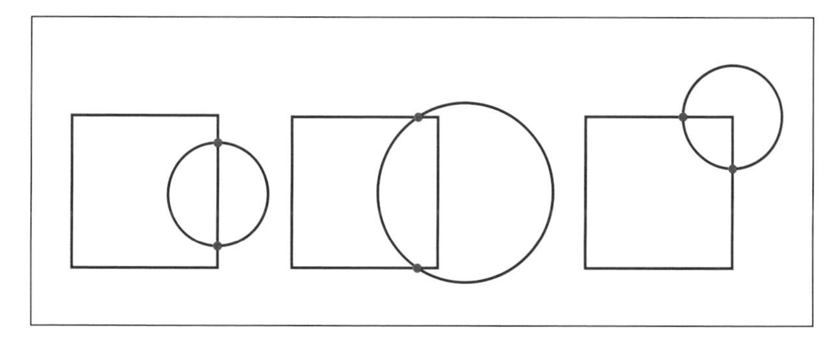

Fig. 19.3. Three configurations classified as having two crossing intersection points

Indeed, as they generated and compared different sketches, the students often noted the difficulty of ignoring attributes that were not part of their agreed-on defining properties. They began to view the subject more formally, by considering consequences of their stipulated definitions. This perspective has been characterized as an important aspect of advanced mathematical thinking (Tall 1992).

What Relationships Do You Notice in Your Classification?

As students worked to classify the different ways a circle and square can intersect according to their definitions, we encouraged them to look for connections and relationships. In one class of preservice and in-service teachers, several students noticed that the total number of crossing intersection points seemed to be even. When asked to share their ideas in a whole-class discussion, the group offered this relationship as a conjecture that they believed would always be true. The entire class was then asked to consider this relationship. Was it always true? Could we explain why or why not? After a few moments, several students found counterexamples to the conjecture, as shown in figure 19.4. They found that by having the circle "cross" at a vertex, the number of crossing points, which they had defined as crossing between endpoints of a side, could be odd.

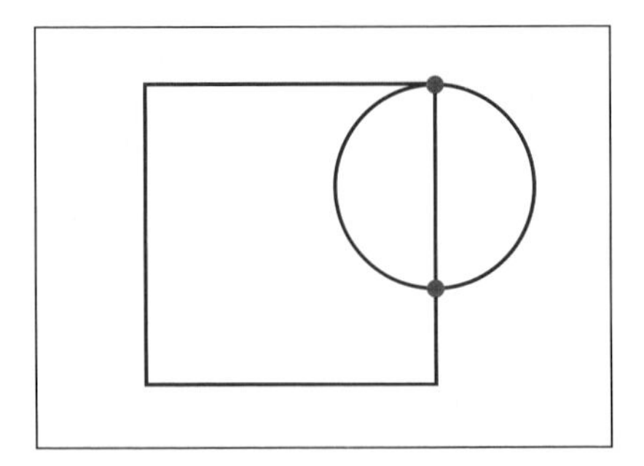

Fig. 19.4. A counterexample to the "even number of crossing points" conjecture because one of the intersection points crosses at a vertex

What happened next highlights just how this investigation really engaged the students in doing geometry. The students decided that the relationship, namely, that configurations always have an even number of crossing points, was really true but that their definitions and classification scheme were inadequate to support this relationship. At this point the instructors encouraged the students to discuss with one another which particular features were important and not to be afraid to "think outside the box." Several students noted that not all the "vertex" points were the same; whether the circle "crossed over" the related segment (an aspect they had already deemed relevant for side points) also made a difference. After some debate, the students decided that they needed to change their definitions for this relationship to be true. They decided that the type of intersections really exhibited two independent characteristics; *where* they occurred (at a vertex or along the interior of a side) and *how* they occurred (tangent to the point or crossing through the point). Thus, they created a classification that yielded the four combinations of these characteristics, as shown in figure 19.5.

With their new classification scheme, students were able to justify why every configuration must have an even number of crossing points. They also agreed that the new scheme was more elegant than their original scheme. We, as instructors, were filled with a sense of joy and amazement by the fact that our students' journey had truly become one of mathematical creation. Lakatos (1975), in his seminal book *Proof and Refutations*, noted that mathematics grows not in the order presented in textbooks but rather through a recurring cycle in which relationships are discovered, proofs are proposed and refuted, and then relationships are ultimately proved by refining appropriate arguments and definitions (Clements 2003). Our students had engaged in this cycle, from making initial definitions based on experimentation, to noticing an important relationship, to the final refinement of their scheme that enabled them to prove the result they felt should be true. Their agreed-on final proof used the fact that because the square is a simple, closed

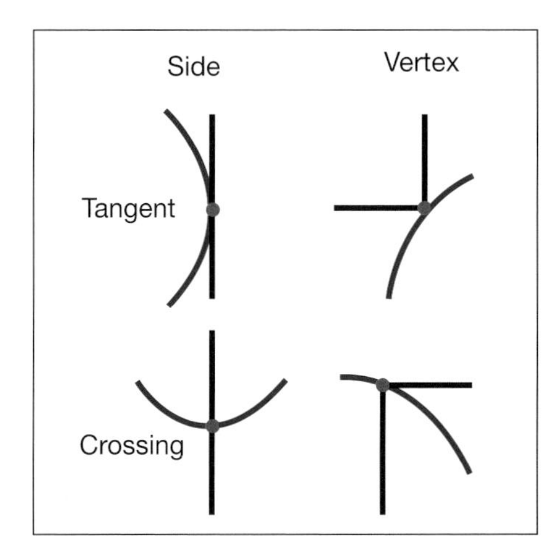

Fig. 19.5. Students' final classification scheme

curve, it partitioned the plane into points interior to, on, and exterior to the square. They argued that by starting at a nonintersection point on the circle and moving clockwise around it one time, they would move through each intersection point once. As they did so their location "state" from interior to exterior, or vice versa, would change only once at crossing intersection points, which is why they had to include crossing vertex points! Hence, because the state change toggled between two options (interior and exterior) and the final state had to be the same as the original state, they concluded that the number of crossing points had to be even.

Once the students were satisfied with their proof, we probed their understanding by trying to uncover some hidden assumptions concerning continuity and finiteness of the total number of intersection points. They argued why they knew the total number of intersection points had to be finite, but did not really understand the need for continuity. We noted that they were in good company, that in fact Euclid made a similar assumption in Proposition 1, Book 1 of *Elements,* and that only relatively recently Hilbert and others saw the need for a formal axiom of continuity (Hilbert 2001). The students were excited to know that their mathematical activity paralleled that of historical figures. After some discussion they agreed on the need to guarantee that curves are continuous to avoid the situation of one curve failing to intersect another by passing through a "microscopic hole." It also made sense to them that we, following in the footsteps of Hilbert, should take this property as an axiom. This follow-up discussion had the additional benefit of helping students articulate the difference between a definition and an axiom, two concepts sometimes viewed identically as "things taken to be true" even though they have very different roles in the process of doing geometry.

Once students settled on their classification scheme, they made a systematic search for all the different configurations for a given number of intersection points. Indeed, their theorem concerning an even number of crossing points became useful at this point because they could use it to narrow down the list. For example, for five total intersection points, zero, two, or four of them must be crossing points. Zero can be quickly eliminated as impossible, leaving only two and four. From the remaining possibilities, students determined that four configurations are possible, as shown in figure 19.6. Note that the first pair of configurations in figure 19.6 has two crossing points and the second pair has four crossing points.

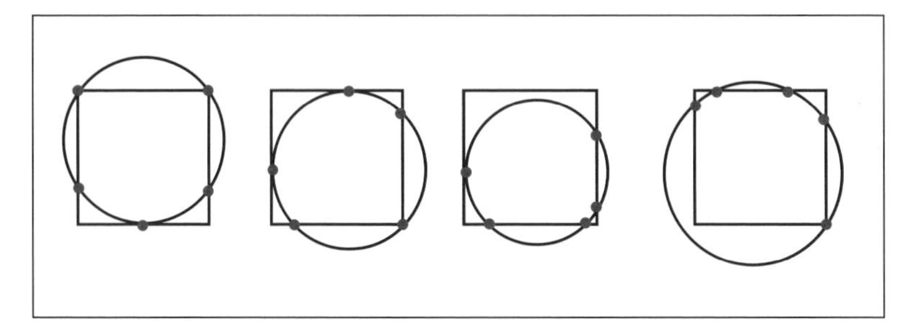

Fig. 19.6. Possible five-point intersections

How Can You Construct a Given Configuration?

As students worked to classify all possible configurations, they continued to identify and explore interesting relationships and soon believed that their classification was complete. When they reached this juncture, we shifted the investigation one more time and asked students how to construct (as in the Euclidean sense, using the equivalent of a straightedge and compass) some of the configurations they had found. This extension was not completely out of the blue for students, because they had already been trying to represent different configurations using a variety of construction tools. The problem for most students, however, lies in moving from using the tools to make an approximate drawing to using them in a coordinated way that logically results in a precise construction (Clements 2003). We asked students for an explicit sequence of steps, that is, an algorithm, that would produce the configuration. Such statements as "… and then play with the circle until it fits" were not allowed (Canada and Blair 2006/2007). Our reason for this assignment was to bridge the gap between finding a configuration and formally proving that it actually exists; we intended for students' constructions to lead to constructive proofs (de Villiers 2003).

The students found these constructions to be both interesting and challenging. As they worked on a variety of constructions, students began to notice and use different relationships about circles, such as the fact that the perpendicular bisector of a chord intersects the center of a circle. Teachers tend to assume that if students are aware of a given theorem, they will be able to use it effectively, but we found that our students struggled to see how a theorem might apply to a particular situation. As they tried to construct different configurations, however, students got quite creative. By using probing questions, we were able to help them draw out and identify more-general relationships. For example, one student discovered that he could construct a "corner" of a square within a given circle by connecting a point on the circle to the endpoints of any diameter (fig. 19.7). This technique was useful for creating several configurations and was quickly adopted by many students. When asked why the strategy worked, the class discussed it for more than ten minutes before they recognized it as a special case of the inscribed angle theorem, namely, that the measure of an inscribed angle of a circle equals one-half the measure of its corresponding central angle.

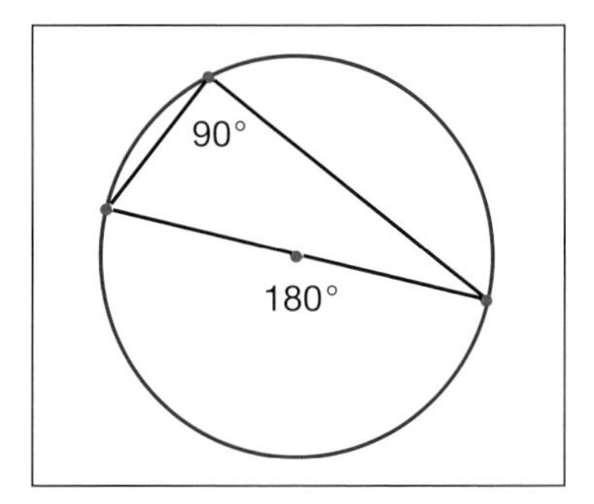

Fig. 19.7. Constructing a right angle within a circle as a useful application of the inscribed angle theorem

An interesting aspect of having students construct different configurations is the variety of ways in which they may approach the task. When a theorem or construction is presented in a textbook, it is often justified in only one way. Thus, students tend to think that a theorem has only one proof and furthermore, that someone has already found it. The constructions in this investigation were different in that students could not copy them from a book. Instead, they tended to find several ways to construct and justify a given configuration. Even more surprising was the fact that the students' methods were correct but often substantially different from what we had anticipated they would find. For example, consider the five-point configuration with one tangent-side, two crossing-side, and two tangent-vertex intersection points (shown in fig. 19.8). Starting with a square, we correctly anticipated that students would first construct the perpendicular bisector of one side, *AE,* making the endpoints the two tangent-vertex points and noting that the perpendicular bisector would intersect the opposite side at the tangent-side point (*C*). From that point on, however, none of the students finished the construction in the anticipated way. We had expected students to construct the chord from the tangent-side point (*C*) to one of the two tangent-vertex points (*E*) and then construct its perpendicular bisector. Since perpendicular bisectors of chords of a circle contain the center of the circle, we can determine the needed center by constructing the intersection of the perpendicular bisectors. Once we have located the center, we can construct the circle with that center through any of the other three points, as shown in figure 19.8.

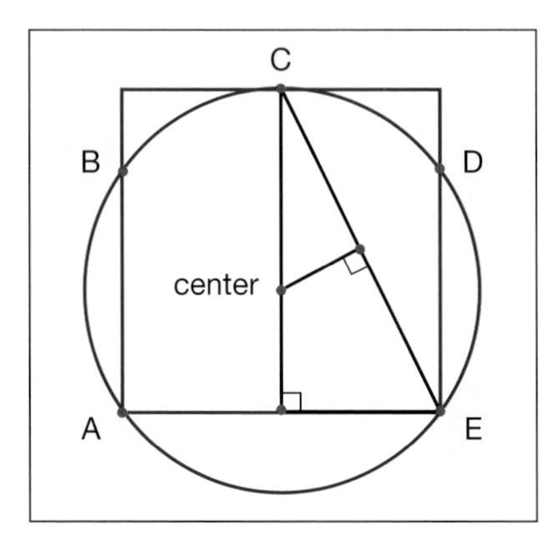

Fig. 19.8. An anticipated construction

The students' actual approaches, however, were more complex and brought out some interesting relationships. One group used cyclic quadrilaterals, a topic we had recently explored. They constructed the segment from one of the tangent-vertex points *A* to the tangent-side point *C* and noticed that if that were connected to the appropriate point *D,* one of the desired crossing-side points, then a cyclic quadrilateral *ACDE* could be formed, as shown in figure 19.9. Furthermore, they remembered that opposite angles of a cyclic quadrilateral are supplementary: if one angle is 90 degrees, the one opposite is also 90 degrees. Hence, they needed to make angle *ACD* a right angle, so they constructed the perpendicular at *C.* They then reasoned that segment *DA* had to be a diameter using the inscribed angle theorem. The midpoint *M* of this diameter determined the center of the circle.

Another group of students came up with a totally different approach. After an extended amount of exploration using The Geometer's Sketchpad, they created the following directions. First construct the midpoints *C* and *F* of the top and bottom sides of the square. Then find the midpoint *G* of segment *CF.* Next find the midpoint *H* of segment *GF*. Finally, find the midpoint *M* of segment *GH;* this point is the desired center of the circle. Crazy as it seems, their method seemed to work, but they were not sure why (fig. 19.10).

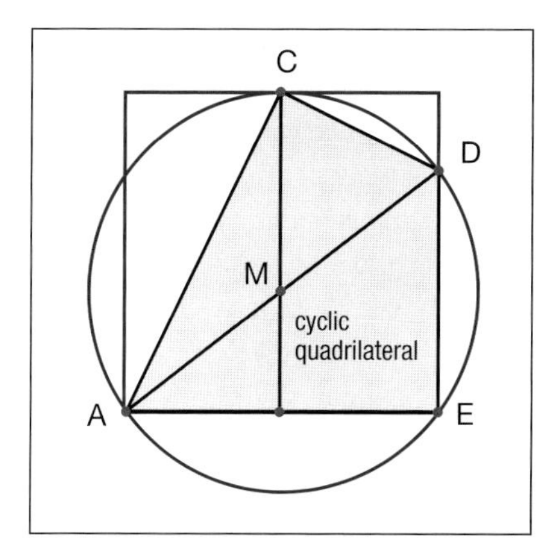

Fig. 19.9. One construction by students of a five-point intersection

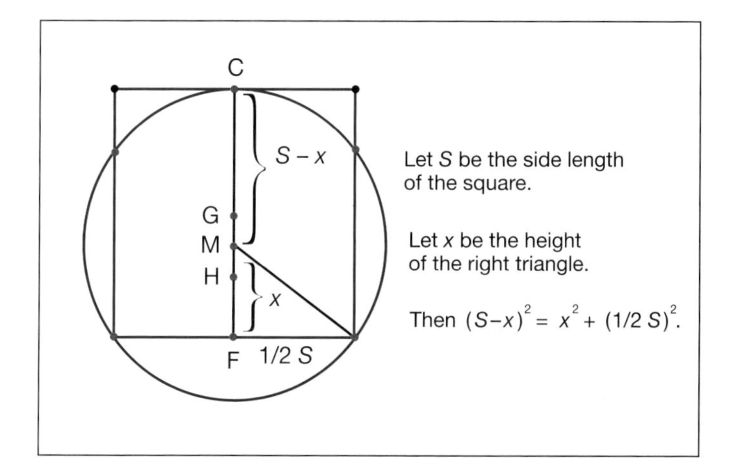

Fig. 19.10. One method by students, using the Pythagorean theorem

After they presented their idea to the whole class, we asked the class to discover why it worked. Before long, several students saw how to use the Pythagorean theorem to prove that point *M* did have the desired property (i.e., that it was equidistant from the intersection points), but they still did not have a good sense of why it worked.

We were intrigued to observe that these students started with a "black box" method that just worked for no known reason; then moved to an analytic method that, in their opinion, failed to provide a clear explanation; and finally produced an intuitive combination of both.

Finally, one student who had been using the "show grid" feature of The Geometer's Sketchpad reported with excitement that she had found a "3-4-5" right triangle, as shown in figure 19.11. On seeing this, the students realized why this method works. Assuming a unit square, because the lower side is split in half, each half is four-eighths of a unit long, whereas segment *CF* is divided such that point *M* is three-eighths of a unit from the bottom. This construction creates a right triangle with legs of three- and four-eighths, so the hypotenuse is five-eighths, matching the length of segment *CM*.

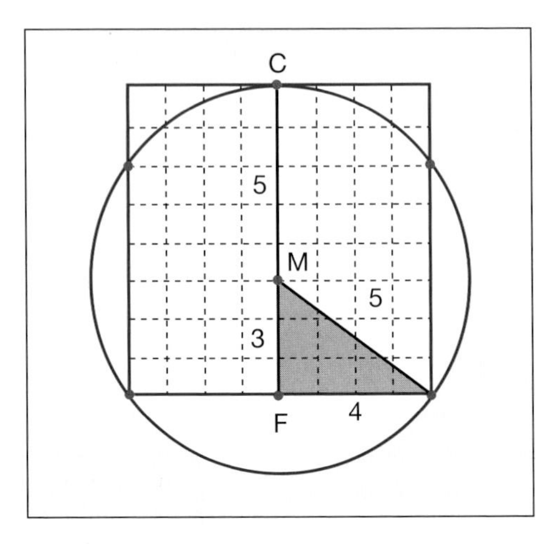

Fig. 19.11. Another method by students, using the Pythagorean theorem and showing an underlying 3-4-5 right triangle

Concluding Remarks

The investigation of intersections of a circle and a square served as a rich context for our students to engage in the process of doing geometry, and we encourage you to try it with your students. More important than using this particular task, however, is the opportunity it presents to help us create other rich tasks. We did not design the task to present or apply a particular concept or theorem, but rather as a means to guide students though the processes of representing, defining, classifying, constructing, and proving. The general structure of the investigation followed a *launch, explore,* and *extend* format, which transfers to other strands of mathematics education besides geometry.

We *launched* the investigation by beginning with a simple question and imprecise representational tools. From the beginning, students were able to take ownership of the investigation because the problem was posed in an open-ended way. They were able to generate a wealth of configurations and were encouraged to decide for themselves whether any two configurations were the same. The initial question was deliberately imprecise in that we did not tell them what we meant by "a different way." Consequently, our students had to engage in the processes of defining and classifying as they explored the situation.

Secondly, as our students *explored* the situation more deeply, we expanded the initial investigation by encouraging them to attend to relationships and make conjectures concerning the situation. Exploring these conjectures led some students to see shortcomings in their initial definitions, which resulted in their decision to refine their classification schemes. They also began to use more-precise representational tools, such as The Geometer's Sketchpad. Doing so helped them connect the relationships they found with other concepts, such as the inscribed angle theorem. Thus, the problem, although initially presented in an intuitive manner, was open to more-precise analysis and connected with meaningful mathematics.

Furthermore, creating their own classification system for the circle-and-square intersection situation also helped our students see the need for more-formal justifications of their results. Our students were willing to *extend* the investigation by examining how the configurations they discovered could be constructed, and they were able to do so in multiple ways. The students were able to prove their constructions through other relationships, such as the Pythagorean theorem. Their proofs not only verified their results but also established connections with different concepts, an important role for proof in geometry (de Villiers 2003). When we designed the circle-and-square investigation, we intended it to begin informally and lead to formal constructions and proofs in a natural way.

Variations on this problem can lead to similarly rich geometric investigations. One might explore how other planar shapes, such as a square and an ellipse, intersect. Another would be to explore how a cube and a plane intersect in space. The amount of interesting mathematics that can be generated from such seemingly simple questions is surprising. Indeed, the theory of conics is historically related to a similar question concerning the intersection of a plane and a double cone. In sum, if instructors are willing to embark on a mission of exploration with their students with rich tasks such as these, then they are inviting students to "do geometry" with them.

REFERENCES

Atiyah, Michael. "What Is Geometry?" In *The Changing Shape of Geometry,* edited by Chris Pritchard, pp. 24–29. Cambridge, U.K.: Cambridge University Press, 2003.

Canada, Daniel, and Stephen Blair. "Intersections of a Circle and Square: An Investigation." *Mathematics Teacher* 100, no. 5 (December 2006/January 2007): 324–28.

Clements, Douglas H. "Teaching and Learning Geometry." In *A Research Companion to "Principles and Standards for School Mathematics,"* edited by Jeremy Kilpatrick, W. Gary Martin, and Deborah Schifter, pp. 151–78. Reston, Va.: National Council of Teachers of Mathematics, 2003.

de Villiers, Michael. *Rethinking Proof with The Geometer's Sketchpad.* Emeryville, Calif.: Key Curriculum Press, 2003.

de Villiers, Michael, Rajendran Govender, and Nikita Patterson. "Defining in Geometry," In *Understanding Geometry for a Changing World,* Seventy-first Yearbook of the National Council of Teachers of Mathematics (NCTM), edited by Timothy V. Craine, pp. 189–204. Reston, Va.: NCTM, 2009.

Driscoll, Mark, Michael Egan, Rachel Wing DiMatteo, and Johannah Nikula. "Fostering Geometric Thinking in the Middle Grades: Professional Development for Teachers in Grades 5–10." In *Understanding Geometry for a Changing World,* Seventy-first Yearbook of the National Council of Teachers of Mathematics (NCTM), edited by Timothy V. Craine, pp. 155–72. Reston, Va.: NCTM, 2009.

Goldenberg, E. Paul, Albert Cuoco, and June Mark. "A Role for Geometry in General Education." In *Designing Learning Environments for Developing Understanding of Geometry and Space,* edited by Richard Lehrer and Daniel Chazan, pp. 3–44. Mahwah: NJ.: Lawrence Erlbaum Associates, 1998.

Hilbert, David. *Foundations of Geometry*. La Salle, Ill.: Open Court Classics, 2001.

Jackiw, Nicholas. The Geometer's Sketchpad. Version 4.0. Software. Emeryville, Calif.: Key Curriculum Press, 2001.

Lakatos, Imre. *Proofs and Refutations.* Cambridge, U.K.: Cambridge University Press, 1975.

National Council of Teachers of Mathematics (NCTM). *Professional Standards for Teaching Mathematics.* Reston, Va.: NCTM, 1991.

———. *Principles and Standards for School Mathematics.* Reston, Va.: NCTM, 2000.

Tall, David. "The Transition to Advanced Mathematical Thinking: Functions, Limits, Infinity and Proof." In *Handbook of Research on Mathematics Teaching and Learning,* edited by Douglas Grouws, pp. 495–511. New York and Reston, Va.: Macmillan Publishing Co. and National Council of Teachers of Mathematics, 1992.

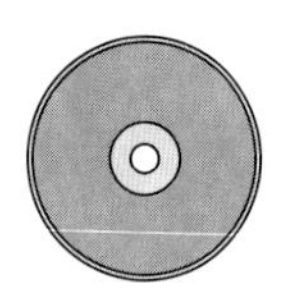

Geometer's Sketchpad files that support this article are found on the CD-ROM disk accompanying this Yearbook.

Area Formulas with Hinged Figures

Alfinio Flores

STUDENTS often learn the formulas for the areas of such figures as rectangles, triangles, parallelograms, trapezoids, and regular polygons without making connections among them. Often, too, students do not realize that the algebraic transformation of one area formula into a different equivalent expression, for example,

$$\frac{bh}{2} = \frac{b}{2}h,$$

can sometimes be interpreted in geometric terms as well as algebraically. The goal of this article and its corresponding Web site is to help students make such connections with the use of interactive, hinged figures that correspond to each of figures in this article. (The interactive figures can be found in the companion disk and at www.math.udel.edu/~alfinio/yearbook09/areahinged.html. Readers are encouraged to try using these interactive figures before or during their reading of this article.) These hinged figures can be manipulated in different ways. Some are transformed by rotating some of their parts to form other shapes that have the same area. Others are rotated as a whole to form, together with the original, a new figure. By comparing the original figures with the new figures, students can establish relationships among different formulas for areas. As they rotate the figures or their parts, students can get a kinesthetic as well as a visual sense of what parts of the original figure correspond to parts of the transformed figure. The guidance of a teacher is crucial in helping students interact with the figures in a way that will maximize their understanding.

We briefly discuss how this interaction can play out when the emphasis is on the learning of geometrical ideas per se and then discuss how the interactive figures and the formulas for areas can provide a bridge to algebraic thinking and notation. Finally, we look at each of the figures and indicate the questions that can be asked to get students to engage with them.

Several of the arguments of this article depend on the fact that when a segment is rotated 180 degrees about its midpoint, it coincides with itself. This result can be demonstrated by letting students experiment with concrete, straight objects such as toothpicks or ice-pop sticks to represent segments that can be rotated about their midpoints. The interactive figures then provide an additional experiential setting where students can see what properties are preserved when figures or their parts are rotated in other ways. They can see, for example, that a segment rotated 180 degrees around a point not on the segment will give another segment that is congruent and parallel to the first. They can also see relationships that emerge from the way the figures are partitioned—for example, that the segment connecting the midpoints of two sides of a triangle is parallel to the third side. For older students who are learning to develop deductive arguments in geometry, the interactive figures can suggest conjectures and offer leads to understanding why they are true. Students can then prove the relationships between the original and the transformed figures by using traditional arguments of congruence or by basing their arguments on properties of rotations previously proved or accepted. Students can also see how the formulas for areas can be organized into a deductive system in which some formulas are derived from others.

When the emphasis is on bridges to algebra, students can work through a sequence of geometric figures with the same area and find corresponding algebraic expressions for each figure and its parts. The different parts of the algebraic expression correspond to parts of the figure in each instance. Alternatively, teachers can give students a sequence of algebraic expressions that are equivalent and ask students to find geometric interpretations for each term or factor in the different expressions. Area formulas can thus provide a context for meaningful manipulation of algebraic expressions. On one hand, students have the opportunity to interpret geometrically the parts of the different formulas and see why they are equivalent. On the other hand, students can use algebraic principles such as the associative, commutative, and distributive properties to transform one algebraic expression into another. Whereas often students simply learn to substitute specific numerical values into the formulas, here the purpose is to treat the different elements in a formula as mathematical objects per se. Some instances require subtleties in the algebraic reasoning, and the teacher may want to make them explicit. Students with a wide range of mathematical backgrounds can learn about areas and geometrical relationships by interacting with the figures. Teachers can offer help and guidance according to the mathematical maturity of the students to develop the more subtle algebraic aspects.

Questions are posted on the CD and Web site for each of the interactive figures to help students focus their attention on relevant relationships. The teacher still needs to play an active role, however, guiding students so they realize the area formulas that are being illustrated in each interactive figure. Appropriate and timely prompts from the teacher support students in seeing the relationships, describing them in their own words, and using mathematical notation.

The formula for area on which all the other formulas in this article are based is that for the area of a rectangle. To make the connection more explicit, instead of using length and width to describe the rectangle, we say that the rectangle has base b and height h, so that the area (A) formula is given by $A = bh$.

Triangle

For the area of triangles, the simplest case is the right triangle. Figure 20.1 shows one example. The hinge or pivot is at the midpoint of the hypotenuse. The triangle is rotated 180 degrees around the hinge. Teachers can help students make explicit the mathematical relationships by asking questions to focus their attention on properties and relations of different parts.

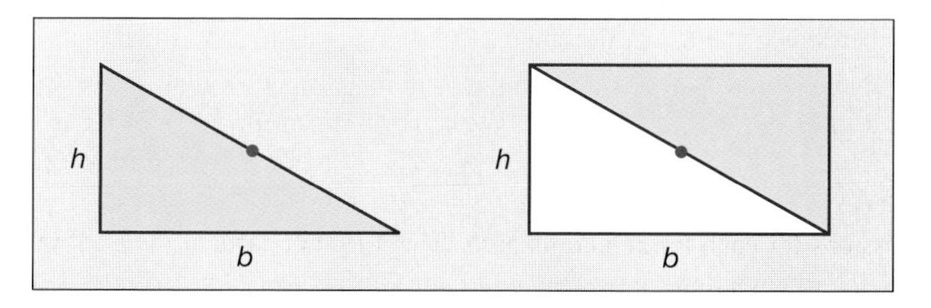

Fig. 20.1. Right triangle and rectangle

- What is the area of the original triangle compared with that of the rotated triangle?
- What shape do the original right triangle and the rotated triangle form together? How can you justify your claim?
- What is the relationship between the area of the triangle and the area of the rectangle?
- If the base of the rectangle (its length) is b and its height (width) is h, what is the formula for the area of the rectangle?

- What should be the formula for the area of the right triangle?

The goal is for students to express the area of the right triangle as

$$A = \frac{1}{2}bh.$$

The rectangle can also be used to derive the formula for more-general triangles. In the next case, the altitude of the triangle falls in the triangle's interior (fig. 20.2). In figure 20.2, the hinges are at the midpoints of two sides of the triangle. The segment with measure h is perpendicular to the base of the triangle and forms two right triangles. As students interact with this figure, teachers can again ask questions so that they can make explicit the relationships among parts.

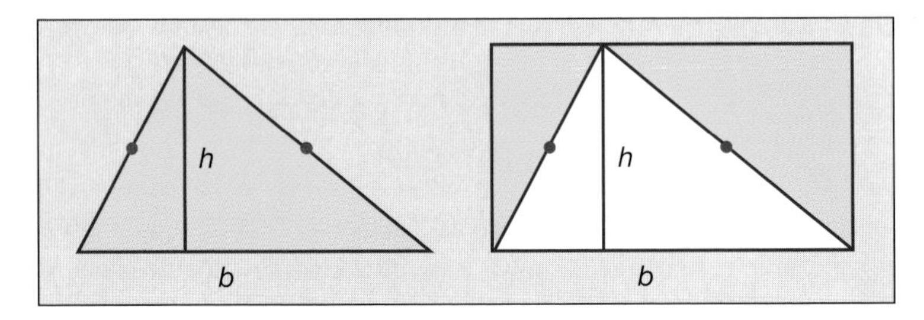

Fig. 20.2. Triangle with inside altitude

- How are the areas of these two triangles together related to the area of the original triangle?
- What figure do the two rotated parts form together with the original triangles?
- How can we justify that the new figure is, indeed, a rectangle?
- What is the area of the original triangle compared with the area of this rectangle?
- What is the base of this rectangle in terms of the original triangle?
- What is the height of this rectangle compared with the height of the original triangle?
- Using your own words, express the relationship between the areas of the original triangle and the rectangle verbally.
- If the area of the rectangle is given by *bh,* write a formula for the area of the triangle.

The goal is for students to express the area of the triangle as

$$A = \frac{1}{2}bh.$$

Of course, students can also say that the area of the triangle is equal to the area of the rectangle divided by 2, and write it as

$$A = \frac{bh}{2}.$$

Students can also describe how the triangle is formed by two right triangles and can use letters to label the parts and algebraic notation to express the relationships. If an altitude of measure h cuts the base of the triangle into parts q and p (fig. 20.3), what are the bases of the two right triangles? What is the sum of the bases of the two right triangles compared with the base of the original triangle? ($q + p = b$.) Students can express the area of the triangle as the sum of the areas of the two right triangles and can simplify the algebraic expression using the distributive and commutative properties:

$$A = \frac{1}{2}qh + \frac{1}{2}ph = \frac{1}{2}h(q + p) = \frac{1}{2}hb.$$

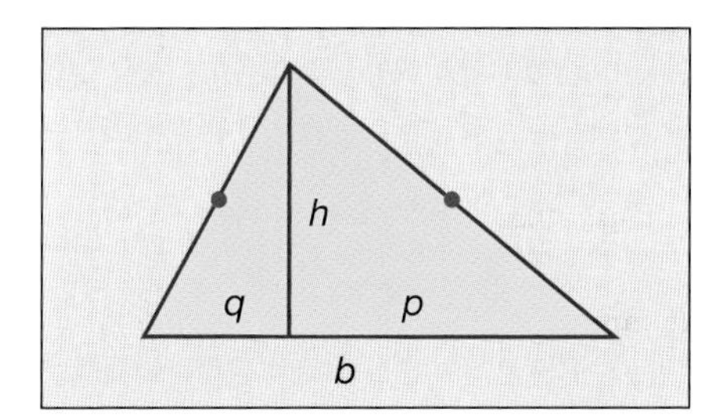

Fig. 20.3. Triangle area as sum of areas of right triangles

Every triangle has three altitudes, and the area can be found with any side as base and the corresponding altitude. In figure 20.4, three copies of a triangle are shown to illustrate how the argument above could be made using any of the three altitudes. Working with these copies may help students overcome the misconception that a triangle has only one altitude, and that it must be associated with a side that appears horizontally.

A different approach is to rotate parts of each of the two right triangles. In figure 20.5, the hinges are again at the midpoints of two sides. Students need to realize that the line connecting the midpoints is parallel to the base, is perpendicular to the altitude with measure h, and cuts the altitude in half. The teacher may want to ask students to justify each of these facts. Here are some other questions teachers can ask to make the relationships explicit as students interact with the figures.

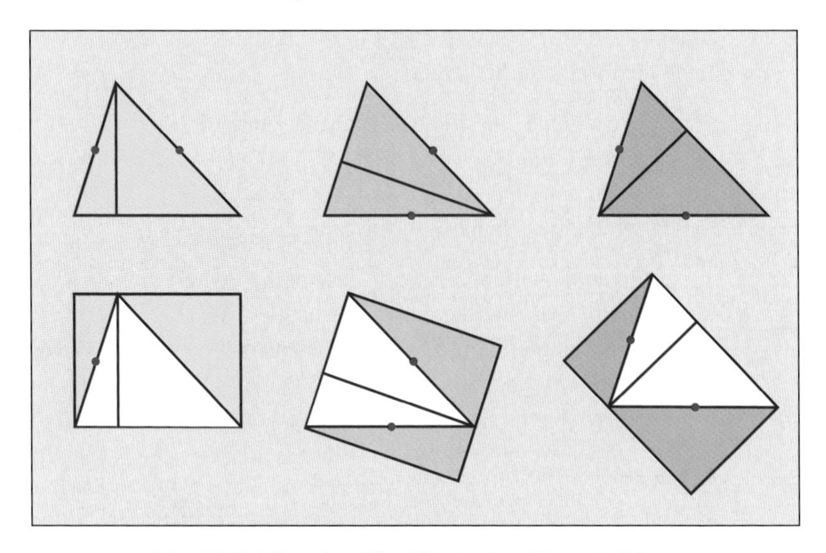

Fig. 20.4. Triangle with altitudes to different sides

- What kind of shape do the two rotated triangles form together with the section of the original triangle that did not move? How can you justify your claim?
- How does the base of this rectangle compare with the base b of the original triangle?
- How does the height of the rectangle compare with the height h of the original triangle?
- If we write $\frac{h}{2}$ for the height of the rectangle, what would be a formula for the area of the rectangle?

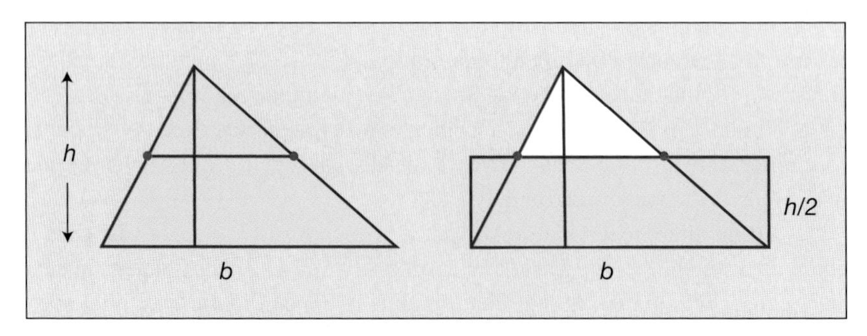

Fig. 20.5. Rectangle with half the height of triangle

The goal is for students to see that the area of a triangle is equal to the area of a rectangle with the same base and half the height,

$$A = b \times \frac{h}{2}.$$

In figure 20.6, the hinges are again at the midpoints of the sides, and two small right triangles are formed by segments through the midpoints perpendicular to the base of the triangle. The segments through these points are thus parallel to the altitude and will cut each of the segments of the base of length p and q in half. Teachers might ask their students why this result is true. Teachers can guide students to express the relationship between these half-segments and b using algebraic notation,

$$\frac{p}{2} + \frac{q}{2} = \frac{(p+q)}{2} = \frac{b}{2}.$$

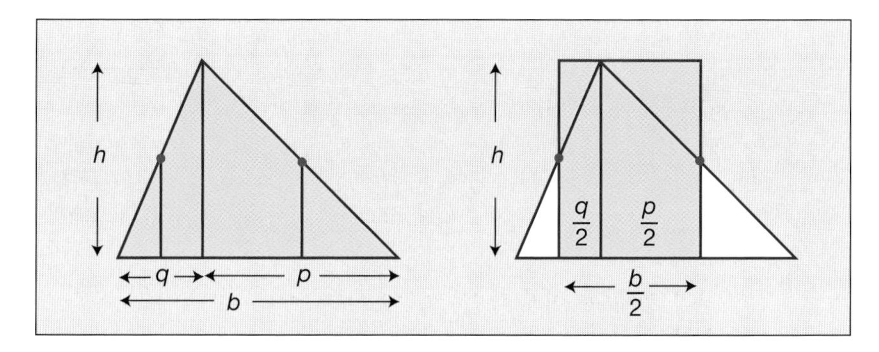

Fig. 20.6. Rectangle with half the base of triangle

The teacher may also have students explain why the new shape is a rectangle. Students can thus see that the area of a triangle is equal to the area of a rectangle with the same height and half the base,

$$A = \frac{b}{2} \times h.$$

We thus have different algebraic expressions for the area of the triangle,

$$\frac{1}{2}bh = \frac{bh}{2} = b \times \frac{h}{2} = \frac{b}{2} \times h.$$

The first two express essentially the same geometric fact; each of the last two has a different geometrical interpretation. The teacher may also want to make explicit how the commutative and associative properties are used in transforming one formula into the other, for example,

$$b \times \left(\frac{1}{2} \times h\right) = \left(b \times \frac{1}{2}\right) \times h$$

by the associative property, and

$$b \times \frac{1}{2} = \frac{1}{2} \times b$$

by the commutative property.

Rectangle and Parallelogram

In figure 20.7, the hinges are at the midpoints of the sides of the parallelogram. When the small triangles are rotated, a rectangle is formed. (Why?) The base of the parallelogram has the same length as the base of the rectangle formed by rotating the two triangles; students see this relationship visually or by reasoning as follows. If c is the base of the small triangle, the base of the rectangle will be $b - c + c = b$. The area of the rectangle is thus bh, and therefore the area of a parallelogram is given by $A = bh$. This approach works for parallelograms in which the projections of the midpoint intersect the bases. An approach that works with all parallelograms, including the "long skinny ones," is to use the connection with the formula of a triangle, as shown in the next section.

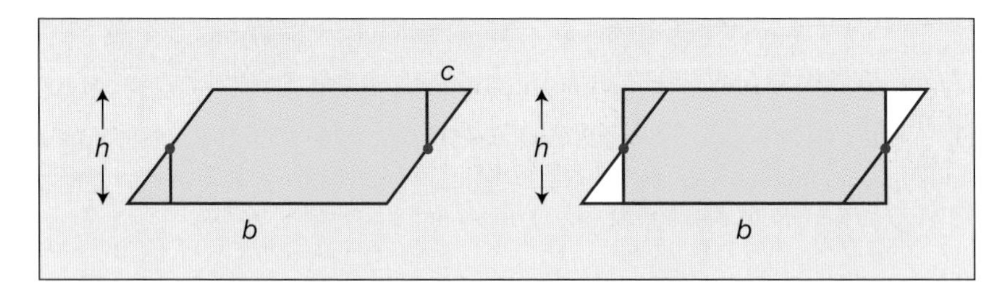

Fig. 20.7. Rectangle with the same base and same height as parallelogram

Triangle and Parallelogram

We can find another route to the formula of the parallelogram by using the formula for the area of a triangle. Figure 20.8 shows that two copies of an acute triangle make a parallelogram. Teachers can ask questions so that students explicitly understand that, indeed, the rotated triangle together with the original form a parallelogram, that the base of the triangle is the same as the base of the parallelogram, and that the height of the triangle is also the same as the height of the parallelogram. Students can thus see that the area of the parallelogram is two times the area of the triangle with the same base and the same height. The teacher can also ask students to express the relationship between the area of the triangle and the parallelogram using algebraic notation,

$$A = 2 \times \frac{1}{2} bh = bh.$$

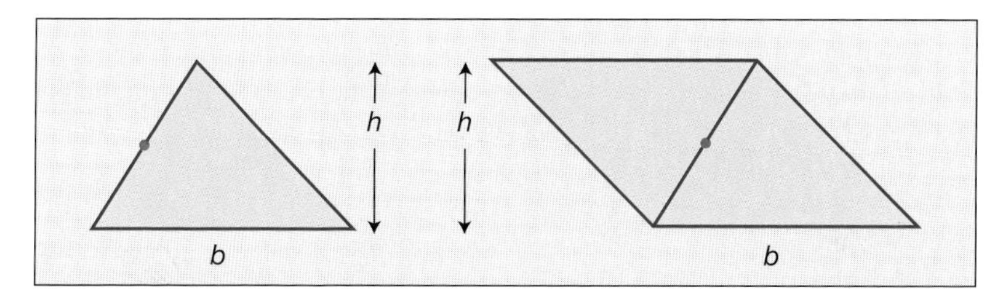

Fig. 20.8. Parallelogram with the same base and same height as triangle

Figure 20.9 shows another connection between the area of a triangle and a parallelogram. The hinge is at the midpoint of a side of the triangle, and a small triangle is formed by a segment through the midpoint parallel to the base. Students can be guided to see and justify why (1) the rotated small triangle together with the part of the triangle that did not move form a parallelogram, (2) the area of the triangle is the same as the area of the parallelogram, (3) the base of the triangle is the same as the base of the parallelogram, and (4) the height of the parallelogram is half the height of the original triangle. If we let students express the height of the parallelogram as h, they will express the height of the triangle as $2h$. They can then express the area of the triangle algebraically and can simplify the expression to obtain a formula for the area of a parallelogram,

$$A = \frac{1}{2} b \times 2h = bh.$$

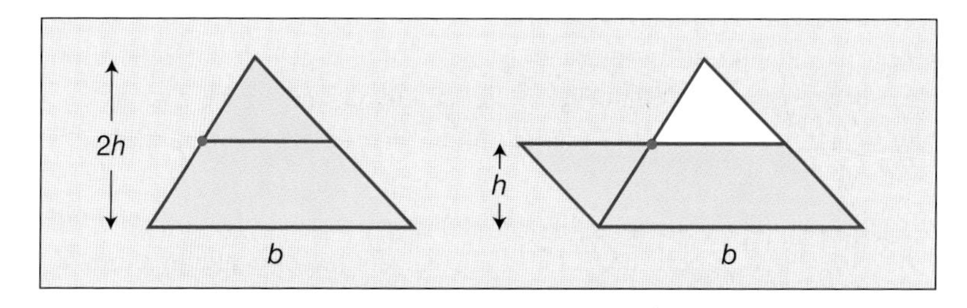

Fig. 20.9. Parallelogram with half the height of triangle

The area formula for triangles was developed in figures 20.1 through 20.6 only for right and acute triangles. Once we know the formula for the area of a parallelogram, we can find that the formula for the area of a triangle works for all kinds of triangles, even when the obtuse angle is at the base, that is, when the altitude lies in the triangle's exterior. In figure 20.10, the hinge is at the midpoint of the longest side of the triangle, the side opposite the obtuse angle. Students need to make explicit (1) that the rotated triangle together with the original form a parallelogram, (2) that the base of the triangle is the same as the base of the parallelogram, and (3) that their heights are the same. Students can thus view the obtuse triangle as half of a parallelogram with the same base and the same height. The area of the triangle with its altitude in the exterior is therefore also given by

$$A = \frac{1}{2}bh.$$

Of course, students can see that the area formula works for obtuse triangles using other methods, for example, expressing the area of the obtuse triangle as the difference of the areas of two right triangles.

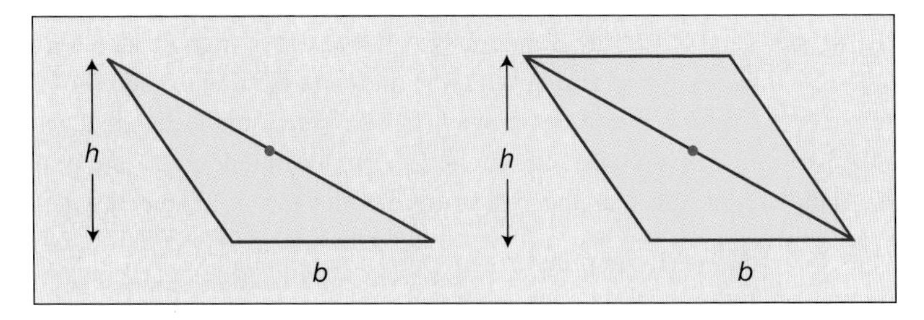

Fig. 20.10. Parallelogram from triangle with altitude outside

Trapezoid and Parallelogram

The formula for the area of a trapezoid can be related to the formula for the area of a parallelogram. In figure 20.11, the hinge is at the midpoint of one of the legs of the trapezoid. The teacher can guide students to see and justify why (1) the rotated trapezoid together with the original form a parallelogram, (2) the base of the parallelogram is equal to the sum of the lengths of the bases of the trapezoid, and (3) the parallelogram and trapezoid have the same height. Students can thus see that a trapezoid has half the area of a large parallelogram with the same height and whose base is the sum of the bases of the trapezoid. If we denote the measure of one base of the trapezoid as a and the other as b, then students can express the area of the large parallelogram as $(a + b)h$. Students can therefore write the area of the trapezoid as

$$A = \frac{1}{2}(a+b)h.$$

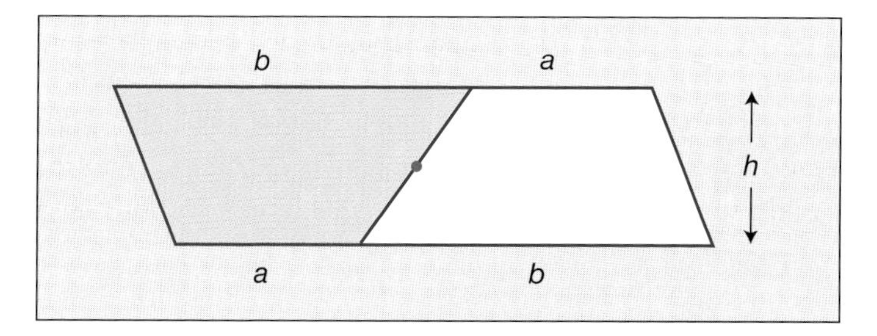

Fig. 20.11. Trapezoid and parallelogram with the same height

Students can also see that the segment joining the midpoints of the legs of a trapezoid is parallel to the bases and has a length equal to the arithmetic average of the two bases, $(a + b)/2$ (fig. 20.12), because two times this segment is equal to the sum of the bases. By writing the formula as

$$A = \frac{(a+b)}{2}h,$$

students can see another interpretation for the area of the trapezoid as the product of the average of the bases times the height.

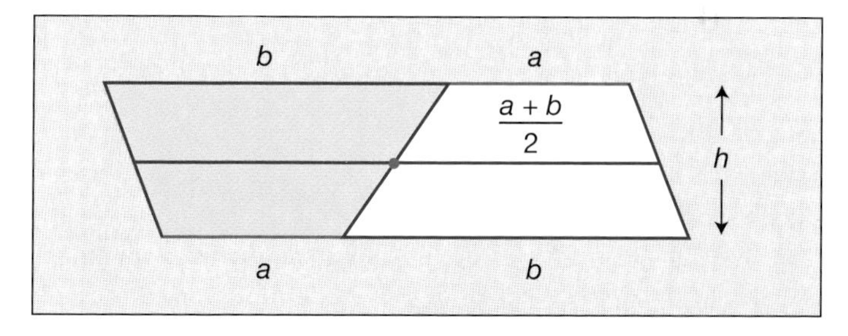

Fig. 20.12. Trapezoid showing the average of the bases

In figure 20.13, teachers can direct the attention of the students to the distance from the segment through the midpoints of the legs to each of the bases. Students can realize and justify why the line is equidistant from both bases and why the distance is half the height of the original trapezoid. Thus, when the upper part of the trapezoid is rotated, a parallelogram is formed. The teacher can ask questions so that students make explicit that, indeed, the new shape is a parallelogram, that its height is half the height of the original trapezoid, and that the base of the long parallelogram is the sum of the bases of the trapezoid. Students can state that a trapezoid has the same area as a parallelogram with half the height and whose base is equal to the sum of the bases of the trapezoid,

$$A = \frac{h}{2}\left(a + b\right).$$

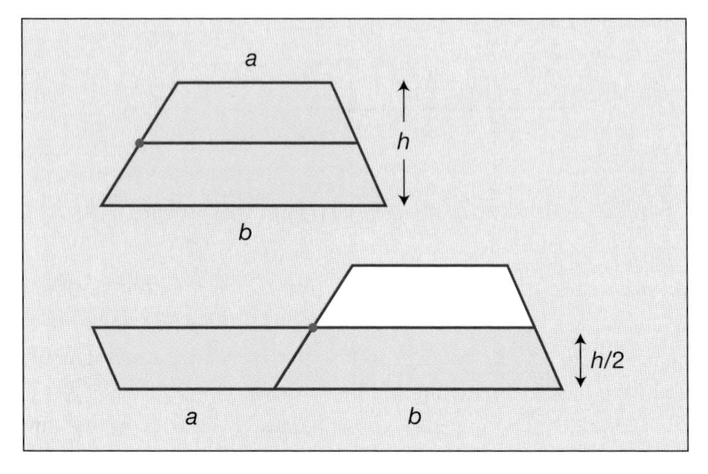

Fig. 20.13. Parallelogram with half the height of trapezoid

We have thus three equivalent formulas for the area of the trapezoid,

$$\frac{1}{2}\left(a + b\right)h = \frac{\left(a + b\right)}{2}h = \left(a + b\right)\frac{h}{2},$$

each having a different geometrical interpretation. Students should be able to explain when we use associative and commutative properties to transform one formula algebraically into another.

We can also transform a trapezoid into a parallelogram by rotating one small triangular part. In figure 20.14, the hinge is at the midpoint of one of the legs of the trapezoid. The small triangle has one side parallel to the other leg of the trapezoid. The teacher can ask students to justify that indeed the new shape is a parallelogram.

How to find the length of the base of this parallelogram may not be immediately obvious, however. Students might observe that the base is equal to the segment through the hinge and parallel to the bases of the trapezoid. Alternatively, if

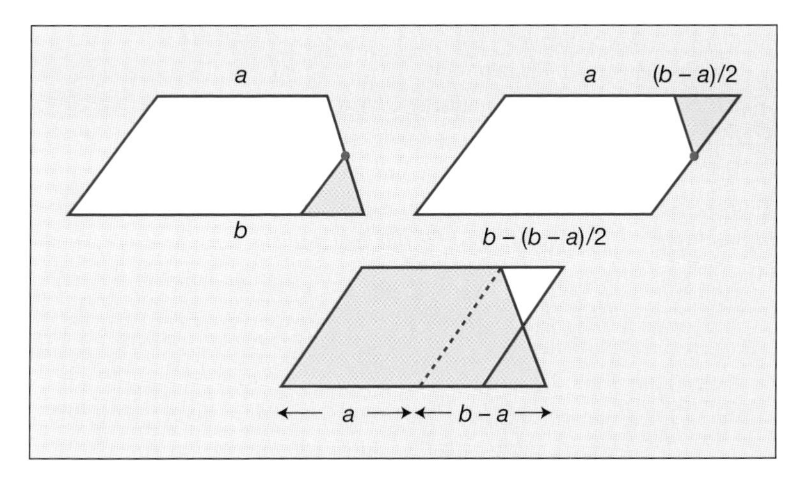

Fig. 20.14. The base of the parallelogram is the average of the bases of the trapezoid.

the length of the larger base is b and the shorter base is a, students can see that the length of the base of the parallelogram in figure 20.14 is $(a + b)/2$ in the following way. Let a and b be two numbers with $a \le b$, and let $b - a$ be their difference. Students can show that the average of a and b, $(a + b)/2$, can be expressed as $a + (b - a)/2$ or as $b - (b - a)/2$. They can represent a and b on the number line and give a geometrical interpretation of the average, $(a + b)/2$, as the midpoint of segment on the number line from a to b (fig. 20.15).

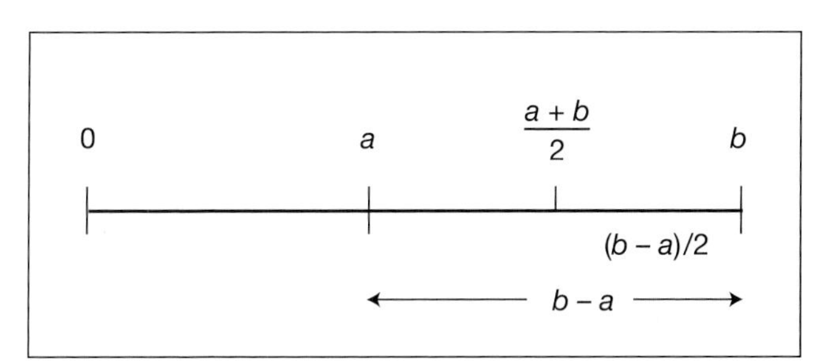

Fig. 20.15. The average as the midpoint of *a* and *b*

In either instance, as they saw before in figure 20.12, the length of this segment is the average of the lengths of the bases. Thus students can see in another way that a trapezoid has the same area as a parallelogram with the same height and whose base is the average of the bases of the trapezoid,

$$A = h\frac{a+b}{2}.$$

Trapezoid and Rectangle

A trapezoid can also be transformed into a rectangle by rotating two small triangles. In figure 20.16, the hinges are at the midpoints of the legs of the trapezoid. Students can remember that the line segment connecting these midpoints will be parallel to the bases of the trapezoid, and its length will be the average of the bases. Figure 20.16 shows a trapezoid that has the same area as a rectangle with the same height and whose base is the average of the bases of the trapezoid. Students can also verify algebraically that, indeed, the base of the rectangle is the average of the bases of the trapezoid in the following way. Let q be the length of the projection of one leg onto the longer base, and let p be the length of the projection of the other leg (fig. 20.17). Then $q + p = b - a$. The bases of the small triangles are $q/2$ and $p/2$. When the triangles are rotated up (fig. 20.18), students can see that the base of the rectangle is $a + q/2 + p/2 = a + (q + p)/2 = a + (b - a)/2 = (a + b)/2$. Here again, the teacher may want to help students recognize the use of the distributive property in the first equation and of the associative and commutative properties in the second.

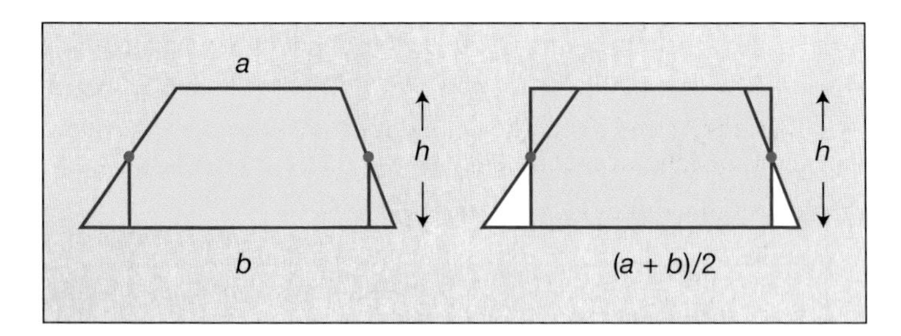

Fig. 20.16. The base of the rectangle is the average of the bases.

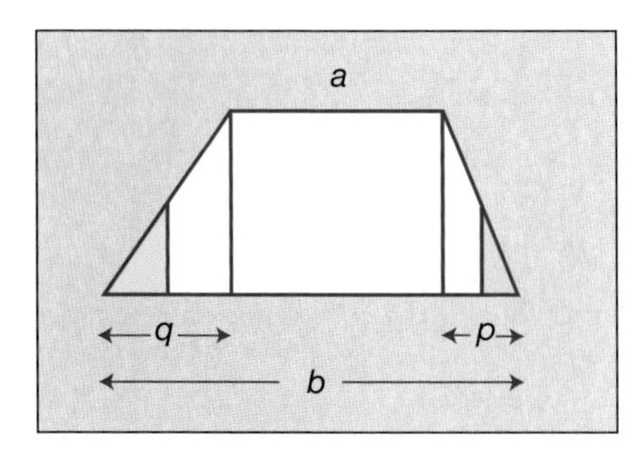

Fig. 20.17. $q + p = b - a$.

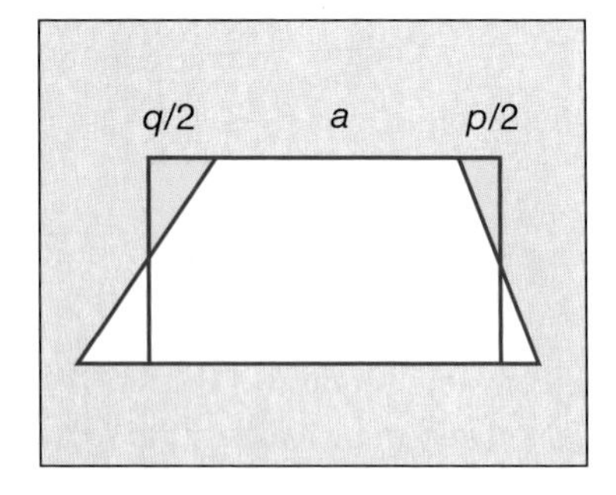

Fig. 20.18. Base of the rectangle using midpoints of trapezoid sides

Kite

In a kite, the two diagonals are perpendicular. In figure 20.19 the hinges are at the midpoints of the sides of the kite. Four right triangles form the kite. The four rotated triangles together with the original kite form a rectangle. (Why?) The teacher can ask questions to focus the attention of the students on the relationship between the lengths of the diagonals of the kite and the base and height of the rectangle. Students can see and justify why the area of a kite is half the area of a rectangle with height equal to one diagonal d of the kite and base equal to the other diagonal c of the kite. The area of the kite is thus

$$A = \frac{1}{2}dc.$$

By focusing on one-half of the rectangle, students can also find a geometrical interpretation for the expression

$$A = d\frac{c}{2}.$$

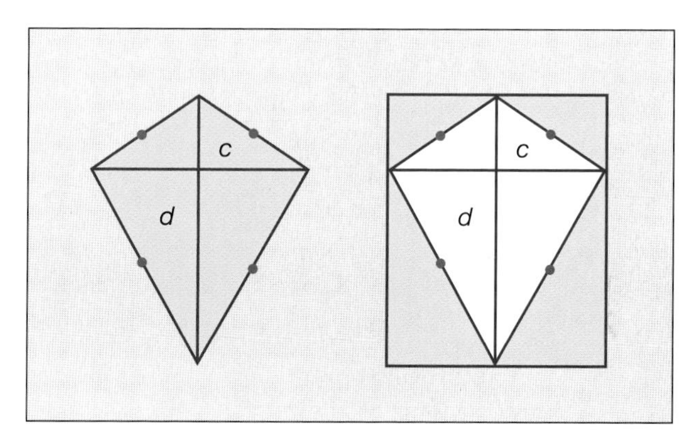

Fig. 20.19. Kite and rectangle

Regular Polygons

Figure 20.20 shows a connection between the area of a parallelogram and the area of a regular polygon with an even number of sides. Triangles of two different colors each form half the regular polygon. Therefore, the length of the parallelogram's base is the sum of the lengths of half the number of sides of the polygon, and the base of the parallelogram is equal to half the perimeter of the regular polygon, that is,

$$b = \frac{1}{2}p.$$

The apothem a of a regular polygon is the distance from its center to the midpoint of one of its sides. We can see that the height of the parallelogram is equal to the apothem a of the regular polygon. Thus, the area of the parallelogram is given by

$$A = \frac{1}{2}pa,$$

or

$$A = \frac{pa}{2}.$$

The area of the regular polygon will thus be

$$A = \frac{pa}{2}.$$

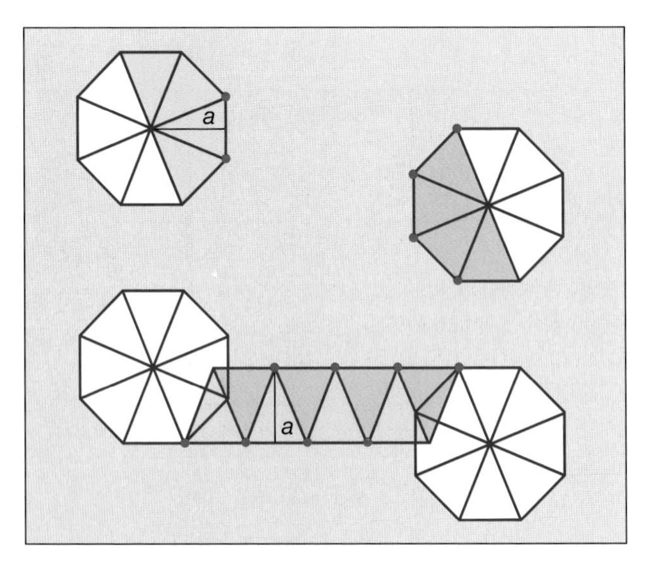

Fig. 20.20. Regular polygon (even number of sides) and parallelogram

Figure 20.21 shows the case of a regular polygon with an odd number of sides. In this case, a rectangle is formed, but the reasoning is the same to show that the area of a regular polygon with an odd number of sides is also

$$A = \frac{pa}{2}.$$

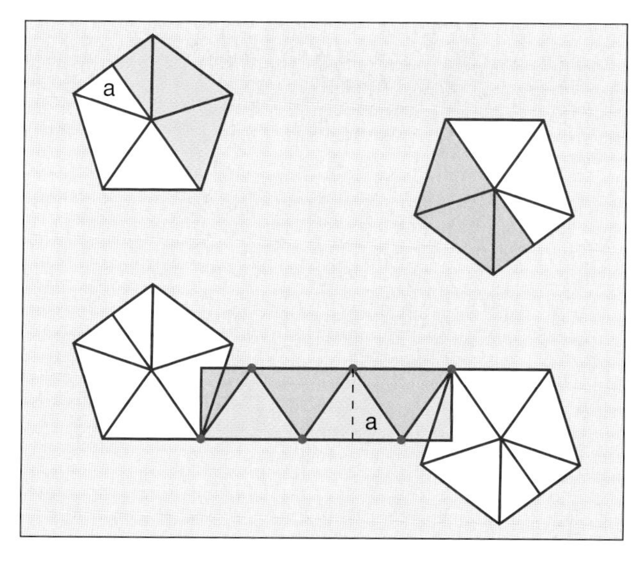

Fig. 20.21. Regular polygon (odd number of sides) and rectangle

Concluding Remarks

These interactive figures, with appropriate guidance from the teacher, give students the opportunity to connect formulas for the areas of the rectangle, right triangle, arbitrary triangle, parallelogram, trapezoid, kite, and regular polygon. Rather than have a collection of isolated formulas, students can see how the formulas form a network of relationships and how some of them can be derived from other formulas or from special cases. The interactive figures can also serve as an experimental laboratory to facilitate the discovery of properties of shapes and transformations to be proved later, as well as to provide opportunities for conjecturing and practicing deductive thinking as students justify their observations. Further, the activities can be used to develop the algebraic thinking of students. Students need to give meaning to algebraic expressions and to see why certain operations on algebraic expressions will indeed give equivalent expressions. When students are still developing their understanding of algebraic operations, contexts arising outside algebra can help them develop meanings. They may need to form specific images for the algebraic expressions, as well as to be able to explain them verbally or by using pictorial representations. Formulas for areas can thus serve as a vehicle to give meaning to, and practice for, algebraic operations.

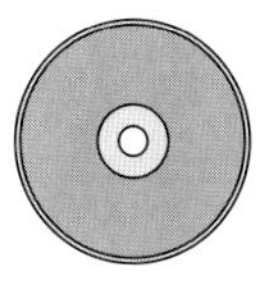

Interactive figures files that support this article are found on the CD-ROM disk accompanying this Yearbook.

An Integrated Approach to Teaching and Learning Geometry

David Wilson

THE NOTION of an integrated course in geometry is certainly not novel. John Swenson's (1936) geometry textbook *Integrated Mathematics with Special Application to Geometry* sought to distinguish his approach from the traditional separation of algebra and geometry, in part through his introduction of the coordinate plane into the traditional Euclidean geometry curriculum. Although many geometry textbooks since have employed the term *integrated,* the "text" presented in this article refers to an approach that weaves topics in algebra, geometry, and introductory trigonometry together seamlessly—that is, without chapters or other separations into content strands. This particular approach is problem-based and emphasizes coordinate geometry as a foundation for the development of many fundamental geometric concepts and as a means to develop students' reasoning and proof abilities. Technology is embedded, and students are expected to use technology as a tool to explore relationships and develop conjectures.

The integrated approach provides continual opportunities for students to make connections among mathematical topics and is central to students' development of geometric knowledge and understanding. The materials wonderfully illustrate recommendations in *Principles and Standards for School Mathematics* (National Council of Teachers of Mathematics 2000 [NCTM], p. 289) for—

> instructional materials that are intentionally designed to weave together different content strands … [and] to make sure that courses oriented toward any

> particular content area (such as algebra or geometry) contain many integrative problems—problems that draw on a variety of aspects of mathematics, that are solvable using a variety of methods, and that students can access in different ways.

The integrated geometry materials, titled simply Mathematics 2, are written by the mathematics faculty at Phillips Exeter Academy (PEA 2006a) and are available in downloadable PDF files from its Web site. These materials, explored in this article, have the study of plane geometry at their core but integrate several other content strands as well.

Phillips Exeter Academy has a long history of producing curricular materials extending back to the latter part of the nineteenth and early twentieth centuries. The father-and-son faculty members George A. and George Wentworth authored a series of algebra and geometry textbooks that dominated the market at that time. The popularity of the Wentworth geometry textbook was attributed in part to "the abundance of 'original' exercises (proofs left to students' analysis and ingenuity) as opposed to 'book proofs' (full demonstrations to be memorized for reproduction)" (Austin [1919], quoted in Donoghue [2003, p. 335].) The philosophy of leaving much to students' analysis and ingenuity has lasted at Exeter and is evident in the present course.

The materials emphasize students' engagement in problem solving with reflection and communication as essential components of students' learning. These aspects of the materials are reminiscent of ideals promoted in Harold Fawcett's (1938) study of students learning geometry in an environment that provided them with opportunities to explore, derive, and document their own findings in the form of a geometry text. Although the PEA materials do not expect students to produce a geometry "textbook," they do promote students' participation in the exploration, discovery, and justification of significant geometric relationships. The problems help the students develop thinking strategies and reasoning and proof abilities throughout and parallel what Fawcett noted should be a fundamental consideration in teaching geometry:

> If the real purpose of teaching demonstrative geometry is to give the pupil an understanding of the nature of proof, the emphasis should not be placed on the conclusions reached, but rather on the kind of thinking used in reaching these conclusions. (Fawcett [1938, p. 466], quoted in González and Herbst [2006])

The authors (PEA 2006b, Specific comments, para. 1) describe their text as a "mathematical whole" and in their overview emphasize several important characteristics and expectations.

> There is no Chapter 5, nor is there a section on tangents to circles. The curriculum is problem-centered, rather than topic-centered. Techniques and theorems will become apparent as you work through the problems, and you will need to

> keep appropriate notes for your records—there are no boxes containing important theorems. There is no index as such, but the reference section that starts on page 201 should help you recall the meanings of key words that are defined in the problems.

For students, this paragraph is brought into reality as they peruse the Math 2 "text" (PEA 2006a)—nearly eighty pages containing ten to fifteen problems on every page. Diagrams accompany many problems, but students see that the text has no introductory paragraphs and no sample problems—nothing but problem after problem. This format can be slightly intimidating for both students and teachers, particularly since the latter try to imagine how such a text might be implemented and how the day-to-day instruction would occur. To generate some thoughts on how this implementation might be done, the following pages provide a general overview of the Math 2 materials and discuss how some content from a traditional geometry course is developed in the PEA text. In addition, this article discusses the substantive changes in the classroom environment that accompany the implementation of these materials and some personal reflections from the author's experience teaching with them.

Foundations of the Curriculum

A Spiral Approach

The curriculum follows a spiral approach to learning on several levels. On one level it is the embodiment of Jerome Bruner's (1975) reflections on teaching and of a spiral curriculum:

> [S]uccessful efforts to teach highly structured bodies of knowledge like mathematics ...often took the form of metaphoric spiral in which at some simple level a set of ideas or operations were introduced in a rather intuitive way and, once mastered in that spirit, were then revisited and reconstrued in a more formal or operational way.

The PEA text begins as Bruner describes—with an intuitive approach. Frequently this approach takes the form of a problem presented in the coordinate plane so that the student has an accurate figure to work from and has available a variety of tools that allow for calculations of slope, distance, intersections, angle measure, and so on. Problems are then revisited with an increasing level of difficulty and gradually embedded in related contexts to allow for deeper understanding to develop and connections to become apparent. For example, several problems (from page 2 in the text) shown in figure 21.1 are designed to develop the notion of a locus of points that satisfy a given condition.

The first two problems permit access for students of varying ability levels to begin intuitively exploring the problem in the coordinate plane. The Pythagorean

1. Two different points on the line $y = 2$ are each exactly 13 units from the point (7, 14). Draw a picture of this situation, and then find the coordinates of these points.

2. Give an example of a point that is the same distance from (3, 0) as it is from (7, 0). Find lots of examples. Describe the configuration of all such points. In particular, how does this configuration relate to the two given points?

3. Verify that $P = (1, -1)$ is the same distance from $A = (5, 1)$ as it is from $B = (-1, 3)$. It is customary to say that P is *equidistant* from A and B. Find three more points that are equidistant from A and B. By the way, to "find" a point means to find its *coordinates*. Can points equidistant from A and B be found in every *quadrant*?

Fig. 21.1. An intuitive beginning. Produced by members of the mathematics department at Phillips Exeter Academy. Used by permission.

theorem is employed in these problems as students develop the notion of distance in the plane through right triangles. Later, problems require the more formalized distance formula, but throughout the text the student is prompted to think of distance as the hypotenuse of a right triangle.

Problem 3 in figure 21.1 has several interesting aspects. First, it is extending problems 1 and 2 to a more challenging level by involving points that are neither horizontally nor vertically aligned. Second, the authors use a scaffolding approach as they give an example of one point that the student has to verify is equidistant from the given points prior to being asked to generate additional points. This type of scaffolding is common throughout the text. Another aspect of problem 3 is the introduction of formal terminology in the problem. Although terminology is frequently defined in problems, the reference section (also downloadable) to the textbook includes a glossary of all the italicized words from the text.

Figure 21.2 displays problems that allow students to revisit the topic of locus and set the stage for further discussions of how to describe the locus of points both verbally and algebraically as well as connect that understanding with related ideas.

Problem 3 occurs a bit later in the text and prompts students to extend their thinking about the perpendicular bisector as a locus of points to the reasoning behind the circumcenter of a triangle. Similarly, problem 4 extends and connects the themes of distance and locus of points with given conditions to parabolas. The many related problems that are completed prior to problem 4 allow students to see its solution as a natural extension of the previous work.

The van Hiele Model

The authors' use of the spiral to provide an intuitive approach, and to revisit

1. Let $A = (1, 5)$ and $B = (3, -1)$. Verify that $P = (8, 4)$ is equidistant from A and B.Find at least two more points that are equidistant from A and B. Describe all such points.

2. Write a formula for the distance from $A = (-1, 5)$ to $P = (x, y)$, and another formula for the distance from $P = (x, y)$ to $B = (5, 2)$. Then write an equation that says that P is equidistant from A and B. Simplify your equation to linear form. This line is called the *perpendicular bisector of AB*. Verify this by calculating two slopes and one midpoint.

3. Let $A = (3, 4)$, $B = (0, -5)$, and $C = (4, -3)$. Find equations for the perpendicular bisectors of segments AB and BC, and coordinates for their common point K. Calculate lengths KA, KB, and KC. Why is K also on the perpendicular bisector of segment CA?

4. Let $F = (0, 4)$. Find coordinates for three points that are equidistant from F and the x-axis. Write an equation that says that $P = (x, y)$ is equidistant from F and the x-axis.

Fig. 21.2. Revisiting the topic. Produced by members of the mathematics department at Phillips Exeter Academy. Used by permission.

and extend topics accompanied by appropriate scaffolding, aligns the text's approach with teaching recommendations based on the van Hiele levels of geometric thinking and phases of learning. The van Hiele levels are theorized levels of geometric knowledge and understanding that students' progress through sequentially; they constitute a framework for the types of tasks that are appropriate for students to engage in as they study geometry.

The PEA text is appropriately designed for students to transition from a late level 2 (descriptive-analytic) and early level 3 (relational-inferential) to level 4 (formal deductive). Furthermore, although possibly not the explicit intent of the authors, the problems in the PEA text are written in ways that align well with the van Hiele's phases of learning and thus support progress within the levels. Mary Crowley (1987) described the five sequential phases of learning that Pierre and Dina van Hiele proposed would assist students in progressing from one level to the next. In Phase 1: Inquiry, Crowley suggests that—

> observations are made, questions are raised, and level-specific vocabulary is introduced. (P. 5)

Phase 2: Directed Orientation should feature carefully sequenced explorations that lead to Phase 3: Explication, where—

> students express and exchange their emerging views about structures that have been observed. (P. 5)

Phase 4: Free Orientation involves—

> more complex tasks—tasks with many steps, tasks that can be completed in several ways, and open-ended tasks. (P. 5)

In the last phase, Phase 5: Integration, the students review, summarize, and form an—

> overview of the new network of objects and relations. (P. 6)

The problems in figures 21.1 and 21.2 discussed above illustrate the alignment to a great extent, but an additional sample of problems that look at congruence will further illustrate the style of questions and the challenges posed to students as they deepen their understanding. The notion of congruence is introduced informally early in the text through an interesting problem involving the shape shown in figure 21.3. The problem offers an entry point for students of varying levels of understanding and allows for multiple responses that promote discussion and allow for the introduction of relevant terminology.

1. Some terminology: Figures that have exactly the same shape and size are called *congruent*. Dissect the region shown at right into two congruent parts. How many different ways of doing this can you find?

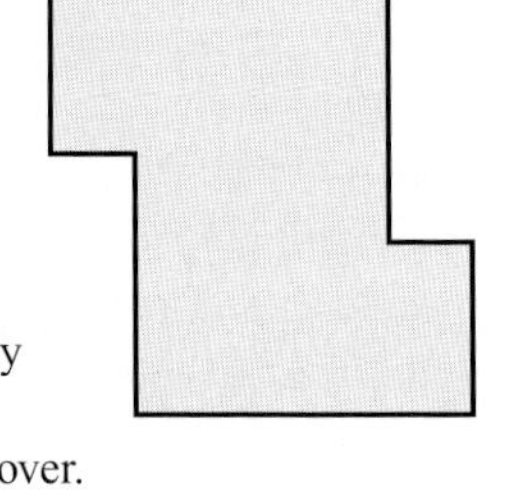

2. Let $A = (2, 4)$, $B = (4, 5)$, $C = (6, 1)$, $T = (7, 3)$, $U = (9, 4)$, and $V = (11, 0)$. Triangles ABC and TUV are specially related to each other. Make calculations to clarify this statement, and write a few words to describe what you discover.

Fig. 21.3. Congruence—van Hiele Phase 1 (Inquiry). Produced by members of the mathematics department at Phillips Exeter Academy. Used by permission.

Schettino (2003) discussed her students' experience with this problem as she implemented the PEA Math 2 materials and the varied topics that arose, including rotational symmetry and isometries. The second problem shown in figure 21.3 continues Phase 1 as opportunities for exploration and further observations are made possible and additional terminology such as *corresponding* can be introduced. Figure 21.4 displays problems that are presented a bit later in the text and provide additional terminology and structure for the students to explore some implications of congruence. The problems in figure 21.4 are representative of Phase 2: Directed Orientation, and students are guided toward recognizing that corresponding parts of congruent triangles are congruent.

Figure 21.5 presents problems that provide some opportunity for Phase 3: Explication. In the first problem, students are given a set of points that do not

1. Let $A = (-5, 0)$, $B = (5, 0)$, and $C = (2, 6)$; let $K = (5, -2)$, $L = (13, 4)$, and $M = (7, 7)$. Verify that the length of each side of triangle *ABC* matches the length of a side of triangle *KLM*. Because of [these] data, it is natural to regard the triangles as being in some sense equivalent. It is customary to call the triangles *congruent*. The basis used for this judgment is called the *side-side-side* criterion. What can you say about the sizes of angles *ACB* and *KML*? What is your reasoning? What about the other angles?

2. Let $A = (0, 0)$, $B = (2, -1)$, $C = (-1, 3)$, $P = (8, 2)$, $Q = (10, 3)$, and $R = (5, 3)$. Plot these points. Angles *BAC* and *QPR* should look like they are the same size. Find evidence to support this conclusion.

Fig. 21.4.Congruence—van Hiele Phase 2 (Directed Orientation). Produced by members of the mathematics department at Phillips Exeter Academy. Used by permission.

1. Let $A = (0, 0)$, $B = (1, 2)$, $C = (6, 2)$, $D = (2, -1)$, and $E = (1, -3)$. Show that angle *CAB* is the same size as angle *EAD*.

2. Using a ruler and protractor, draw a triangle that has an 8-cm side and a 6-cm side, which make a 30-degree angle. This is a *side-angle-side* description. Cut out the figure so that you can compare triangles with your classmates. Will your triangles be congruent?

3. With the aid of a ruler and protractor, draw and cut out three [noncongruent] triangles, each of which has a 40-degree angle, a 60-degree angle, and an 8-cm side. One of your triangles should have an angle-side-angle description, while the other two have angle-angle-side descriptions. What happens when you compare your triangles with those of your classmates?

Fig. 21.5. Congruence—van Hiele Phase 3 (Explication). Produced by members of the mathematics department at Phillips Exeter Academy. Used by permission.

clearly create two congruent triangles, but through the introduction of midpoint *F* of *AC* and drawing lines to form triangles *ABF* and *ADE,* students are able to apply their knowledge creatively in a novel way and realize the value of the side-side-side (SSS) congruence theorem. As is typical of many problems, students are asked to furnish evidence for their conclusions and discuss their reasoning. This aspect of the tasks gives students the opportunity to share their developing ideas and understanding of the topic. In the second and third problems, students are asked to consider alternative conditions for congruence and to extend and refine their understanding of criteria for congruence. This request represents one aspect of the spiraling nature of the content, but also a brief return to Phase 2: Directed Orientation, as students are specifically focused on completing prescribed tasks.

Two additional problems shown in figure 21.6 illustrate the last two phases (Free Orientation and Integration). In these problems students are challenged to think broadly about other criteria for congruence and then to refine their thinking once again as they explore a figure that poses problems for using side-side-angle (SSA) as a congruence criterion. As these problems are completed, students can share and summarize their developing knowledge of congruence as it relates to triangles and can set the foundation for further study and application.

1. A triangle has six principal parts—three sides and three angles. The SSS criterion states that three of these items (the sides) determine the other three (the angles). Are there other combinations of three parts that determine the remaining three? In other words, if the class is given three measurements with which to draw and cut out a triangle, which three measurements will guarantee that everyone's triangles will be congruent?

2. Use the diagram to help you explain why SSA evidence is not by itself sufficient to justify the congruence of triangles. The tick marks designate segments that have the same length.

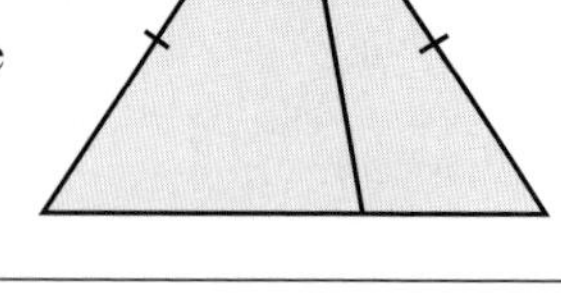

Fig. 21.6. Congruence—van Hiele Phases 4 and 5 (Free Orientation and Integration). Produced by members of the mathematics department at Phillips Exeter Academy. Used by permission.

The PEA Math 2 text engages students in problem solving on multiple topics simultaneously. Thus, as students explore congruence they are also introduced to the concept of a vector and then vector translations and other isometries. Problems within the context of transformations provide opportunities for students to deepen their understanding of congruence. The problems presented in figure 21.7 offer an illustration of this aspect of the content development that again reflects opportunities for students to engage in discussions and Phase 5: Integration activities.

Technology, Cooperative Learning, Reasoning, and Proof

Several other foundations of the PEA materials are the integration of technology, the expectation of students' working together in a cooperative environment,

1. Plot points $K = (0, 0)$, $L = (7, -1)$, $M = (9, 3)$, $P = (6, 7)$, $Q = (10, 5)$, and $R = (1, 2)$. Show that the triangles *KLM* and *RPQ* are congruent. Show also that neither triangle is a vector translation of the other. Describe how one triangle has been transformed into the other.

2. Plot points $K = (-4, -3)$, $L = (-3, 4)$, $M = (-6, 3)$, $X = (0, -5)$, $Y = (6, -3)$, and $Z = (5, 0)$. Show that triangle *KLM* is congruent to triangle *XZY*. Describe a transformation that transforms *KLM* onto *XZY*. Where does this transformation send the point $(-5, 0)$?

Fig. 21.7. van Hiele Phase 5: Integration. Produced by members of the mathematics department at Phillips Exeter Academy. Used by permission.

and embedding reasoning and proof in the problems. Many of the previous problems illustrate these features. Several of the problems are excellent candidates for explorations through such interactive geometry programs as The Geometer's Sketchpad (GSP) (Jackiw 2001). Such an approach would assist students in making initial conjectures and testing to verify or contradict their conjectures. The PEA Math 2 problems vary in their expectation for the use of technology, in that explicit statements may be made regarding technology or the decision about its use may be left to the student. Figure 21.8 presents problems illustrating each of these formats. Problem 1 is from the first pages of the text and clearly relays the

1. Consider the linear equation $y = 3.62(x - 1.35) + 2.74$.
   (a) What is the slope of this line?
   (b) What is the value of y when $x = 1.35$?
   (c) This equation is written in *point-slope* form. Explain the terminology.
   (d) Use your calculator to graph this line.
   (e) Find an equation for the line through $(4.23, -2.58)$ that is parallel to this line.
   (f) Describe how to use your calculator to graph a line that has slope -1.25 and that goes through the point $(-3.75, 8.64)$.

2. The sides of the triangle at right are formed by the graphs of $3x + 2y = 1$, $y = x - 2$, and $-4x + 9y = 22$. Is the triangle isosceles? How do you know?

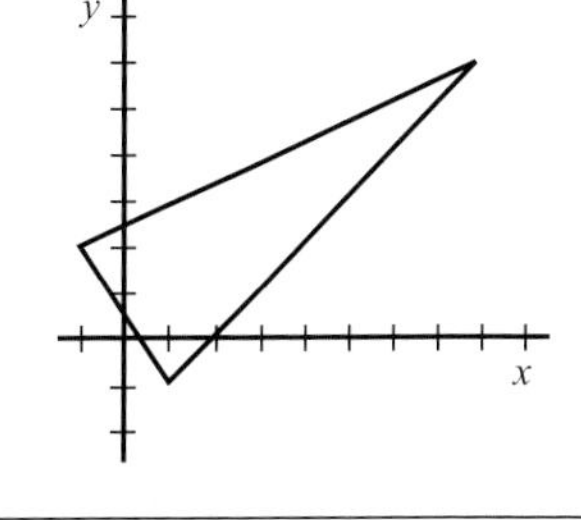

Fig. 21.8. Technology integration. Produced by members of the mathematics department at Phillips Exeter Academy. Used by permission.

expectation of the use of a graphing calculator, whereas problem 2 presents an opportunity for students to decide whether to use a calculator to verify coordinates of vertices.

Little support is found for how to use the graphing calculator in the materials, as that skill is assumed to be prior knowledge for PEA Math 2 students. The PEA authors (2006b) explicitly describe their expectations for technology use and for students' written responses:

> Many of the problems in this book require the use of technology (graphing calculators or computer software) in order to solve them. Moreover, you are encouraged to use technology to explore, and to formulate and test conjectures. Keep the following guidelines in mind: write before you calculate, so that you will have a clear record of what you have done; store intermediate answers in your calculator for later use in your solution; pay attention to the degree of accuracy requested; refer to your calculator's manual when needed; and be prepared to explain your method to your classmates. (Specific comments, para. 4)

Another foundation of the PEA text is that of group learning. The small-group discussions that occur around a table at Exeter can be effectively emulated in larger classes through small-group work and employing aspects of cooperative learning. This approach should promote an environment similar to that at Exeter, where students are—

> exposed to problem solving in a very student-centered, discussion-based classroom. Students are held accountable for attempting solutions to homework problems and the class as a whole decides on correct solutions. (Phillips Exeter Academy 2006b, para. 1)

Students are challenged to make and verify conjectures throughout the text. The development of students' problem-solving abilities and reasoning-and-proof abilities are natural outcomes of engagement with the PEA materials. Problems require students to make calculations to verify conjectures and frequently ask for written explanations to justify their conclusions, as shown in many of the problems presented thus far.

The process of justifying responses is connected with, and extended to, the notion of proof about one-quarter of the way through the Math 2 materials. The text presents two formats of proof. The format presented in figure 21.9a, a paragraph format, allows students to demonstrate their ability to construct a cohesive argument with sentences, as opposed to the second, more traditional format of a statement-reason, two-column proof that is presented simultaneously in figure 21.9b.

For the most part, students are not directed toward one form of proof or the other. The introduction of formal proof is accompanied by an increased emphasis on synthetic geometry—that is, outside the coordinate plane in more generalized

a. Here are two examples of proofs that do not use coordinates. Both proofs show how specific *given* information can be used to logically deduce *new* information. Each example concerns a kite $ABCD$, for which $AB = AD$ and $BC = DC$ is the given information. The first proof, which consists of simple text, shows that diagonal AC creates angles BAC and DAC of the same size.

PROOF 1. Because $AB = AD$ and $BC = DC$, and because the segment AC is shared by the triangles ABC and ADC, it follows from the SSS criterion that these triangles are congruent. Thus it is safe to assume that all the corresponding parts of these triangles are congruent as well (often abbreviated to CPCTC, as in proof 2 below). In particular, angles BAC and DAC are the same size.

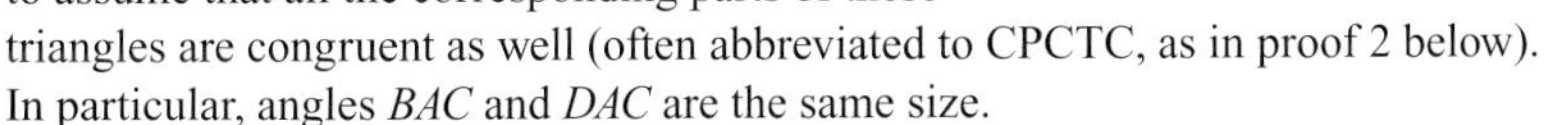

b. Now let E mark the intersection of diagonals AC and BD. The second proof, which is an example of a two-column proof, is written symbolically in outline form. It shows that the diagonals intersect perpendicularly. This proof builds on the first proof, which thus reappears as the first five lines.

	Statement	Reason
PROOF 2.	$AB = AD$	given
	$BC = DC$	given
	$AC = AC$	shared side
	$\Delta ABC \cong \Delta ADC$	SSS
	$\angle BAC = \angle DAC$	CPCTC
	E = intersection of AC and BD	
	$AB = AD$	given
	$\angle BAE = \angle DAE$	preceding CPCTC
	$AE = AE$	shared side
	$\Delta ABE \cong \Delta ADE$	SAS
	$\angle BEA = \angle DEA$	CPCTC
	$\angle BEA$ and $\angle DEA$ supplementary	E is on BD
	$\angle BEA$ is right	definition of right angle

Fig. 21.9. Formal proof. Produced by members of the mathematics department at Phillips Exeter Academy. Used by permission.

terms. For example, problems such as those shown in figure 21.10 are mixed among problems that continue to use the coordinate plane as an avenue for the development of concepts.

1. In quadrilateral $ABCD$, it is given that $AB = CD$ and $BC = DA$. Prove that angles ACD and CAB are the same size. N.B. If a polygon has more than three vertices, the *labeling convention* is to place the letters around the polygon in the order that they are listed. Thus AC should be one of the diagonals of $ABCD$.

2. If the diagonals of a quadrilateral bisect each other, then any two nonadjacent sides of the figure must have the same length. Prove that this is so.

3. Triangle ABC is isosceles, with AB congruent to AC. *Extend* segment BA to a point T (in other words, A should be between B and T). Prove that angle TAC must be twice the size of angle ABC. Angle TAC is called one of the *exterior angles* of triangle ABC.

Fig. 21.10. Synthetic geometry. Produced by members of the mathematics department at Phillips Exeter Academy. Used by permission.

The PEA Math 2 emphasis on reasoning and proof, the focus on independent and cooperative learning opportunities, and the rich connections that arise in this problem-based approach offer great potential for providing students with a very strong background to build on. However, as noted previously, some fairly large impediments arise when it comes to implementing such a text, and those warrant discussion here as well.

Classroom Implementation

The challenge for the teacher implementing the PEA materials is multidimensional. To a great extent, previous experiences for students and teachers with a student-centered instructional format will ease the transition. Many schools that have implemented *Standards*-based curricula will already have struggled with some of the same issues, although the *Standards*-based curricula tend to have themes or topics that focus the students' work and are often defined by chapters or units. The lack of any such separations in the PEA materials presents diverse challenges for the teacher. Issues of how classroom time is structured, how students' learning is assessed, and the student's role and responsibility all require time and effort to develop and coordinate into a successful teaching and learning environment. Furthermore, districts must allot adequate time, support, and professional development opportunities to accompany any implementation effort.

The Classroom Environment

The problem-based nature of the PEA materials, and the interwoven strands of content, may elicit some discomfort from students and teachers as they adjust their expectations for what the teaching and learning environment entails. Students may need support in bringing to bear the necessary skills and confidence to approach problems. Careful selection of items for classroom work, as well as for homework assignments, will help facilitate the development of such skills and confidence. The first problem in the Math 2 text is displayed in figure 21.11 and may be challenging for many students if approached as it is posed.

1. A 5×5 square and a 3×3 square can be cut into pieces that will fit together to form a third square.

   (a) Find the length of a side of the third square.

   (b) In the diagram at right, mark P on segment DC so that $PD = 3$, then draw segments PA and PF. Calculate the lengths of these segments.

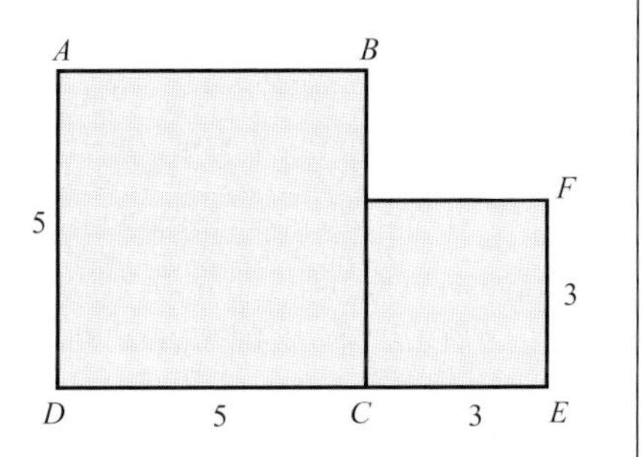

   (c) Segments PA and PF divide the squares into pieces. Arrange the pieces to form the third square.

2. (Continuation) Change the sizes of the squares to $AD = 8$ and $EF = 4$, and redraw the diagram. Where should point P be marked this time? Form the third square again.

3. (Continuation) Will the preceding method *always* produce pieces that form a new square? If your answer is *yes*, prepare a written explanation. If your answer is *no*, provide a counterexample—two specific squares that can *not* be converted to a single square.

Fig. 21.11. First problem. Produced by members of the mathematics department at Phillips Exeter Academy. Used by permission.

Bill Campbell (1997), an instructor from Phillips Exeter Academy, shared an approach he used with this problem to allow greater initial access. The problem can be posed as a puzzle task for which the figure is drawn on a grid as shown in figure 21.12. Students cut along the lines connecting A to P and P to F, as well as on the outline of the squares, to form five pieces. The example shown uses side lengths of 12 and 5 for the two squares rather than 5 and 3. Students are challenged to arrange the five pieces into a square.

The puzzle can be presented without the original problem and thus provides all students with a place to begin. Students' initial attempts to complete the puzzle frequently yield a "solution" of the 12×14 rectangle shown in figure 21.13. The

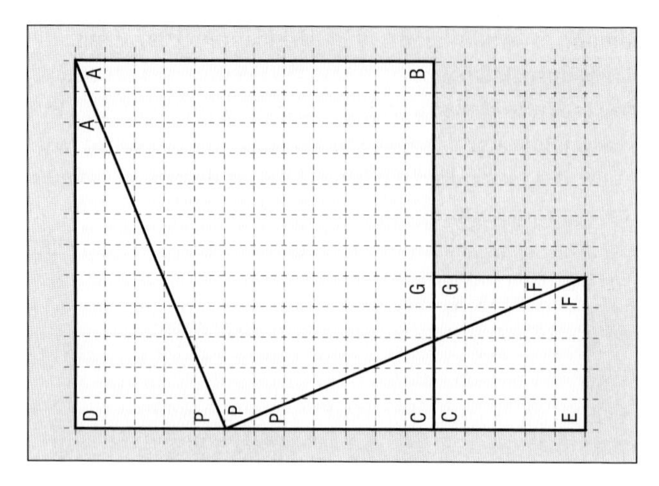

Fig. 21.12. An alternative approach. Produced by members of the mathematics department at Phillips Exeter Academy. Used by permission.

rectangle provides an opportunity to direct students' attention to the areas of the two original squares, and the convenient choice of dimensions of 5 and 12 allows many students to see what the dimensions of the solution square must be. Thus, the approach to the problem through the puzzle offers an opportunity to use part (a) of the problem as a hint toward the solution. Since the puzzle involves a grid, the students are able to look at distance easily, and once they realize that the side length must be 13, they are able to consider which pieces have such a length.

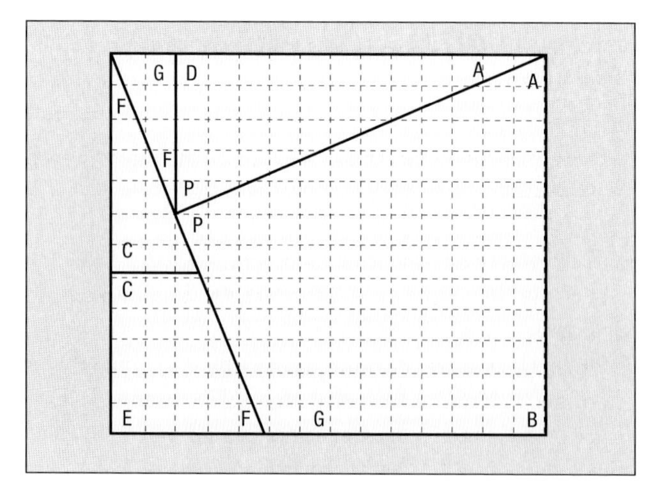

Fig. 21.13. An incorrect attempt. Produced by members of the mathematics department at Phillips Exeter Academy. Used by permission.

The solution shown in figure 21.14 provides opportunities for multiple solution paths to be discussed, including transformations of triangles *ADP* and *PEF,* and of course provides students with a visual connection to the Pythagorean theorem as they respond to problem 3. The shifting of the problem to a more familiar form—that of a puzzle task—engineers a comfort zone for the students and allows them to engage in the task. The original problem can be returned to as a follow-up to the puzzle task, with students now having the requisite knowledge and understanding to proceed.

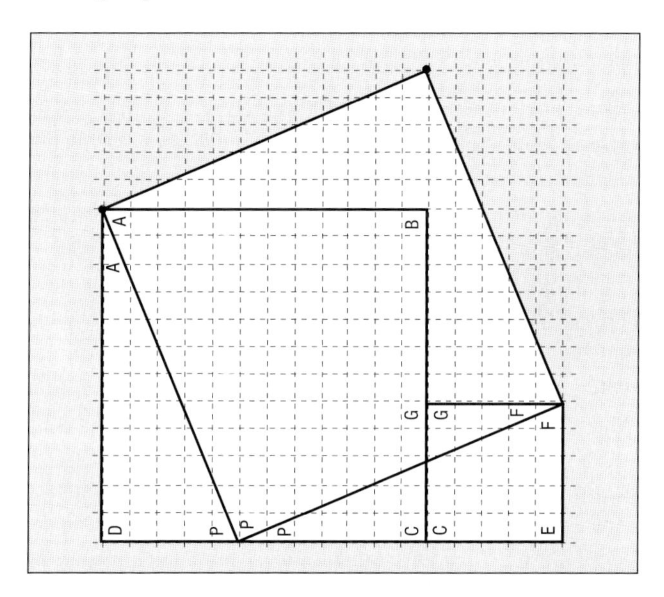

Fig. 21.14. A solution. Produced by members of the mathematics department at Phillips Exeter Academy. Used by permission.

Students will become comfortable with an environment in which they are expected to solve novel problems on a regular basis over time, but the teacher plays a vital role in this adjustment phase as he or she provides appropriate scaffolding for the students' problem-solving efforts and establishes a new "contract" with the students that shifts their role to that of active thinkers with an increased responsibility for learning mathematics.

The notion of a "contract" between the teacher and the students that defines their respective roles is helpful. The metaphor offers a way of anticipating some of the feelings of apprehension and uneasiness that may manifest themselves early on. Patricio Herbst (2006) wrote of one teacher's classroom experiences as she attempted to implement a problem-centered approach to teaching a geometry unit and the "negotiations in the contract" that took place. The teacher was attempting to use a problem-based approach to help students develop the ability to reason and make and test conjectures. Herbst documented the struggles and adjustments that were made as the shifts in the expectations for the learning environment were taking place. He noted that part of the difficulty may have stemmed from limited opportunities for the students to engage in deductive reasoning previously in the year. He concluded that (p. 344)—

> if in order to afford students an opportunity to grasp the meaning of mathematical ideas we recognize the need to afford access to how those ideas connect to and warrant each other, students should be exposed to tasks that require them to reason deductively to find things out … even if the product is not yet fully written as a mathematical proof.

The PEA Math 2 content is certainly well-aligned with such a vision. The difficulties in implementing a shift in practice can be limited in part by beginning the school year with new "contract" expectations that are aligned with the text and by fostering a sense of ownership of the learning. The PEA materials provide a sound foundation for this change.

The classroom environment will also change regarding assessment practices. The students' work during class time, the problems they complete outside class, and their contributions during cooperative learning and during whole-class summary discussions are integral parts of the learning and assessment processes. The traditional quiz and test are harder to place in the schedule because the materials do not have the conventional chapter separations.

Lastly, appropriate structure of the cooperative process can greatly enhance the success of implementing this problem-based material. The fundamental aspects of cooperative learning—including group selection, positive interdependence, individual accountability, social skills, and group processing—are elements that enhance the learning process.

Successful use of the PEA materials is tied to many factors. Substantial time, effort, and commitment, accompanied by appropriate professional development, contribute to successful implementation. The type of implementation will vary and will largely determine the need for the supportive elements mentioned above. For example, one or two teachers implementing the PEA materials for an honors group will require less support than a broad-based adoption for all geometry classes. Either one requires extensive time for the teachers to work through the problems to gain a sense of the how content is developed and how previous work is revisited and built on to introduce new themes and topics. Invitations to serve as coaches may be appropriate to extend to individuals with experience and expertise in implementing problem-based curricula and cooperative learning or in fostering student-centered learning environments. Perhaps an invitation to an Exeter faculty member should be considered as well. The PEA materials are available to be used in any way a teacher sees appropriate and are a wonderful resource, but the full implementation of the text requires extensive planning and support.

The following section provides a glimpse of challenges and rewards the author had experienced during implementation of the PEA Math 2 materials and offers a few additional suggestions.

Some Personal Reflections on Implementation

The PEA text comes with neither a solutions manual nor any guidelines for implementation, although it does include an introduction for students written by former students. My initial experience in using the material was as a source of weekly problem sets for my geometry students that were presented each Monday and due on Friday. Several times each week, students would share ideas and ask questions that would help them progress in their work. The discussions surrounding the problems gradually consumed more and more time, and I found the students' work on the problems to be more beneficial than many of my lectures and other assignments. I began to get a sense of the spiral that was embedded in the PEA text and how work on early problems was often laying the foundation for the more formal development and discussion to come later. I also watched my students' abilities and confidence in problem solving grow dramatically.

During several years of using the problems as a supplemental source, my instructional approach underwent significant changes as I attempted to make my classroom more student-centered and align my practices with the vision put forth in the various NCTM *Standards* documents (NCTM 1989, 1991, 1995, 2000). Part of this change involved my decision to use the PEA Math 2 text as the primary source of material with an honors geometry class. This choice posed a significant challenge to my students as well as to me. Many of the students preferred listening to lectures because their high ability had always allowed for "successful" experiences in that environment. Having to come to class each day and work on problems was not what they had come to expect, and this format evinced some resistance. The unstructured nature of the text bothered some students because they wanted to know what topic they were studying and be able to label notes appropriately. Parents also raised questions regarding the classroom environment and the method of instruction. Most students and parents, however, were quite excited about the challenging work and appreciated the approach. The largest source of questions and anxiety from the students stemmed from the disruption of their expectations of their role in the classroom. For many, this anxiety was resolved by the end of the first quarter.

A second fundamental change involved assessment. Tests became a quarterly event with several quizzes dispersed over the quarter as well, but submitted problems became the primary source of assessment. Part of the students' evolving comfort level stemmed from their ease with the state examination questions that we occasionally attempted. They found those questions to be trivial compared with what they were working on daily, and they became more confident that they were learning much more than they perceived. The lack of a sense of learning "content" may have been partly due to the lack of tests that had so often been their measure of success in previous courses. Working on rich problems each day, and reaching solutions to them, did not seem to equate to achieving a high test score.

Accompanying my transition to student-centered, problem-based classroom was an attempt at cooperative learning. I later realized how valuable a few basic practices of cooperative learning were. For example, group assessment and random selection of students to present solutions would have enhanced the small-group experience and fostered a more successful learning environment.

In spite of my novice efforts at shifting instructional practices and the venture into an unconventional text, the end results were highly successful. The students had become such strong problem solvers that as we reviewed for the New York State examination, those questions, as mentioned earlier, seemed simple. The students' scores reflected this success with many 100 percents and certainly the best state examination scores a group of my students had experienced. The nature of the learning experience allowed them to have a much higher retention rate and to achieve a deeper understanding of the content than they had previously experienced.

Conclusions

The many problems contained in this article reflect a small sample of the PEA Math 2 materials. Actively engaging in the problems as you read will help develop a sense of the expectations and spiraling nature of the curriculum. You may wish to return to different sections and work through the problems as you reread to truly gain some insight into how the PEA problems provide an opportunity for students to attain a rich and deep understanding of geometry.

REFERENCES

Austin, Charles M. "A History of Plane Geometry as a School Study in the United States." Masters thesis, University of Chicago, 1919.

Bruner, Jerome S. "Entry into Early Language: A Spiral Curriculum." Paper read at the Charles Gittins Memorial Lecture at University College of Swansea, U. K., 1975.

Campbell, Bill. "Why Are Manhole Covers Round?" Presentation at the Annual Meeting of the National Council of Teachers of Mathematics, Minneapolis–Saint Paul, April 1997.

Crowley, Mary L. "The van Hiele Model of the Development of Geometric Thought." In *Learning and Teaching Geometry, K–12,* 1987 Yearbook of the National Council of

Teachers of Mathematics (NCTM), edited by Mary M. Lindquist, pp. 1–16. Reston, Va.: NCTM, 1987.

Donoghue, Eileen F. "Algebra and Geometry Textbooks in Twentieth-Century America." In *A History of School Mathematics*, vol. 1, edited by George M. A. Stanic and Jeremy Kilpatrick, pp. 329–98. Reston, Va.: National Council of Teachers of Mathematics, 2003.

Fawcett, Harold P. *The Nature of Proof.* Thirteenth Yearbook of the National Council of Teachers of Mathematics. New York: Bureau of Publications, Teachers College, Columbia University, 1938.

González, Gloriana, and Patricio G. Herbst. "Competing Arguments for the Geometry Course: Why Were American High School Students Supposed to Study Geometry in the Twentieth Century?" *International Journal for the History of Mathematics Education* 1, no. 1 (2006): 7–33.

Herbst, Patricio G. "Teaching Geometry with Problems: Negotiating Instructional Situations and Mathematical Tasks." *Journal for Research in Mathematics Education* 37, no. 4 (July 2006): 313–47.

Jackiw, Nicholas. The Geometer's Sketchpad. Version 4.0. Software. Emeryville, Calif.: Key Curriculum Press, 2001.

National Council of Teachers of Mathematics (NCTM). *Curriculum and Evaluation Standards for School Mathematics*. Reston, Va.: NCTM, 1989.

———. *Professional Standards for Teaching Mathematics*. Reston, Va.: NCTM, 1991.

———. *Assessment Standards for School Mathematics*. Reston, Va.: NCTM, 1995.

———. *Principles and Standards for School Mathematics*. Reston, Va.: NCTM, 2000.

Phillips Exeter Academy. *Math 2*. Available from www.exeter.edu/academics/84_9408 .aspx, June 2006a.

———. *Teaching Mathematics with the Harkness Method*. Available from math.exeter .edu/dept/harkness/index.html, 2006b.

Schettino, Carmel. "Transition to a Problem-Solving Curriculum." *Mathematics Teacher* 96, no. 8 (November 2003): 534–37.

Swenson, John A. *Integrated Mathematics with Special Application to Geometry*. Ann Arbor, Mich.: Edwards Brothers, 1936.

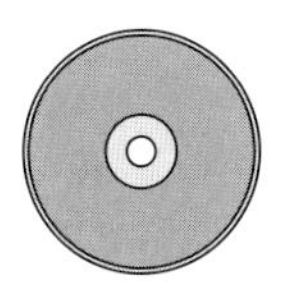

A PDF file of the Phillips Exeter Academy's Mathematics 2 course and a link to Phillips Exeter Academy's Web site are found on the CD-ROM disk accompanying this Yearbook.

Redesigning a Traditional Geometry Lesson as an Investigative Activity

Jon Davis

Principles and Standards for School Mathematics (*Principles and Standards*) (NCTM 2000) describes geometry as "a rich arena in which students can discover patterns and formulate conjectures" (p. 309). Today, increasing numbers of textbooks portray mathematics in just such a way. For example, in *Discovering Geometry: An Investigative Approach* (Serra 2008) students frequently conduct investigations using such tools as patty paper or The Geometer's Sketchpad (Jackiw 2001), construct conjectures based on those explorations, and develop arguments to verify the validity of their conjectures as well as understand why they are true. However, still in use in classrooms across the United States are many textbooks that depict geometry as a collection of theorems, postulates, and axioms developed by unknown persons for unknown reasons to be passively consumed by the student. This article suggests a general model by which teachers can critique and redesign lessons to unlock their potential. The model is used to transform a passive textbook lesson involving the sum of the interior angles of convex polygons into a more dynamic, hands-on, reasoning-based experience for students. The examples used to illustrate this model are derived from the author's experience teaching a course for preservice teachers but could apply to classroom teachers as well.

The Teacher-Curriculum Relationship

Ben-Peretz (1990) describes three roles that curriculum developers envision teachers playing in the classroom with regard to a particular textbook. First, developers may assume that teachers do not actively interpret and alter a curriculum when they teach it. Instead teachers are viewed as the "ether" through which the curriculum developers' ideas are transmitted to students. Second, they may view teachers as interpreters of curriculum and seek to standardize interpretations by designing workshops for teachers to help them understand the philosophy and intent of the intended curriculum. The third perspective, taken in this article, is that teachers are both users of a particular textbook as well as curriculum developers. They may use their professional judgment to adapt a curriculum to the needs of a particular group of students, or they may go beyond the goals of a particular curriculum by using it as the starting point from which a new set of activities not envisioned by the developers can be created. In this respect, the teacher is tapping a curriculum's potential (Ben-Peretz 1990).

A Textbook Lesson

A group of preservice secondary school teachers were given a photocopy from a student textbook of a lesson involving the sum of the angle measures of a triangle, as well as supporting material found in the teacher's edition. In the original lesson, students were asked to measure the three angles of a given triangle with a protractor and calculate their sum. On the following page, the textbook states that the sum of the measures of a triangle is 180 degrees. A paragraph proof of this theorem follows immediately. Students are told that the sum of the measures of a triangle in most non-Euclidean geometries is not 180 degrees. Later, the textbook states and proves the quadrilateral-sum theorem: that the measures of the angles of a quadrilateral sum to 360 degrees. Last, students are given an argument justifying the formula for the sum of the interior measures of the angles of any convex polygon with n sides, $(n - 2)180$.

Principles and Standards as a Framework to Examine Lessons

The Learning Principle of *Principles and Standards* casts students as autonomous learners who are capable of monitoring their progress, struggling with appropriate tasks, and persevering. As students engage in learning mathematics actively, they may be more inclined to explore mathematical ideas on their own. The preservice teachers evaluated the described lesson in view of the Learning Principle. They noticed that glimpses of this more active viewpoint of mathematics were found at the beginning of the lesson, where students measure the interior angles of a triangle to determine their sum, but that by stating the triangle-sum theorem instead of giving students an opportunity to conjecture this relationship, the textbook's approach undermined this perspective.

Moreover, the Learning Principle of *Principles and Standards* states that students should be actively engaged in lessons that develop conceptual understanding, build facility with procedures and factual knowledge, as well as nurture connections between them. The textbook lesson supported some conceptual knowledge and held opportunities to develop students' facility with procedures, but it *presented* this information to students instead of giving them a chance to investigate the concepts and develop connections between them in a supportive environment.

The Learning Principle also highlights the importance of challenging students with appropriate tasks. The textbook lesson supplied students with examples of how to solve problems using procedures that were stated in the textbook or provided students with various proofs. Students had no opportunities to interact with other students while engaging in the challenging task of developing procedures or following up conjectures with proofs. Consequently, students were denied opportunities to reflect on their own or others' thought processes, monitor their work for mistakes, and experience the satisfaction of solving a difficult mathematical problem. In conclusion, *Principles and Standards* gave the preservice teachers an alternative framework that helped them step back from their more traditional mathematical experiences as high school students, supplied them with a lens to examine the lesson, and gave them a means through which the lesson could be reimagined.

The preservice teachers used the alignment between the lesson and the Learning Principle of *Principles and Standards* as a steppingstone from which a more active lesson could be created. The teachers designed new activities by drawing on their own creativity, optional activities in the teacher's edition, their own beliefs about students, past experiences in college mathematics courses, and past experiences in mathematics education courses. When a teacher draws on his or her own personal resources, however, care should be taken not to contradict the framework being used. Thus, as will be demonstrated in this instance, teachers should move back and forth between their own experiences and beliefs and an external framework, such as *Principles and Standards,* as they redesign and evaluate lesson components. As they do so, they may begin to reveal a curriculum's potential (Ben-Peretz 1990).

Identifying Alignment

Some preservice teachers thought that the opening activity in the textbook, in which students measure the interior angles of a triangle and find their sum as 180 degrees, was aligned with the Learning Principle of *Principles and Standards*. As it was depicted, however, the activity demonstrated only that the sum of the measures of one specific triangle was 180 degrees. The preservice teachers thought that this opening activity had the *potential* to help students see that the sum of the measures of the angles of *any* triangle is 180 degrees. As one preservice teacher said, "I guess we kind of thought it would be a way for them to come up with the theorem themselves before they do the proof" (Kevin[1] 10-23-06).

To support students in making a conjecture about the angle sum in all triangles, the preservice teachers increased the number and type of triangles that students investigated. Some groups of teachers gave students a worksheet involving five different triangles for each student to measure, as seen in figure 22.1.

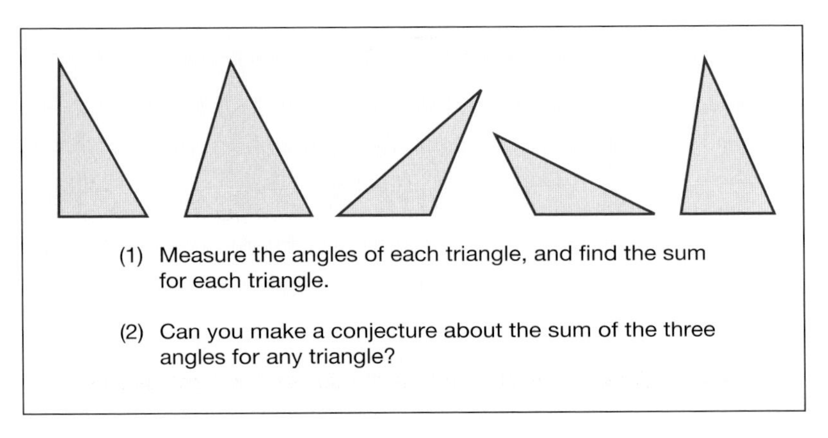

Fig. 22.1. One group's beginning worksheet

The preservice teachers made the decision to provide the triangles because they thought students would draw triangles that were very similar to one another's. This belief about students led the preservice teachers to redesign the lesson so that different examples would be investigated. This element broadened the goals of the lesson to include the skill of searching for and testing a wide range of examples in verifying a conjecture and illustrates the preservice teachers' act of tapping the potential of the curriculum.

1. All names used in this article are pseudonyms.

Proving the Triangle-Sum Theorem

Although some aspects of the lesson, such as discovering the triangle-sum theorem, lent themselves to an investigative lesson, other aspects—such as its proof—did not. One group decided that they would return to a proof of the triangle-sum theorem the next day, and two groups simply allowed students to discover the triangle-sum theorem inductively with examples instead of proving its validity. One group, however, did make the proof of the triangle-sum theorem more investigative.

The proof in the textbook begins with triangle *ABC*. An auxiliary line is drawn through vertex *B* parallel to the base of the triangle as shown in figure 22.2.

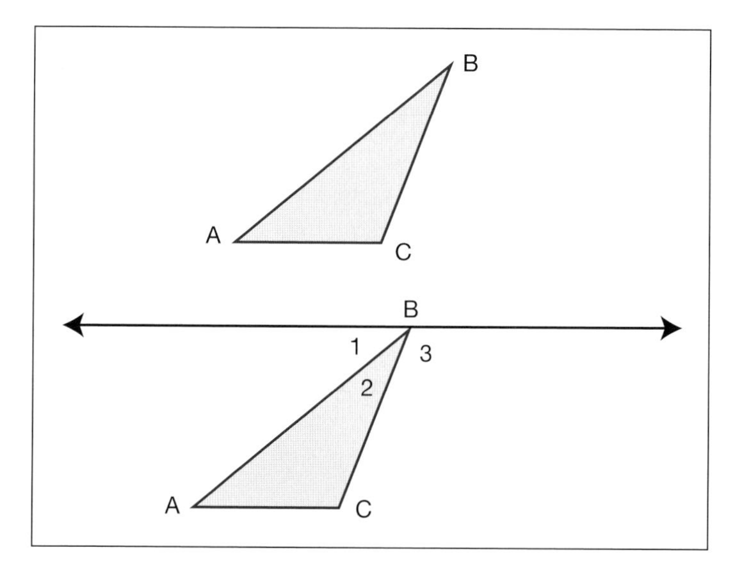

Fig. 22.2. Drawing the auxiliary line through vertex *B* and parallel to side *AC* of triangle *ABC,* forming angles 1 and 3

The proof uses the fact that because the line through *B* is parallel to side *AC,* angle 1 is congruent to angle *A* and angle 3 is congruent to angle *C* by alternate interior angles. The preservice teachers in this group saw the connection between the triangle with the auxiliary line drawn with angles 1, 2, and 3 and the optional activity (described in the teacher's edition), in which students tear the vertices off of a triangle and put them together to form a line. Consequently, they wanted students to cut out angles *A* and *C* and determine whether those matched angles 1 and 3. However, the preservice teachers in this group were not content with students' simply understanding *that* these angles were congruent; they wanted the students to determine *why* angle *A* matched angle 1 and angle *C* matched angle 3.

At this point, though, the preservice teachers were not sure how they could support students in making this discovery.

Instructor: How can you help students see that these are alternate interior angles?

Kevin: State that they're parallel lines.

Instructor: What other information can I give the students besides those two lines are parallel that can help them figure out that alternate interior angles are at work here?

Mark: That these [two sides of the triangle] are transversals.

Instructor: That's almost like a key for kids. As soon as they know that these are parallel and this is a transversal, it's like a light goes on. Oh, my gosh, those are alternate interior angles [angle 1 and angle *A*].

Kevin: We could have this line [line at the top of the triangle] and this line [line at the base of the triangle] and state that they're parallel. Then extend these two lines [pointing to *AB* and *CB*] or at least one of them and state that it's a transversal. Give them that much?

Instructor: It's up to you, however much you are going to put into the worksheet versus how much are you going to have in your mind to be able to tell these kids when they get stuck.

Kevin: So exactly what are the questions we want to ask?

Instructor: What do you want students to get out of this?

Kevin: Well, I basically want them to ... if we cut these angles out, how do we know this angle will fit here and this angle will fit here? Hopefully we want them to come up with alternate interior angles and generalize from there, and this is always 180 degrees here. [*Instructor points to angles 1, 2, and 3 in figure 22.2.*]

Mark: Say you cut off angles *A* and *C* and see if they can fit them into the missing pieces at the top to form a straight angle with angle 2. Is that too trivial, or no? Then you could say, Can you think why this might always work? Remember that these are parallel lines and these are transversals.

Thus the preservice teachers began to understand that most students may not see why the base angles might be congruent to the angles 1 and 3. As curriculum developers, they needed to support the students in seeing why the base angles of the triangle along with the third angle form 180 degrees. Pointing out to students that the base of the triangle and the line through vertex B *were* parallel, along with stating that the sides of the triangle were transversals, seemed to make the task manageable for students without significantly decreasing the demands of the task. The preservice teachers also believed that students would have a chance to engage in the practice of mathematics by cutting out the angles at the base of the triangle and constructing the main argument themselves, as opposed to simply reading the proof in the textbook. This conversation resulted in the worksheet shown in figure 22.3.

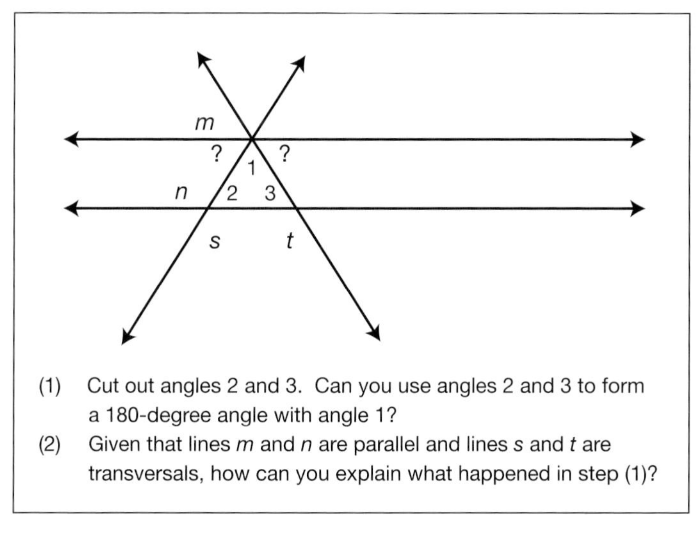

Fig. 22.3. A worksheet that preservice teachers designed to help students construct the proof as depicted in the textbook. Note that a scalene triangle would have been a better way to show the general case.

The preservice teachers' activity to prove the triangle-sum theorem did not end with this worksheet. They wanted students to think about the generality of the steps of the proof, making sure students understood why the argument would work for *any* triangle, taking into consideration the tendency of students to attribute a proof to a specific diagram instead of to a class of objects (Chazan 1993).

The preservice teachers also wanted students to understand the generality of the auxiliary line. That is, can a line be drawn through one of the vertices of any triangle so that it is parallel to the opposite side? If so, why? Such a conversation could lead to why the sum of the angles of a triangle in non-Euclidean geometry does not equal 180 degrees. Such an approach contrasts sharply with how the textbook treats this idea.

The teachers once again began with an activity in the book and used the Learning Principle of *Principles and Standards* to make the lesson more exploratory. However, because the preservice teachers did not have much teaching experience on which to draw, they had to turn to other resources, such as the teacher support materials. They saw an activity in the teacher support materials as a bridge between using the static figure in the textbook as an aid in developing the proof and a more active exploration. They also had to consult their own understandings of proof once again, however, because the exploration focused on a specific figure and they wanted students to generalize their work to *any* type of triangle.

The redesign of the lesson also presented new problems for the preservice teachers because they had to consider the level of scaffolding needed for students to engage actively in constructing a proof. They also had to confront the dilemma that many curriculum developers encounter about what should go into the student textbook (i.e., the worksheet) and what should go into the teacher support materials (e.g., the questions to ask the students depending on their actions).

Quadrilateral-Sum Theorem

The textbook lesson on the quadrilateral-sum theorem presented students with a proof in paragraph form demonstrating that the sum of the measures of the interior angles of a quadrilateral is 360 degrees. The argument relied on showing that any quadrilateral can be cut into two triangles. The triangle-sum theorem was then invoked to demonstrate that the sum of the angle measures of a quadrilateral is 360 degrees. In revising this lesson, many of the preservice teachers, however, wanted students to number the vertices of the quadrilateral, cut off these vertices, and place them together. They expected students to notice that instead of forming a straight line as they did with the triangle, the angles now filled the space around a point. Other preservice teachers provided students with a set of triangles that could be used like pieces from a tangram set to build larger figures. That is, two triangles with one pair of congruent sides could be placed together to form a quadrilateral. A pentagon could be formed by attaching a triangle to one side of the quadrilateral.

Once again, the teachers engaged in a cycle of evaluating the new activity against their own knowledge and beliefs. The teachers made sure that new lessons took students' existing knowledge into account, which in this instance involved knowing that the sum of the angle measures about a point is 360 degrees. In both instances, the new lesson began with current lesson activities and used a more active approach to reconfigure the lesson. The preservice teachers' work with regard to the quadrilateral-sum theorem illustrated two different uses of current textbook activities in creating more active lessons. First, students' active engagement, as seen in one activity (triangle-sum theorem), was appropriated and used in a different textbook activity (quadrilateral-sum theorem), thus making the connection between the two lessons more coherent. In the second instance, the teachers examined existing textbook activities for their implicit action and then thought about how this action could be made more explicit and consequently, engage students.

I encouraged the preservice teachers to think about the productiveness of both of those techniques to help students discover the quadrilateral-sum theorem. With the first technique, students could see that the sum of the angles of a quadrilateral is 360 degrees. That is, it helped to show that the quadrilateral-sum theorem *is* true but not *why* it is true. In this respect, the second technique of building polygons by attaching triangles was more valuable in helping students understand why the theorem is true and leading them to a proof.

Generating the Polygon-Sum Theorem

The textbook gives students a completed table consisting of the name of the polygon along with the sum of its interior angles. The first entry is for a 6-gon with angle sum $4 \cdot 180$ degrees, whereas the last entry involves an n-gon with angle sum $(n - 2) \cdot 180$ degrees. In reimagining earlier textbook activities, the teachers used more active lesson elements as a starting point, but in this final example they were able to bring to bear their previous experiences (developed in previous college mathematics education courses) of translating between mathematical and real-world contexts and tabular representations in altering the lesson.

One group of preservice teachers created the worksheet in figure 22.4.

Number of sides (*n*-gon)	**Sum of Interior Angles**
3	180°
4	
5	
6	

What is the sum of interior angles of a 25-gon? A 100-gon?

What would the sum of interior angles of an *n*-gon be?

Fig. 22.4. A worksheet that preservice teachers designed to help students create the formula for the sum of the interior angles of a convex polygon

This group assumed that students would be able to draw the different figures and calculate the sum of the interior angles. However, they thought that the tabular representation would make the increase in the sum of interior angles by 180 degrees for each increase in the number of sides much more transparent. This insight, they believed, might lead students to develop an argument for this relationship based on the use of triangles to construct figures of $n + 1$ sides from figures with n sides. They also thought that the table had the potential to spark students' generation of the method of triangulating polygons as shown in figure 22.5.

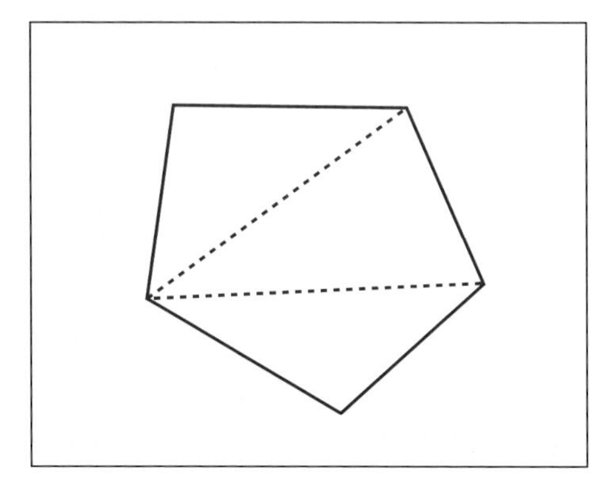

Fig. 22.5. Triangulating a convex pentagon from a vertex

The textbook presented students with the method above of triangulating from a vertex of the figure; however, leaving the generation of this method up to the students left open the possibility that they might divide the polygon into triangles using another method, as shown in figure 22.6.

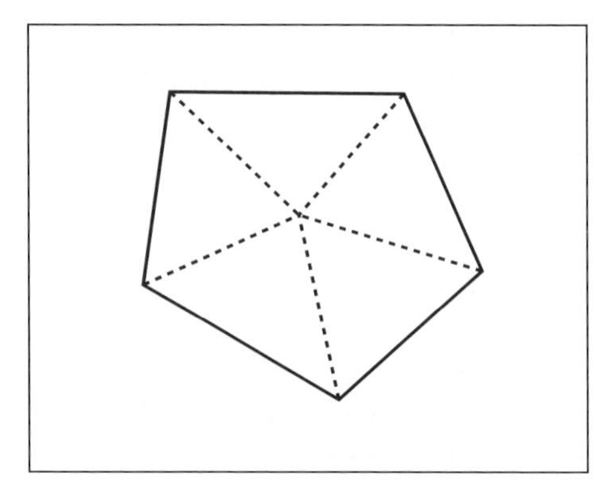

Fig. 22.6. Triangulating a convex pentagon from a point inside the figure

The preservice teachers and I had a conversation about this second method and the formula it might generate.

Cheryl: We thought that students might divide the figure into triangles by using the center [*sic,* i.e., an interior point].

Instructor: What is the formula if they put the center there?

Cheryl: Well, you have n triangles because there is one for each side of the figure. And since there is 180 degrees in a triangle, you have $180n$, but you have to consider the angles in the center, which are a circle, 360 degrees. So the formula here is $180n - 360$.

By providing spaces in which students could approach the lesson in a number of different ways, the preservice teachers began thinking about other ways that students might break a polygon into triangles. In the exchange above, I pushed the preservice teachers to think about how students might use this method to develop a formula for the sum of the angles of a convex polygon. This method led to another discussion.

Instructor: What if I place my point on one side and form triangles from that point [shown in fig. 22.7]?

Instructor: How many triangles do I get from a five-sided figure when I do that?

Mark: Four triangles.

Instructor: Right, I get four triangles, and I know that the sum of the measures of the angles of each of those triangles is how much?

Martha: One hundred eighty.

Instructor: And that's times $n - 1$. [*Teacher writes* $180(n - 1)$ *on the board.*] Now how do I adjust that?

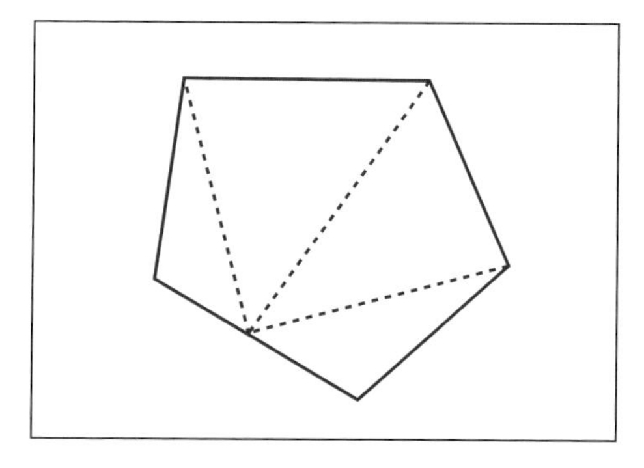

Fig. 22.7. Triangulating a convex pentagon from a point on a side

Mark: Subtract 180.

Instructor: Why?

Mark: Because you have three angles at the bottom that add up to 180, but they aren't part of the angles of the figure. [*Teacher writes on the board:* $180(n-1)-180$.]

Together the preservice teachers and I further expanded the potential of this activity to include multiple solution methods and to connect geometry with algebra by asking students to explain the origin of their formulas. Moreover, the activity could be used to further develop students' symbolic manipulation capabilities. This last goal could be accomplished by asking students to show that the three expressions $180(n-2)$, $180n-360$, and $180(n-1)-180$ for the sum of the interior angles of a convex polygon are equivalent.

The preservice teachers had worked with recursive formulas in previous college courses. This background helped prepare them for further expanding the potential of this lesson. The preservice teachers envisioned that students might develop a recursive formula for the polygon-sum theorem because students begin their explorations with a triangle and later move to the quadrilateral. Furthermore, they thought that technology might be useful in developing a recursive formula for the polygon-sum theorem and understanding why this formula makes sense. Involving technology could be accomplished by asking students to use The Geometer's Sketchpad (Jackiw 2001) to determine a method for creating a figure with $n+1$ sides from a figure with n sides.[2] They might do so by breaking a side in two and opening it up to form two sides as shown in figure 22.8.

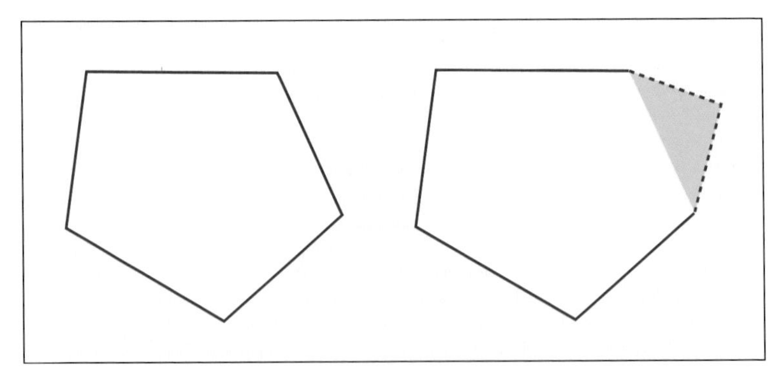

Fig. 22.8. Creating a figure with $n+1$ sides from a figure with n sides

2. An activity similar to this one, but without the use of technology, is presented in the teacher's edition. However, the preservice teachers had not read this activity when developing the lesson.

Reflecting on the Teachers' Lessons

Although the activities the teachers created improved on the textbook lesson in many ways, the new activities also had several shortcomings. First, the triangles that the teachers created in figure 22.1 all have a base parallel to the horizon. Research suggests that students who often see right triangles presented in this way have a difficult time identifying them if their base is not parallel to the horizon (Clements and Battista 1992). Therefore, some of these triangles should be rotated. The triangle the teachers constructed to be used in the proof of the triangle-sum theorem in figure 22.3 is equilateral. Students may believe that this proof applies only to equilateral triangles (Battista 2007). Accordingly, a more general figure here might be a scalene triangle.

Second, only one group of preservice teachers out of four incorporated an activity involving proof of the triangle-sum theorem into their lessons: however, as *Principles and Standards* argues, reasoning and proof are "fundamental aspects of mathematics" (NCTM 2000, p. 342). Consequently, they all should have crafted an activity requiring students to develop a proof of the triangle-sum theorem. This pursuit would not only have given students an opportunity to engage in practices more aligned with the work of mathematicians but may have also helped them understand *why* the theorem is true.

Third, the teachers who had engaged in reimagining this textbook activity had themselves learned geometry in a course that drew extensively on The Geometer's Sketchpad (Jackiw 2001); few of the teachers, however, incorporated this software into their lessons. The use of this technology could have quickly facilitated the testing of numerous cases, either to generate conjectures for or to verify the triangle-sum, quadrilateral-sum, and polygon-sum theorems.

Conclusion

The process of realizing the curriculum potential of a textbook lesson begins with its critique through an external framework (e.g., *Principles and Standards*), the teacher's personal beliefs and experiences, or in this instance, both. The goal of this critique is to search for alignment between the framework or the teacher's knowledge and beliefs and the activities in either the student or teacher's edition of the textbook. The teacher then uses these areas of alignment to redesign a textbook lesson. The teacher must continually evaluate these newly designed activities against either the framework or her or his knowledge and beliefs, and may need to re-evaluate the alignment several times to refine the activity.

The teacher revisits this process once the redesigned lesson has been taught, as it can then be viewed in the context of the actual experiences with students. Redesigning lessons was a learning experience for the preservice teachers. For example, they learned about the importance of, and subtleties associated with, auxiliary lines. They also learned about how diagrams may promote as well as restrict students' generalizations. Furthermore, they learned aspects of the curriculum development process itself when deciding what should be placed in the student textbook and what needed to be reserved for the teacher's edition.

Principles and Standards describes an ambitious vision for school mathematics. Creating completely new tasks that embody this vision can be a daunting task for beginning as well as veteran teachers. The activity described in this article provided preservice teachers with a concrete example of how they might begin the journey of transforming a traditional mathematics lesson into one more aligned with the vision described in *Principles and Standards*.

REFERENCES

Battista, Michael T. "The Development of Geometric and Spatial Thinking." In *Second Handbook of Research on Mathematics Teaching and Learning,* edited by Frank K. Lester, pp. 843–908. Charlotte, N.C., and Reston, Va.: Information Age Publishing and National Council of Teachers of Mathematics, 2007.

Ben-Peretz, Miriam. *The Teacher-Curriculum Encounter: Freeing Teachers from the Tyranny of Texts.* Albany: State University of New York Press, 1990.

Chazan, Daniel. "High School Geometry Students' Justification for Their Views of Empirical Evidence and Mathematical Proof." *Educational Studies in Mathematics* 24, no. 4 (December 1993): 359–87.

Clements, Douglas H., and Michael T. Battista. "Geometry and Spatial Reasoning." In *Handbook of Research on Mathematics Teaching and Learning*, edited by Douglas A. Grouws, pp. 420–64. Reston, Va., and New York: National Council of Teachers of Mathematics and Macmillan Publishing Co., 1992.

Jackiw, Nicholas R. The Geometer's Sketchpad. Software. Emeryville, Calif.: Key Curriculum Press, 2001.

National Council of Teachers of Mathematics (NCTM). *Principles and Standards for School Mathematics*. Reston, Va.: NCTM, 2000.

Serra, Michael. *Discovering Geometry: An Investigative Approach*. 4th ed. Emeryville, Calif.: Key Curriculum Press, 2008.

Looking Forward to Interactive Symbolic Geometry

Philip H. Todd

INTERACTIVE geometry software has features for facilitating discovery-based learning. It has been widely adopted in the past twenty years, and a vast amount of creativity has been brought to bear on applying it to the educational process (King and Schattschneider 1997; Gawlick 2002). This is not to say that it is without its limitations.

Interactive geometry systems are constructive rather than constraint-based. In a construction-based system, for example, to create an incircle of a triangle, one constructs first the intersection of the angle bisectors, then the perpendicular from one side through this point, then the circle centered at the intersection point whose circumference goes through the foot of the perpendicular. In a constraint-based system, by contrast, one sketches a circle inside the triangle, then constrains the circle to be tangent to each side of the triangle. A construction-based system models the process of geometry as practiced a couple of millennia ago by the ancient Greeks, whereas a constraint-based system models the process of geometry practiced today in computer-aided design systems. Further, interactive geometry programs are purely numeric and thus are limited in their ability to reinforce the important connections between geometry and algebra. In this article, we imagine the potential of an interactive symbolic geometry system that addresses both these limitations and provides an interesting complement to existing interactive geometry software programs and computer algebra systems (CAS).

Geometry in a Classroom of the Future

Constraint-based symbolic geometry software has the potential to make an impact on mathematics education in the next twenty years. With the reader's indulgence, then, we are going to embark on a little time travel into a high school of the future, let us say to 2019. We assume that the students of the future will have access both to some form of symbolic geometry software and to CAS in much the same way that the students of today have access to graphing calculators. The examples in this article use Maple (Maplesoft 2005) and Geometry Expressions (Todd 2007).

Exploring Heron's Formula

Our first stop will be a sophomore class. The students all have computers: some are laptops, some are tablets, some look like calculators. All have symbolic geometry software and CAS. The students are investigating relationships between the radii of incircles and excircles of a triangle. As a warm-up, they create a formula for the area of a triangle. The students first sketch a triangle *ABC,* then add length constraints to each side, specifying that *AB* is length *c, BC* is length *a,* and *AC* is length *b*. They select the triangle and designate its interior as a polygonal region, then have the software display the area of the polygon. Their symbolic geometry program automatically creates the expression for the area shown in figure 23.1. The shaded red pentagon is an icon representing the fact that the quantity displayed is an area.

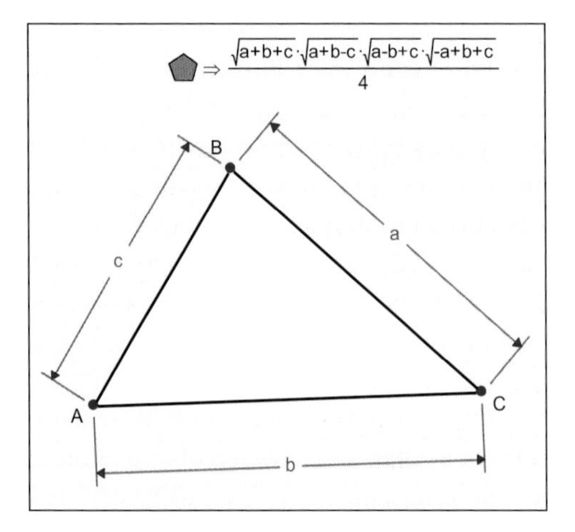

Fig. 23.1. Symbolic geometry software computes a formula for the area of a triangle (identified by a red pentagon) given the lengths of three sides.

Ms. Johnson, the teacher, informs her class that this beautiful formula is due to Heron of Alexandria, and that it frequently appears in textbooks in the following form:

$$\sqrt{s(s-a)(s-b)(s-c)} \text{ where } s=\frac{a+b+c}{2}. \tag{1}$$

She asks the class to verify that expression (1) is indeed equivalent to the equation generated by their symbolic geometry system.

The students have Heron's formula now in two different formats, but how could they prove the formula? Ms. Johnson prompts her class to constrain a triangle relative to the altitude and the two nonbase sides (fig. 23.2) and then to show the formula for the length of the base.

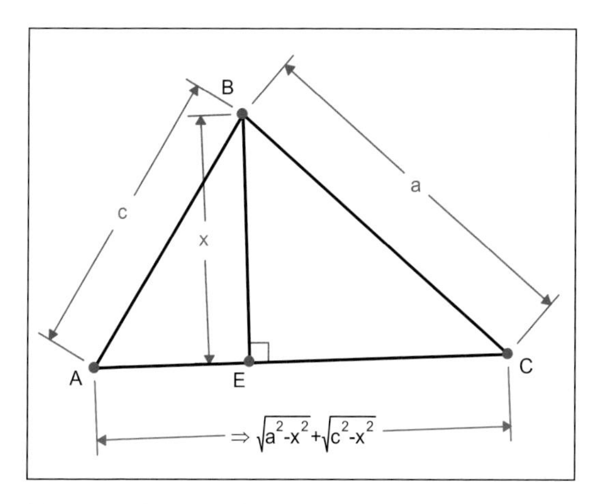

Fig. 23.2. Length of the base relative to the other sides of the triangle and its altitude

The students easily see how they could have derived this expression using the Pythagorean theorem. They agree that if they equate this quantity with b, then solve for x, they will have a formula for the altitude relative to the side lengths. Strategically, they have finished the problem. The only issue is the matter of solving the equation. They entrust this task to an algebra system, as shown below.

solve(sqrt(a^2–x^2)+sqrt(c^2-x^2)=b,x);

$$\frac{\sqrt{2a^2b^2-a^4-c^4+2c^2a^2+2c^2b^2-b^4}}{2b}, -\frac{\sqrt{2a^2b^2-a^4-c^4+2c^2a^2+2c^2b^2-b^4}}{2b}$$

Two solutions are given, but the second is negative. The expression under the square root is in expanded form, so it is difficult to compare with Heron's formula. The students use the **factor()** command in the CAS to straighten this out:

>factor(%);

$$\frac{\sqrt{-(b+a-c)(b+a+c)(-b+a+c)(-b+a-c)}}{2b}$$

With this expression for the altitude, the students immediately multiply by half the base ($b/2$) and verify Heron's formula.

The class's next stop is the *incircle,* or *inscribed circle*. The students are asked to find its radius. Again, their symbolic geometry system automatically gives them a formula relative to the lengths of the sides of the triangle (fig. 23.3).

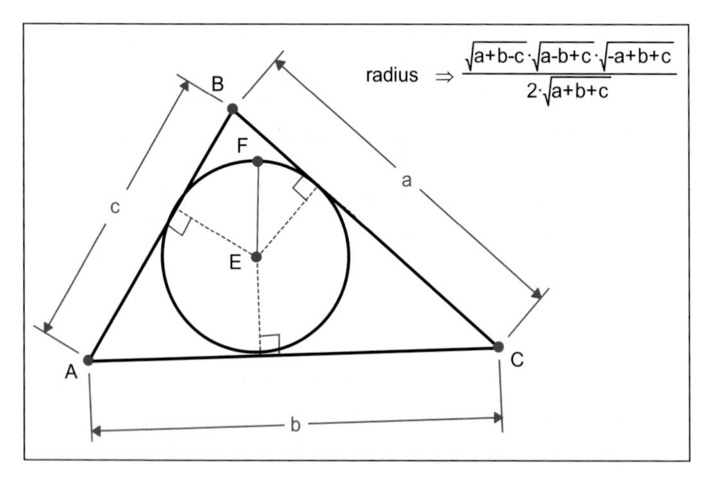

Fig. 23.3. Software-generated expression for the radius of the incircle

Some discussion follows, and one student notices that the formula is symmetric in *a, b,* and *c*. Another points out that it is quite similar to Heron's formula. That comment is exactly the cue Ms. Johnson has been waiting for: she challenges the students to work out how to use the area formula to get the radius. Some get to work in their CAS, copying the two expressions from the symbolic geometry system. Others work directly in the geometry system (fig. 23.4).

Ms Johnson strolls through the aisle and offers gentle prompts:

"The area formula has a 4 in the denominator, and the radius has a 2; how could you change the 4 to a 2?"

"Now the area has $\sqrt{a+b+c}$ in the numerator, which is not there in the radius. How would you get rid of it from the numerator? That's right, divide by it. But now the radius has this same radical term in the denominator, so you'd need to divide by it again."

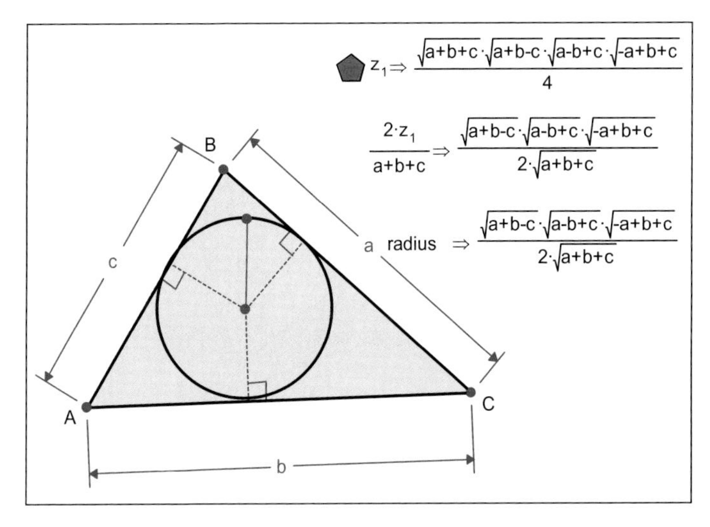

Fig. 23.4. Relationship of the incircle radius to the triangle area

"So you've divided by $\sqrt{a+b+c}$ twice. What is that equivalent to? That's right, dividing by $(a + b + c)$."

The students conclude that the expression for the radius of the incircle is twice the area of the triangle divided by the sum of the sides.

"How could this be proved?" their teacher asks. She suggests dividing the triangle into three parts, as in figure 23.5.

The class has a collective "aha" moment: the radius of the incircle is the altitude of each of the smaller triangles. Ms. Johnson suggests finding a formula for the area of the large triangle relative to the radius and the sides by adding the smaller triangles, and the class quickly obtains the following.

(2) $$A = \frac{ar}{2} + \frac{br}{2} + \frac{cr}{2}$$

Some in the class type this expression directly into their CAS to solve for r, but Ms. Johnson gently scolds them, "This is too simple an equation for your CAS. Solve it by hand, please."

Ms. Johnson ends the session by defining an *excircle* (Coxeter and Greitzer 1967; Posamentier 2002), or *escribed circle,* as a circle that is tangent to one of

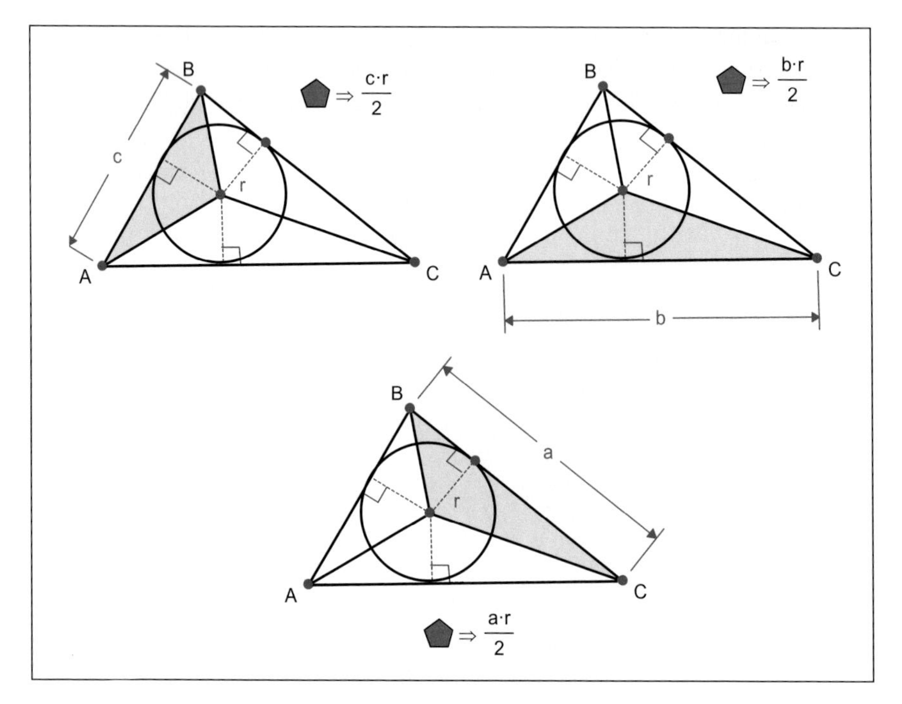

Fig. 23.5. Area-decomposition proof of the incircle radius

the sides of the triangle and to the extensions of the other sides; but whereas the incircle lies inside the triangle, the excircle lies outside. An illustration showing the three excircles of a triangle (fig. 23.6) clears up any confusion. She asks the class to find the radius of one of the excircles. They do this quickly with their symbolic geometry systems. As the class departs, Ms. Johnson suggests they might want to think about how they could prove this result, but we suspect they are more likely thinking about lunch.

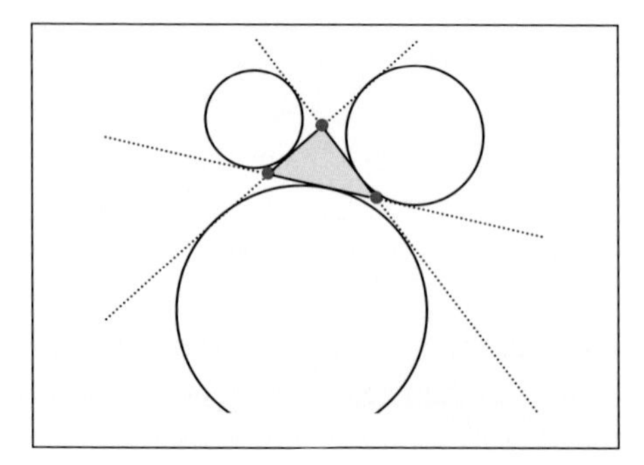

Fig. 23.6. A triangle has three excircles, each tangent to one of its sides and to extensions of the other sides but external to the triangle.

When the class reconvenes the next day, their teacher reminds the students of their result for the radius of the incircle. She splits them into three groups, one for each excircle, and asks them to derive an expression for their excircle in terms of the triangle area. She instructs the groups not to divulge their radii. She then asks the students not only to write down the radius of their own excircle but also to guess the radii of the other groups' excircles.

The team investigating the excircle external to *BC* presents its results (fig. 23.7).

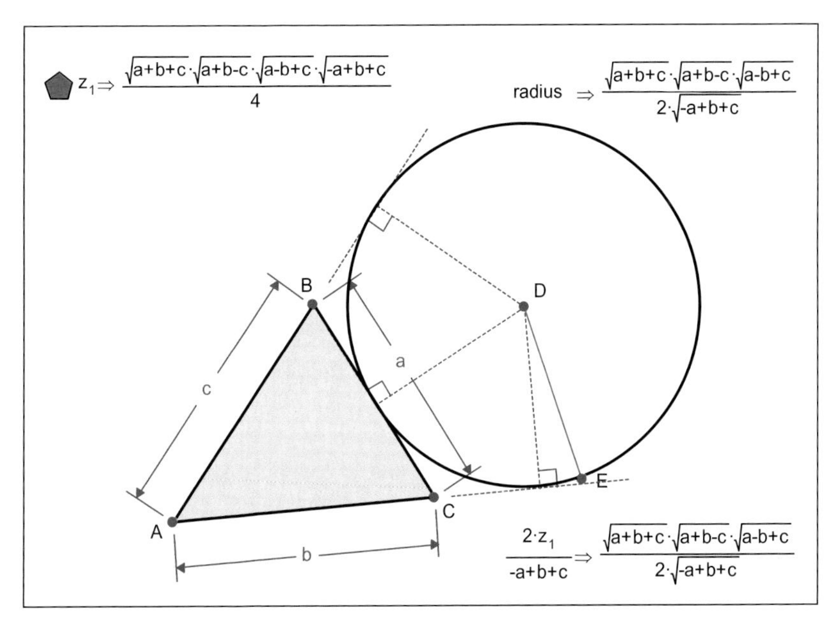

Fig. 23.7. Relationship between the excircle radius and the triangle area

The team's guesses for the other two excircles are accurate:

$$\frac{2A}{a-b+c}, \quad \frac{2A}{a+b-c} \tag{3}$$

Ms. Johnson reminds the class how they proved the incircle radius by dividing the triangle into three smaller triangles whose common point is the center of the incircle. She asks them for suggestions about using a similar strategy with the excircle. Despite some initial confusion as to how to divide a triangle using an external point, with a little guidance from Ms. Johnson the class comes up with the pictures in figure 23.8.

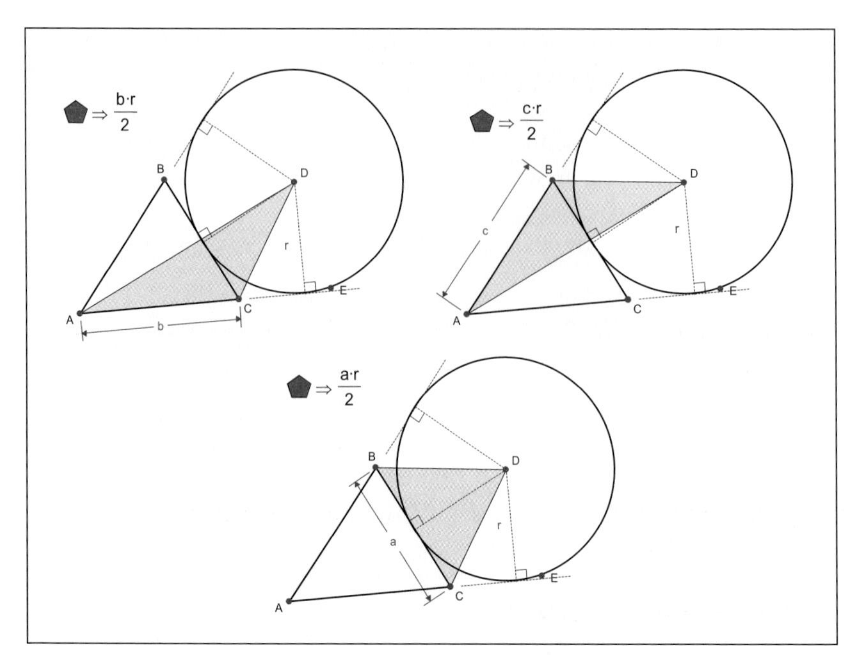

Fig. 23.8. Area-decomposition proof of the excircle radius

The connection between the geometry and the algebra immediately becomes apparent—the area of the original triangle can be found by adding the smaller triangles with bases length b and c and subtracting the triangle with base length a.

$$A = -\frac{ar}{2} + \frac{br}{2} + \frac{cr}{2} \tag{4}$$

This result corresponds to the fact that the term in the denominator of the radius expression is the sum of two lengths minus the third. This time nobody reaches for the CAS; everyone solves for r by hand. Ms. Johnson writes the results of the class investigation on the board, a set of formulas for the radii of the incircle and the three excircles.

She assigns the class homework: to come up with expressions involving radii

$$r_0 = \frac{2A}{a+b+c}, \quad r_1 = \frac{2A}{a+b-c}, \quad r_2 = \frac{2A}{a-b+c}, \quad r_3 = \frac{2A}{-a+b+c} \tag{5}$$

and area, but without the side lengths. "In other words," she asks, "are there any relationships between the radii and area that are true for all triangles?"

Unfortunately we do not get to visit Ms. Johnson's class to see what the students came up with; we are off to Mr. Ford's precalculus class tomorrow. Would anyone recognize that multiplying all the radii would give a denominator closely related to the square of the area?

(6) $$r_0 r_1 r_2 r_3 = \frac{16A^4}{(a+b+c)(a+b-c)(a-b+c)(-a+b+c)}$$

And hence,

(7) $$r_0 r_1 r_2 r_3 = \frac{A^4}{A^2} = A^2.$$

Would someone identify that inverting the radius expressions would make them amenable to addition? Would anyone try adding the reciprocals of the ex-circle radii?

(8) $$\frac{1}{r_1} + \frac{1}{r_2} + \frac{1}{r_3} = \frac{a+b-c}{2A} + \frac{a-b+c}{2A} + \frac{-a+b+c}{2A} = \frac{a+b+c}{2A} = \frac{1}{r_0}$$

Perhaps not, but I am sure Ms. Johnson will have a set of prompts to elicit these results from the class when they next meet.

Ladders around Corners

Mr. Ford in precalculus has posed to his class the problem of moving a ladder around a corner. "What," he asks, "is the longest ladder that you can carry round a right-angled junction between a corridor of width *a* and a second corridor of width *b?*" (Kalman 2007)

He works with the students to set up the problem in their symbolic geometry system. They use the axes to represent one pair of walls. They offset the other walls of the corridor by *a* and *b,* and constrain the length of the ladder to be *L*. Dragging the foot of the ladder along the wall lets the class experiment with different ladder lengths (fig. 23.9).

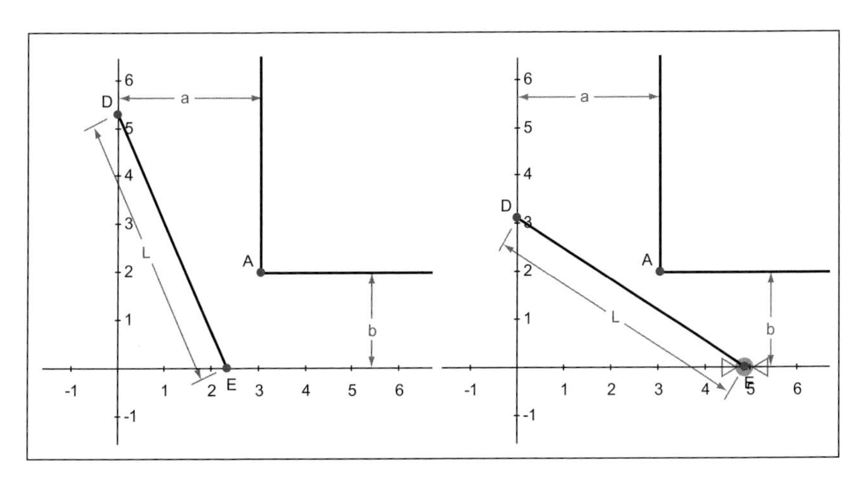

Fig. 23.9. Moving a ladder around a corner

Mr. Ford assigns different values of a and b to different students and asks them to find by trial and error a value for L that just clears the corner. He writes the values in a table on his whiteboard, but the group is unable to guess a general form from the data. He asks the class whether the problem can be simplified in any way. He elicits the answer that he is clearly looking for: what if we make the two corridors the same width?

The students decide that if the ladder only just fits, then its center will hit the corner of the wall. Mr. Ford suggests they create the locus of the center of the ladder with maximum length as it moves so that the ends are touching the outer walls. Asked what kind of curve this is, the class suspects it may be a quarter circle, or part of an ellipse. The symbolic geometry system shows the equation of the locus, which is quickly identified as a circle with radius $L/2$ (fig. 23.10).

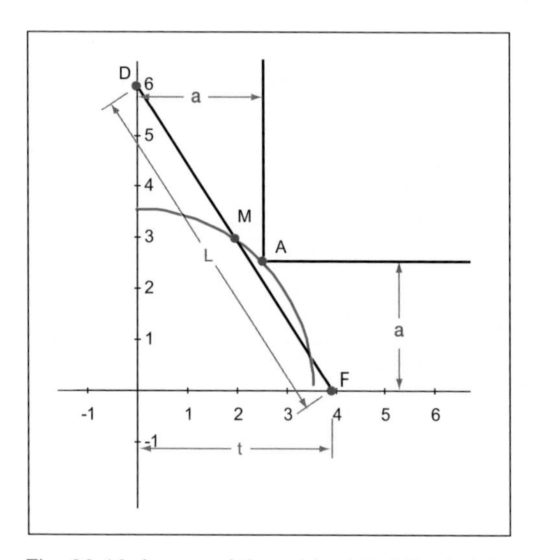

Fig. 23.10. Locus of the midpoint of the ladder

"What," asks Mr. Ford, "is the distance of A from the origin?" The class immediately generates the result with their geometry system, but Mr. Ford asks how they could have worked out this distance without the computer. With a little prompting, the class agrees that it is the diagonal of a square of side a, and is an easy result of the Pythagorean theorem.

What, then, is the length of the longest ladder that will fit around the corner? Several of the students have already put the answer into their geometry system to confirm it by the time he gets the official response:

(9) $$L = 2a\sqrt{2}$$

At the next session, when they reconsider the original problem, Mr. Ford introduces the class to the envelope curve (fig. 23.11). Some experimentation convinces the class that this curve can be used to determine whether the ladder fits.

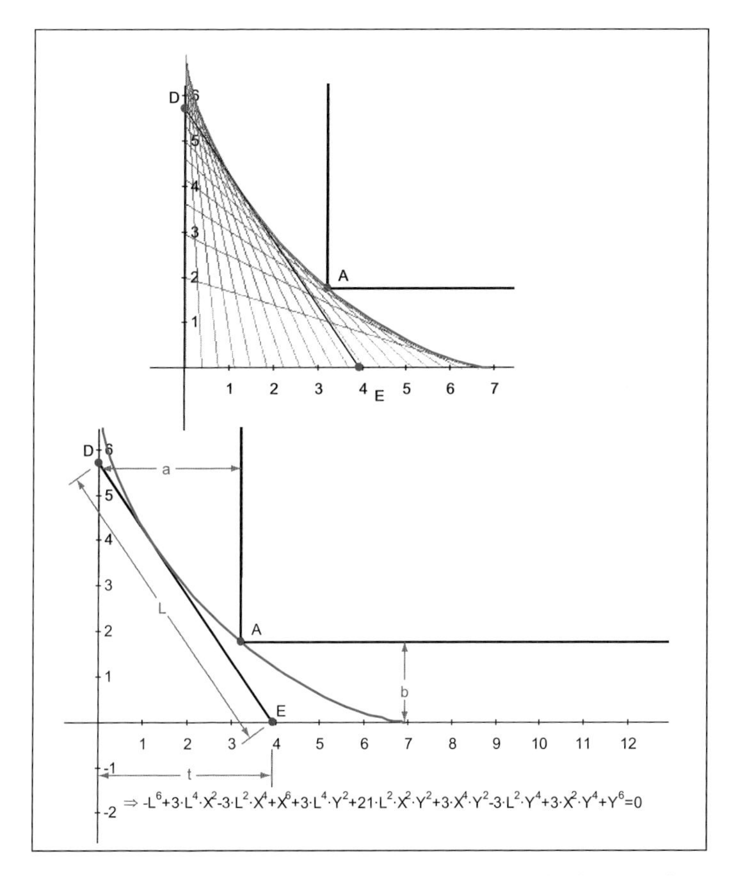

Fig. 23.11. Envelope curve of the ladder, along with its equation

The symbolic geometry system generates the equation of the envelope, and the class is quick to copy this equation into their algebra system and solve for L. They get six rather long and complicated solutions. Mr. Ford points out, however, that one solution is positive real, another is negative real, and the others involve the symbol i, which, he reminds them, is the square root of negative one and not appropriate for a length. Thus only the first solution is considered:

$$\frac{\sqrt{\left(ba^2\right)^{(1/3)}\left[a^2\left(ba^2\right)^{(1/3)}+3b\left(ba^2\right)^{(2/3)}+3ba^2+b^2\left(ba^2\right)^{(1/3)}\right]}}{\left(ba^2\right)^{(1/3)}}.$$

Mr. Ford suggests substituting this value back in for *L*, and they verify that with this length the corner lies on the envelope curve and so this value is the crucial length of the ladder.

He asks the class whether they would expect the result to be symmetric in *a* and *b*. The consensus is yes, it should be, because the maximum length of ladder that can get around a corner from a corridor of width *a* into a corridor of width *b* is the same as the maximum length of a ladder that can go in the other direction. Mr. Ford points out that the expression as written is not obviously symmetric in *a* and *b*—for example, the denominator contains *a* and *b* to different powers. He challenges the students to simplify the expression so that its symmetry is clear.

The first attempt of the class is to apply the **simplify()** command in their CAS, but it returns the same result that was fed in. After a flurry of pencil-and-paper work, a symmetric answer is produced.

(10) $$\sqrt{b^2 + 3a^{\frac{4}{3}}b^{\frac{2}{3}} + 3b^{\frac{4}{3}}a^{\frac{2}{3}} + a^2}$$

Meanwhile one student, lazier than the rest but with a better knowledge of his algebra system, has persuaded it to simplify the expression as shown below by applying assumptions that *a* and *b* are positive real.

$$\left(a^{(2/3)} + b^{(2/3)}\right)^{(3/2)}$$

"You know that what we have to do now," Mr. Ford says, "is to reconcile the solutions" (fig. 23.12). One student points out that the power 3/2 can be regarded as the square root of a cube. We could do the cube and the square root in either order. Cubing first would give us the following:

$$a^2 + 3a^{(4/3)}b^{(2/3)} + 3a^{(2/3)}b^{(4/3)} + b^2.$$

Mr. Ford looks at the clock and sees that five minutes of class time remain. "Verification," he says. "Can anyone give me a known case we can use to verify this result?" The students set $b = a$ in the formula and observe that it simplifies to equation (11).

(11) $$\left(2a^{\frac{2}{3}}\right)^{\frac{3}{2}} = 2^{\frac{3}{2}}a = 2a\sqrt{2}$$

A Rugby Problem

Across the hall we find Ms. Franklin drawing a calculus problem on the board. "The sport of rugby," she says, "is a lot like football here in America,

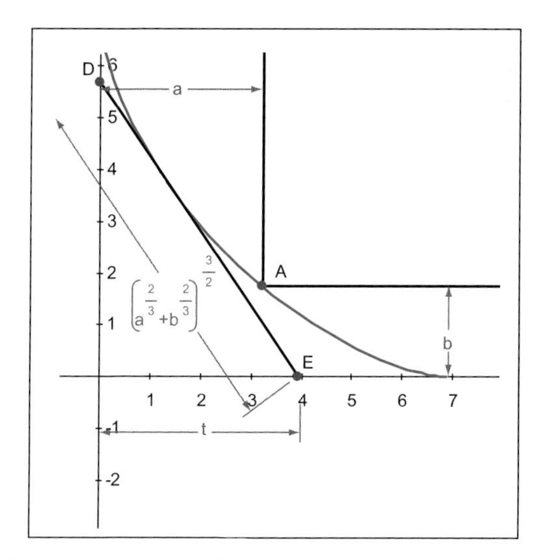

Fig. 23.12. Solution to the longest-ladder problem

but when you score a touchdown—they call it a 'try'—you take the extra-point kick—they call it a 'conversion'—from a place in the field in line with the place the try was scored. So if the try is scored close to the sideline, the kick is taken from somewhere close to the sideline. If it is scored in the middle of the field, the kick can be taken from the middle of the field" (fig. 23.13).

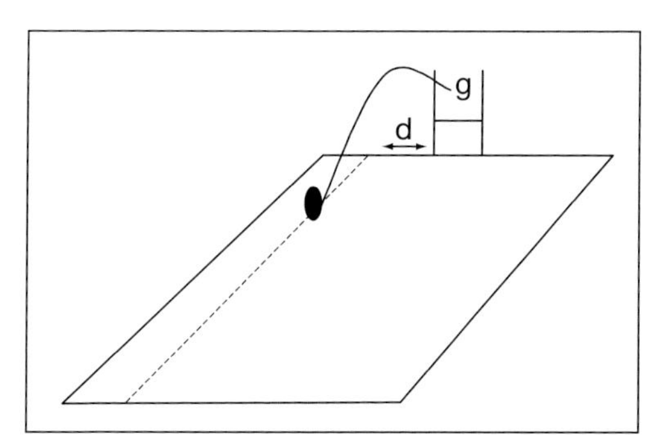

Fig. 23.13. A rugby extra point kick is taken from a position on the field in line with the location of the touchdown. *g* is the width of the goal, and *d* the distance from the nearest post to the location of the touchdown.

"My question is, if the goalposts are width *g*, and the try is scored distance *d* outside the left post, what is the best place to take the kick?" (Jones and Jackson 2001) She asks the class to set up the problem in their geometry system.

The class then discusses how to phrase the question as an optimization problem by asking what is to be optimized. An initial suggestion that they should minimize the length of the kick is disposed of with the realization that the minimum distance would be found on the goal line itself, which would lead to a need to bend the kick to stand any chance of scoring.

Ms. Franklin suggests maximizing the angle made by the goalposts at the point of kick, and the class quickly derives the angle from their geometry system (fig. 23.14).

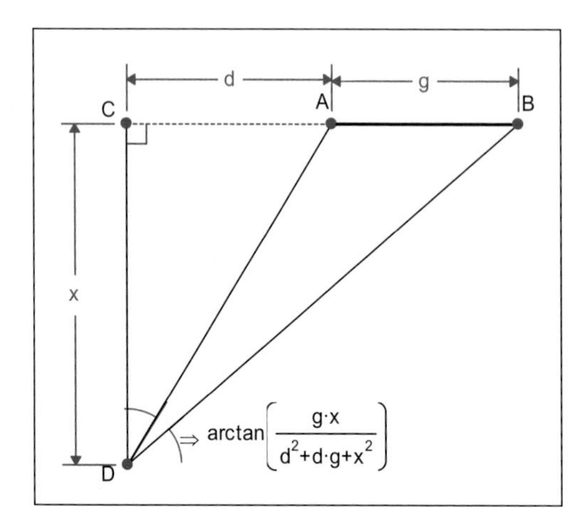

Fig. 23.14. Angle made by the goalposts at the point of the kick

The class agrees that assuming the kicker is able to make the distance, the best place to kick will be where this angle is largest. Their calculus background directs them to set the derivative of the angle (as a function of x) equal to zero.

"We are going to be doing this problem by hand," says Ms. Franklin, "but first, let's say a and b are angles between 0° and 90°, and I tell you that b is greater than a, what can you tell me about tan(b) and tan(a)? Think about the graph of the tangent function. That's right, tan(b) > tan(a). So if I tell you I've found the largest possible value for the angle in our problem, then will that also give the largest possible value of the tangent function?" The class agrees, and is prepared to concede that a maximum angle will occur where the argument of the arctan is maximized. An almost audible relief settles over the class as they realize that they do not need to differentiate the arctan, but instead, just its arguments.

In due course, the derivative of the argument is found to be

$$\frac{-gx^2 + gd^2 + g^2d}{\left(d^2 + dg + x^2\right)^2}. \quad (12)$$

When the derivative is set equal to zero and the equation is solved for x, the result is

$$x = \sqrt{d^2 + dg} \tag{13}$$

"Let's feed equation (13) back into your geometry diagram," suggests Ms. Franklin, "so we can see the locus of the best kick locations for different values of d." Ms. Franklin asks what kind of curve they will get. Parabola and hyperbola are put forward as ideas. The weight of popular opinion is behind hyperbola. "Can we work out the equation of this curve?" Ms. Franklin asks. "We haven't established a coordinate system for our diagram yet. I suggest we make the origin the center of the goalposts and place the y-axis perpendicular to the goal line."

The class feeds $x = (g/2) + d$ and $y = \sqrt{d^2 + dg}$ into the program, and popular opinion is vindicated by the curve's equation (fig. 23.15). "Hyperbolas have asymptotes," says Ms. Franklin. "What is the asymptote of this one?" "When x and y are large," she prompts, "will the size of g matter? So if we ignore g, what is this equation?" The class deduces the asymptote is the line $y = x$ and places it on their diagram.

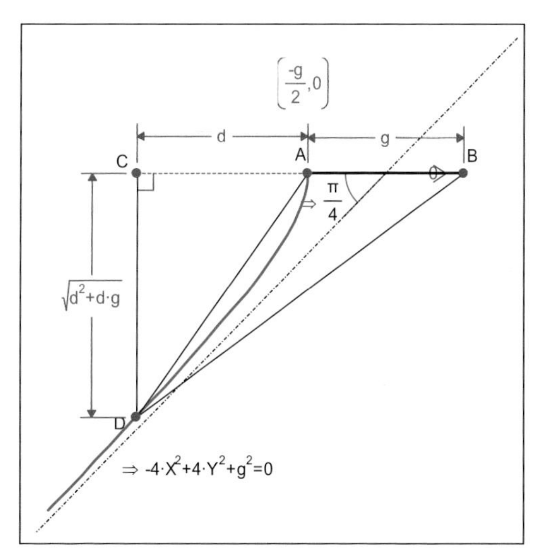

Fig. 23.15. The locus of optimal kick points is a hyperbola whose asymptote is the line $y = x$.

So what practical advice would this class give a rugby kicker? After discussion on how to measure angles on a rugby field, they come up with the statement "Kick from approximately the same distance out from the goal line as the try is from the center of the field."

As the class dissolves we return to our own decade to try and draw conclusions from our excursion into the future.

Conclusion

Reviewing the fictional classroom experiences described in the article, we try to identify the precise points of contact between the algebraic and the geometric components of the problems.

In the incircle-excircle example, the algebraic expression for the radius of the incircle is observed to be closely related to the expression for the area of the triangle. This observation directs the search for a geometric proof through a decomposition of areas. In its turn, the geometry provides context for algebraic manipulation, in which the expression for the area relative to the incircle radius needs to be inverted to give the radius relative to the area. Further, the geometry provides an opportunity to pose an algebraic discovery problem. Strikingly, in this example the students are asked to discover geometric theorems by manipulating algebraic expressions.

Algebraic symmetry is a further point of contact. The expression for the radius of the incircle is symmetric in all three lengths. This property is related to the fact that the geometric definition of the incircle is symmetric with respect to each side of the triangle. An excircle, however, is defined symmetrically with respect to two of the sides but not with respect to the third. The corresponding algebraic expression is symmetric with respect to two of the lengths but not the third.

Symmetry also plays a role in the second example. Geometrically, it is obvious that the longest ladder that can turn a corner from a corridor of width a into a corridor of width b is the same as the longest ladder that can turn a corner in the opposite direction. The initial form of the algebraic result, however, is not symmetric in a and b, and this outcome provides the motivation for an algebraic simplification assignment. Algebraic simplification can seem an academic exercise and somewhat arbitrary, but in this instance, the goal of exposing the inherent symmetry of the expression gives a strong motivation for, and sense of direction to, the manipulation.

An elegant mathematical result for a real problem is achieved through a combination of symbolic geometry, computer algebra, and old-fashioned, by-hand algebraic manipulation. The symbolic geometry is essential to compute the equation of the envelope to the ladder as it moves around the corner. Although the definition of the envelope curve can be grasped intuitively, the mathematics behind computing its equation is beyond students at this level. Likewise, solving a nonfactoring cubic is something best left to a CAS.

Students, however, are not passive participants in the process. To drive the geometrical end of the problem, they need to have a good understanding of loci and of parametric and implicit equations of curves. To interpret the results from the CAS, they need to be familiar with complex numbers and have the mathematical common sense to eliminate geometrically meaningless results. Finally, they need good algebraic manipulation skills of their own to massage the automatic results into a desired form or to identify that different forms of an algebraic expression are equivalent.

In the third example, geometry is initially used in a traditional role in a calculus class: providing the context for an optimization problem. As the symbolic geometry software is able simply to derive the expression for the angle, the student is able to focus on the task at hand: using calculus to solve an optimization problem. Here, the role of the technology is in isolating the component of a problem that is relevant to the current lesson. The algebraic results of the calculus problem are fed back into the geometry software in the form of a curve. The algebraic equation of the curve indicates its form and suggests the use of an asymptote as a linear approximation. The asymptote in turn can be computed from the curve equation. At this point the distinction between algebraic and geometric representation has become blurred—arguably a desirable conclusion.

REFERENCES

Coxeter, Harold S. M., and Samuel L. Greitzer. *Geometry Revisited.* Washington, D.C.: Mathematical Association of America, 1967.

Gawlick, Thomas. "On Dynamic Geometry Software in the Regular Classroom." *International Reviews on Mathematical Education* 34 (2002): 85–92.

Jones, Troy, and Steven Jackson. "Rugby and Mathematics: A Surprising Link among Geometry, the Conics, and Calculus." *Mathematics Teacher* 94, no. 8 (November 2001): 649–54.

Kalman, Dan. "Solving the Ladder Problem on the Back of an Envelope." *Mathematics Magazine* 80, no. 3 (June 2007): 163–82.

King, James, and Doris Schattschneider. *Geometry Turned On! Dynamic Software in Learning, Teaching, and Research*. Washington, D.C.: Mathematical Association of America, 1997.

MapleSoft. Maple. Version 9.5. Software. Waterloo, Ontario: Maplesoft, 2005.

Posamentier, Alfred S. *Advanced Euclidean Geometry; Excursions for Secondary Teachers and Students*. Emeryville, Calif.: Key College Publishing, 2002.

Todd, Philip. Geometry Expressions. Version 1.01. Software. Tigard, Ore.: Saltire Software, 2007.

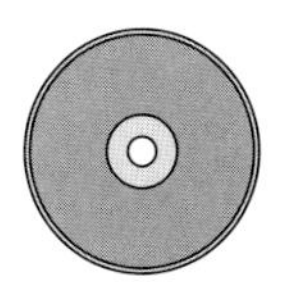

Sample lessons and Geometry Expressions reader files that support this article are found on the CD-ROM disk ccompanying this Yearbook.

Index

A

B

C

D

Q

R

U

V

W

Y